Nanostructured Materials by
High-Pressure Severe Plastic Deformation

NATO Science Series

A Series presenting the results of scientific meetings supported under the NATO Science Programme.

The Series is published by IOS Press, Amsterdam, and Springer (formerly Kluwer Academic Publishers) in conjunction with the NATO Public Diplomacy Division.

Sub-Series

I. Life and Behavioural Sciences	IOS Press
II. Mathematics, Physics and Chemistry	Springer (formerly Kluwer Academic Publishers)
III. Computer and Systems Science	IOS Press
IV. Earth and Environmental Sciences	Springer (formerly Kluwer Academic Publishers)

The NATO Science Series continues the series of books published formerly as the NATO ASI Series.

The NATO Science Programme offers support for collaboration in civil science between scientists of countries of the Euro-Atlantic Partnership Council. The types of scientific meeting generally supported are "Advanced Study Institutes" and "Advanced Research Workshops", and the NATO Science Series collects together the results of these meetings. The meetings are co-organized by scientists from NATO countries and scientists from NATO's Partner countries — countries of the CIS and Central and Eastern Europe.

Advanced Study Institutes are high-level tutorial courses offering in-depth study of latest advances in a field.
Advanced Research Workshops are expert meetings aimed at critical assessment of a field, and identification of directions for future action.

As a consequence of the restructuring of the NATO Science Programme in 1999, the NATO Science Series was re-organized to the four sub-series noted above. Please consult the following web sites for information on previous volumes published in the Series.

http://www.nato.int/science
http://www.springeronline.com
http://www.iospress.nl

Series II: Mathematics, Physics and Chemistry – Vol. 212

Nanostructured Materials by High-Pressure Severe Plastic Deformation

edited by

Yuntian T. Zhu

Los Alamos National Laboratory,
NM, U.S.A.

and

Viktor Varyukhin

Donetsk Physics & Technology Institute,
Ukrainian Academy of Sciences,
Donetsk, Ukraine

Springer

Published in cooperation with NATO Public Diplomacy Division

Proceedings of the NATO Advanced Study Institute on
Nanostructured Materials by High-Pressure Severe Plastic Deformation

A C.I.P. Catalogue record for this book is available from the Library of Congress.

ISBN-10 1-4020-3922-0 (PB)
ISBN-13 978-1-4020-3922-5 (PB)
ISBN-10 1-4020-3921-2 (HB)
ISBN-13 978-1-4020-3921-8 (HB)
ISBN-10 1-4020-3923-9 (e-book)
ISBN-13 978-1-4020-3923-2 (e-book)

Published by Springer,
P.O. Box 17, 3300 AA Dordrecht, The Netherlands.

www.springeronline.com

Printed on acid-free paper

Printed in the Netherlands.

TABLE OF CONTENTS

II. Advances in SPD Technologies

III. Processing, Microstructures, and Mechanical Properties

IV. Mechanical Properties

V. PROPERTIES AT ELEVATED TEMPERATURES

PREFACE

Bulk nanostructured materials are defined as solids with nanoscale (typically 1–100 nm) substructures. In recent years, bulk nanostructured materials have been the subjects of intensive research due to their superior physical and mechanical properties. These include a combination of high strength and ductility at room temperature, high strain rate and low temperature superplasticity, ductility in usually brittle ceramics and intermetallics, and high saturation magnetisation and coercive force in magnetic materials. These superior mechanical and physical properties make nanostructured materials attractive for numerous advanced applications in medical, aerospace, sporting goods, and transportation industries.

The most promising approach for producing nanostructured materials is to refine coarse-grained metal through severe plastic deformation (SPD). The most developed SPD techniques are equal channel angular pressing (ECAP) and high-pressure torsion (HPT). Recently, new SPD techniques have been developed, such as accumulative roll bonding, repetitive corrugation and straightening, and twist extrusion. SPD techniques have the capability of producing bulk nanostructured materials large enough for structural applications. In addition, the nanostructured materials produced by SPD are 100% dense and contamination free. Therefore, the SPD techniques have significant advantages over other techniques in synthesizing bulk nanostructured materials.

Due to their great potential for commercialization, the research and development activities on SPD and SPD-processed materials have increased exponentially in recent years, as evidenced by the increase in the numbers of published papers. Recent publications have indicated that high pressures during SPD significantly affect the microstructural evolution and consequently the mechanical properties of bulk nanostructred materials. For example, the HPT technique, which exerts the highest hydrostatic pressure, can refine grains most effectively and produce the finest nanostructures among all SPD techniques. Another example is ECAP under high back-pressure, which can process materials to larger processing strains without fracture. High pressure is believed to reduce porosity, cracks and other macroscopic defects that are detrimental to mechanical properties.

Recently, it was reported that nanostructured materials processed under high pressure by HPT and ECAP have an extraordinary combination of both high strength and high ductility, which are two desirable, but rarely co-existing properties. These findings indicate that high-pressure is a critical factor that can be employed to process nanostructured materials with superior mechanical, and possibly also physical, properties. However, the reports on high pressure effects are mostly fragmental, and there has been no systematic effort in studying the high-pressure effect on the processing and properties nanostructured materials. Neither has this issue been adequately addressed in recent conferences on nanostructured materials. Recently, the processing of nanostructured materials by SPD under high-pressures has begun to catch the attention of the scientific community and will become one of the major research and development thrusts in the next few years.

The NATO Advanced Research Workshop on *Nanostructured Materials by High-Pressure Severe Plastic Deformation* was held to discuss the above issue and to promote more research in this area. It was opened on September 22, 2004 in the famous Stariy Zamok hotel located on the beautiful bank of Seversky Donets River and surrounded by historical sites including the 17th-century Orthodox Church Monastery. The workshop was later moved to the Donetsk Physics & Technology Institute of National Academy of Science of Ukraine. In addition to serious scientific presentations and discussions, the beautiful Ukrainian folk music and dances at the workshop banquet was one of the highlights that were especially memorable.

During the 3-day workshop, about 60 scientists (see the photo) from 12 countries presented 60 papers and discussed topics including high pressure effect on the nanostructure and properties of SPD-processed materials, fundamentals of nanostructured materials, development of high-pressure SPD technologies for commercial production of nanostructured materials, recent advances of SPD technologies as well as applications and future markets of SPD-processed nanostructured materials. This book collects 42 peer-reviewed papers presented in this workshop.

Participants of the NATO Advanced Research Workshop on *Nanostructured Materials by High-Pressure Severe Plastic Deformation*, Donetsk, Ukraine, September 22-26, 2004.

We would like to thank the following persons in the local organizing committee without whose hard work this workshop would have been impossible:

Dr. Dmitry Orlov, Executive secretary

Mrs. Svetlana Miroshnichenko, Organizing Committee member

Prof. Viktor Beloshenko, Organizing Committee member

Special thanks are extended to Dr. Dimitry Orlov, the secretary of this workshop, who tirelessly and efficiently worked on all logistics, and ran this workshop smoothly and successfully despite of difficulties in securing the conference hotels. Thanks are also extended to the International Advisory Committee, which provided many valuable advices.

We would also like to thank the NATO Science Committee, the Science & Technology Center in Ukraine, and GE Company Global Research Center for their financial support .

Yuntian T. Zhu and Viktor Varyukhin

Workshop Co-directors

This workshop was sponsored by

NATO Physical and
Engineering Science &
Technology Program

GE Company
Global Research Center

Science & Technology Center
in Ukraine

Donetsk Physics and
Technology Institute,
National Academy of Science
of Ukraine

Los Alamos National
Laboratory

I. FUNDAMENTALS OF NANOSTRUCTURED MATERIALS AND SPD PROCESSING

Deformation Twinning in Nanocrystalline fcc Copper and Aluminum

Yuntian T. Zhu*
Materials Science and Technology Division, Los Alamos National Laboratory
Los Alamos, NM 87545, USA

Abstract

Deformation twins have been oberved in nanocrystalline (nc) Al processed by cryogenic ball-milling and in nc Cu processed by high-pressure torsion at room temperature and very low strain rates. They were found formed by partial dislocations emitted from grain boundaries. This paper first reviews experimental evidences and molecular simulation results on deformation twinning and partial dislocation emissions from grain boundaries, and then discusses recent analytical models on the nucleation and growth of deformation twins. These models are compared with experimental results to establish their validity and limitations.

Introduction

Nanocrystalline (nc) materials have been reported to have superior mechanical properties [1-3]. Their mechanical properties are controlled by their unique deformation mechanisms [4-8]. Some deformation mechanisms have been predicted to operate in nc face-centered-cubic (fcc) metals by molecular-dynamics (MD) simulations [7], and experimentally observed [9-13]. This paper reviews deformation twinning and partial dislocation emissions from GBs, and then discusses recent analytical models on the nucleation and growth of deformation twins.

Molecular dynamics simulations

Our understanding of the deformation mechanisms in nc metals were initially largely from the results of MD simulations [4,6,7,14-16]. Such simulations predict nc metals to deform *via* GB sliding and rotation at very fine grain sizes (e.g. 3-10 nm) [4,14]. The GB sliding and rotation were experimentally verified in nc gold [17] and nickel [18]. At larger grain sizes of several tens of nanometers, M D simulations predict that partial dislocation emission from GBs [7,14] is a dominant deformation mechanism. The activation of partial dislocations provides a critical precondition for the formation of deformation twins.

Indeed, deformation twins have been predicted by MD simulations in nc Al [7,15], Ni [6,14] and Cu [19]. Three twinning mechanisms were predicted [7]: 1) homogeneous twinning inside nanosized grains by coincidental overlapping of wide stacking fault ribbons. The stacking fault ribbons were formed by dissociated lattice dislocations, and become very wide due to the effect of small grain size as well as external stress [20]. 2) heterogeneous twinning from the GBs. 3) twinning by GB splitting and migration. Note that MD simulations use very high strain rate (up to 2×10^8), which is much higher than that in a real experiment. Strain rate is

* Corresponding Author. yzhu@lanl.gov

Y.T. Zhu and V. Varyukhin (eds.), Nanostructured Materials by High-Pressure Severe Plastic Deformation, 3-11.

known to affect the twinning tendency in CG materials. Therefore, the MD results need to be taken with caution when compared with experimental results.

Experimental Observations

The partial dislocation emission from GB and deformation twinning in nc Al and Cu have been experimentally verified. Fig. 1 is a high resolution electron microscopy (HRTEM) micrograph that shows direct evidence of partial dislocation emission from GBs in nc Cu processed by high pressure torsion [12]. As shown, there is only one twin boundary at the upper part of Fig. 1 that divides twin domains I and II. However, there are high densities of micro-twins and stacking faults at the lower part of domain II. These micro-twins and stacking faults do not pass across the whole grain but stop in the grain interior with Shockley partial dislocations located at the fronts of the micro-twins and stacking faults. It is obvious that these twins and stacking faults were formed by partial dislocations emitted from the lower GB segment. The micro twins shown were formed by heterogeneous twinning.

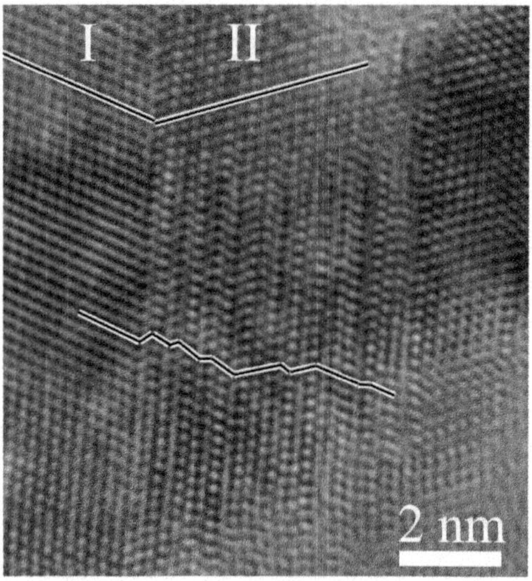

Fig. 1. A HRTEM image showing micro-twins and stacking faults. The lower part of domain II has a lot of micro-twins and stacking faults with one end of the micro-twin/stacking fault stops within the crystallite.

The other two types of twins, formed by homogeneous twinning and GB splitting/ migration, were also recently verified in nc Al [8-10,20] by HRTEM. Al in its coarse- grained (CG) state has never been observed to deform by twining, except at crack tips [21], because of its very high stacking fault energy. CG Cu does not deform by twinning [22,23] except at very high strain rates [24,25] and/or low temperatures [26]. Moreover, in CG Cu smaller grain size was found to impede deformation twinning, which is also true for many other metals [27,28]. In contrast, twinning becomes a major deformation mechanism in nc Cu processed by high pressure torsion (HPT) at room temperature and a low strain rate [12]. In addition, it was also found that under the same HPT condition, twinning only occurred in crystalline domains that are smaller than 50 nm [29]. These results indicate that the nc materials indeed deform via mechanisms not accessible to their coarse-grained counterparts and grain size plays a critical role in the formation of deformation twins.

The heterogeneous twins (see Fig. 1) nucleated on GBs is most frequently observed in nc Al and Cu. However, it is not clear under what conditions they will nucleate and grow. For example, MD simulations do not tell what critical stress is needed for their nucleation and growth. Moreover, MD simulations [4,7,14] usually use extremely high strain rates in the order of 10^6 to 10^8 s^{-1}, which correspond to explosive deformations in a real experiment. It is well known that the strain rate significantly affects the deformation mechanisms of materials [12,19]. This adds complexity to the interpretation of the atomic simulation results. In the following section, I shall review and examine analytical dislocations models on the formation of deformation twins.

Analytical dislocation models

Conventional dislocation models

Two similar models were recently proposed to explain the formation of deformation twins in nc metals [8,11]. In the model by Chen et al [8], the stress needed to activate a lattice dislocation is described as

$$\tau_L = \frac{2\eta G b}{d} \qquad (1)$$

where η is a parameter that reflects the characteristics of the dislocation (η=0.5 for an edge dislocation and η=1.5 for a screw dislocation), G is the shear modulus, and b is the magnitude of the Burgers vector of the lattice dislocation. The stress to activate a partial dislocation is described as

$$\tau_P = \frac{2\eta G b_1}{d} + \frac{\gamma}{b_1} \qquad (2)$$

where b_1 is the magnitude of the Burgers vector of the partial dislocation, and γ is the stacking fault energy. When $\eta = 0.5$, these two equations become identical to those in the work by Liao et al [11].

Because $b > b_1$, τ_p will increase at a slower rate than τ_L, which means that it will be easier to activate a partial dislocation than a lattice dislocation when the grain is below a critical size. These two models seem very straightforward in explaining the activation of partial dislocations, which is a prerequisite of deformation twinning.

Unfortunately, experimental data shows that smaller grain size hinders, not promotes, deformation twinning [27,28], which directly contradicts these two models [8,11]. It has been found that the critical stress for twinning follows a Hall-Petch relationship

$$\sigma_T = \sigma_{T0} + \kappa_T d^{1/2} \qquad (3)$$

where σ_{T0} is a constant and κ_T is the Hall-Petch slope for twinning. Equation 3 is similar to the well-known Hall-Petch relationship for a lattice dislocation:

$$\sigma_L = \sigma_{Lo} + \kappa_L d^{1/2} \tag{4}$$

where σ_{L0} is a constant and κ_L is the Hall-Petch slope for the slip of a lattice dislocation.

Experimental data shows that the Hall-Petch slope for twinning is higher than that for the slip of lattice dislocation for body-centered-cubic (bcc), face-centered-cubic (fcc), and hexagonal close-packed (hcp) metals and alloys [27]. Therefore, the models are not supported by experimental data. Moreover, these two models predict an unrealistically high critical stress for twinning [8], which reflect their problem. The models also made an inexplicit assumption that activation of partial dislocation equals formation of twins, which, as shown later, is not correct.

A recent model based on grain boundary emissions of partial dislocations

A recent analytical dislocation model by Asaro et al [30] utilized the simulation and experimental results that partial dislocations are emitted from GBs of nano-grains. The critical stress needed to move a lattice dislocation is described as

$$\tau_L = \frac{Gb}{d} \tag{5}$$

and the critical stress needed to move a partial dislocation is described as

$$\tau_P \approx \frac{Gb}{3d} + (1-\delta)\frac{\gamma}{Gb} \tag{6}$$

where δ is the ratio of equilibrium stacking fault width to grain size.

This model predicts that below a certain critical grain size partial dislocations from GBs need a lower stress to move than lattice dislocations in nc metals. Most importantly, it predicts a realistic, low twinning stress that is obtainable under experimental conditions such as ball-milling. However, the model does not address two critical issues. First, the emission of a partial dislocation does not guarantee the nucleation of a deformation twin because a trailing partial could easily follow to erase the stacking fault formed by the first partial. Second, random emissions of partial dislocations from a GB would give a deformation twin equal probability to grow or to shrink, which cannot explain the deformation twin growth and the observed large deformation twins in nc Al, Cu and Pd [9-13].

A new model addressing the nucleation and growth of deformation twins

To address the above two issues, we recently developed an analytical model to describe the nucleation and growth of deformation twins in nc Al [31].

The model assumes a grain with a square (111) slip plane, as shown in Fig. 2, similar to that used in previous studies [20,30]. Under an external shear stress $\boldsymbol{\tau}$, a

90° leading Shockley partial, $\mathbf{b}_1$=a/6[11 $\bar{2}$], is emitted from grain boundary AB, depositing two segments of partial dislocation lines (Aa and Bb) on GB. $\boldsymbol{\tau}$ is oriented at an angle α with line ab. A trailing 30° partial, $\mathbf{b}_2$=a/6[2 $\bar{1}$ $\bar{1}$], is also emitted (line Aa'b'B). The two partials ab and a'b' are separated by a stacking fault. The two partials react to form two perfect dislocation segments, Aa' and Bb' at the GB. This dislocation system is called a *60° I system* hereafter.

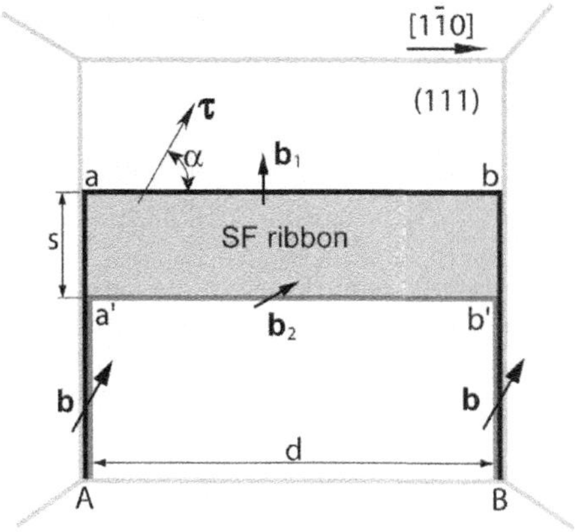

Fig. 2. A schematic illustration of the dislocation model for deformation twin nucleation.

To nucleate a deformation twin, a stacking fault needs to be first created. This can occur via: (1) emission of a 90° partial at a GB, (2) extending the stacking fault ribbon in Fig. 2 across the grain. As shown later, both scenarios may occur depending on the orientation of $\boldsymbol{\tau}$. In Fig. 2, for the partial $\mathbf{b}_1$ to move, $\boldsymbol{\tau}$ has to perform a work to overcome increases in both the stacking fault energy and dislocation energy from lengthening segments Aa and Bb [32,33]. The critical stress for moving partial $\mathbf{b}_1$ can be derived as

$$\tau_p = \frac{1}{\sin\alpha}\left(\frac{\sqrt{6}\gamma}{a} + \frac{Ga}{2\sqrt{6}\pi d}\ln\frac{\sqrt{2}d}{a}\right) \tag{7}$$

where γ is the stacking fault energy, *a* is the lattice parameter, and *d* is the grain size defined in Fig. 2.

The τ needed to move the stacking fault ribbon is equivalent to τ for moving a 60° lattice dislocation. τ has to overcome the work needed to lengthen the lattice dislocation segments Aa' and Bb', and can be derived as:

$$\tau_L = \frac{Ga(4-3\nu)}{8\sqrt{2}\pi(1-\nu)d\cos(\alpha-60°)}\ln\frac{\sqrt{2}d}{a} \tag{8}$$

where ν is the Poissons ratio.

After the stacking fault formation, a twin may nucleate via the emission of a second 90° partial from the grain boundary on a plane adjacent to the stacking fault. On the other hand, a trailing partial may also emit on the stacking fault plane and erase the stacking fault in its path. The critical twin nucleation stress can be derived as:

$$\tau_{twin} = \frac{Ga}{2\sqrt{6}\pi d \sin\alpha} \ln\frac{\sqrt{2}d}{a} \tag{9}$$

The trailing partial requires a stress, τ_{trail}, to move, which can be derived as

$$\tau_{trail} = \frac{\sqrt{6}}{\cos(\alpha - 30°)}\left[\frac{Ga(8-5\nu)}{48\pi(1-\nu)d}\ln\frac{\sqrt{2}d}{a} - \frac{\gamma}{a}\right] \tag{10}$$

Once a twin is nucleated, it may grow via the emission of more 90° twinning partials under stress τ_{twin}. It may also shrink via the emission of a shrinking partial, $\mathbf{b}_2$, on a plane adjacent to the twin boundary but on the twin side. The stress needed to move a shrinking partial can be derived as:

$$\tau_{shrink} = \frac{\sqrt{6}}{\cos(\alpha - 30°)} \frac{Ga(8-5\nu)}{48\pi(1-\nu)d}\ln\frac{\sqrt{2}d}{a} \tag{11}$$

The stresses, τ_P, τ_L, τ_{tiwn}, τ_{trail}, and τ_{shrink}, determine the nucleation and growth of a deformation twin. For example, at $\tau_P < \tau_L$, partial dislocations will be emitted from the GB, and at $\tau_{tiwn} < \tau_{trail}$, a twin may nucleate. In the following, Al is used as a model material to validate the model. For Al, G = 26.5 GPa, ν = 0.345, a = 0.404 nm, and γ = 122 mJ/m^2 [33,34]. In Fig. 3, the stresses, τ_p, τ_L, τ_{twin}, τ_{trail}, and τ_{shrink}, are plotted as a function of grain size d for: (a) α = 90° and (b) α = 135°. The point B in Fig 3a represents the critical grain size (d_B = 5.16 nm) below which a deformation twin nucleates because $\tau_{twin} < \tau_{trail}$. However, a deformation twin can nucleate only after the formation of a stacking fault. As shown, at grain size d_B, $\tau_L < \tau_p$, i.e. the lattice dislocation is operating at τ_L = 0.88 GPa (point B'). The stacking fault width at τ_L is far larger than d_B [31]. This means that a deformation twin nucleates after a stacking fault ribbon spreads across the grain. Interestingly, the critical grain size is larger and the critical stress is lower for twin nucleation (point B') than for the partial dislocation emission (point A), because the stacking fault of a dissociated lattice dislocation would spread across the grain before a partial is emitted.

Fig. 3b shows the stresses vs d at α = 135°.

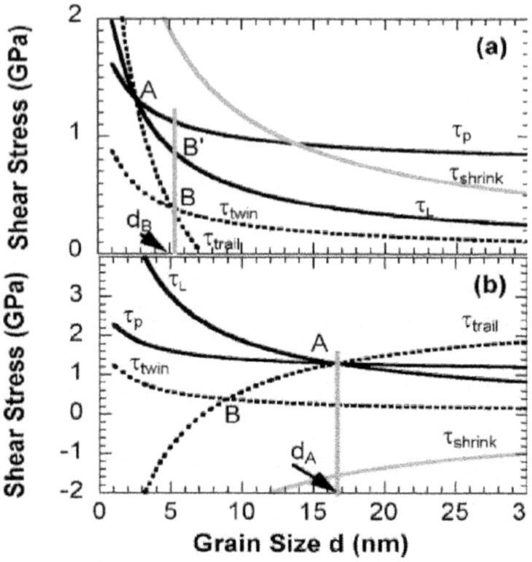

Fig. 3. The critical stresses, τ_p, τ_L, τ_{twin}, τ_{trail}, and τ_{shrink} as a function of nanocrystalline Al grain size d for a given α value of (a) 90° and (b) 135°.

As shown, τ_{trail} or τ_{shrink} is actually the critical stress curve above which the trail or shrink partial is prohibited. At $d < d_A$ (16.66 nm) and $\tau > \tau_p$, deformation twins will nucleate.

The above equations and analysis are for the *60° I* system only. There are two other possible dislocation systems: a *60° II system* with a leading 30° partial and a trailing 90° partial, and a *Screw system* with a leading 30° partial and a trailing 30° partial [20]. These two systems are amenable to the same procedure, and therefore we only present the final results. It is found that the 60° II system does not operate because it requires much higher stress to nucleate a deformation twin. Therefore we shall drop the 60° II system in the following analysis. In a polycrystalline nc sample, grains are likely to orient in all orientations. Therefore, a deformation map linking stresses of deformation twin nucleation and growth with grain size is very useful and desirable. Such a map can be constructed by plotting the critical stresses against the critical grain sizes (see Fig. 4).

The deformation map in Fig. 4 reveals the following four interesting points: (i) The deformation twin *nucleation* curves have a cup and handle geometry. The cup section is from $\boldsymbol{\tau}$ orientations at which all stresses behave like those in Fig. 3a, while the handle section is from $\boldsymbol{\tau}$ orientations at which all stresses behave like those in Fig. 3b; (ii) The optimum grain sizes for deformation twin nucleation (the lowest stress point at cup bottom) are 4.85 nm and 7.25 nm, respectively, for the 60° I and Screw systems; (iii) The 60° I system has a slightly lower critical stress (0.88 GPa) than the Screw system (0.91 GPa) for deformation twin nucleation; (iv) The stress for deformation twin growth is much lower than that for its nucleation.

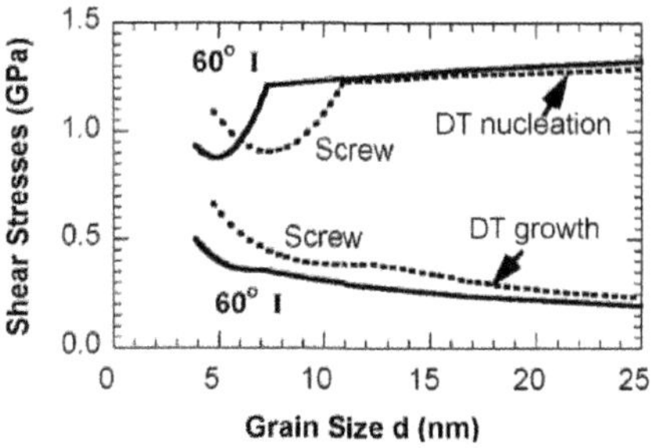

Fig. 4. A deformation map showing the critical stresses for deformation twin nucleation and growth in nanocrystalline Al as a function of grain size for the 60° I and the Screw dislocation systems.

Clearly, the critical stress for deformation twin nucleation is very high (>0.88 GPa). Such a high stress can only be obtained under high strain rates and/or low temperatures, which is consistent with experimental conditions for cryogenically-ball-milled nc Al [9-11]. In addition, the high nucleation stress also explains the low deformation twin density in the nc Al.

On the other hand, the stress for deformation twin growth is low enough to be easily attained during a normal static deformation. This is consistent with a recent MD simulation [35], which shows that deformation twins are difficult to nucleate but easy to grow. This explains why a twin would grow without the traditional pole mechanism. This new twin growth mechanism is defined as the "stress-controlled-growth" mechanism.

Summary

When the grain sizes of nc materials are in the range of 10-50 nm, deformation twins will form in medium to high stacking fault energy fcc metals such as Al and Cu. Several types of deformation twins are predicted by MD simulations and observed experimentally. The most common twins observed experimentally are those nucleated heterogeneously on grain boundaries via the emission of partial dislocations.

Two analytical models based on conventional dislocation theories [8,11] are not supported by experimental data and cannot give a reasonable critical twining stress. Another model [30], which is based on the partial dislocation emission from grain boundaries, gives a reasonable critical stress for partial dislocation activation, but does not address the nucleation and growth of deformation twins. A new model [31], which is also based on the partial dislocation emission from grain boundaries, addresses both the nucleation and growth of deformation twins in nc fcc metals and alloys. It describes a stress-controlled mechanism for the growth of deformation twins, which explains why and how a twin grows without the conventional pole mechanism. The model can successfully explain the experimental observation of twins in nanocrystalline Al processed by cryogenic ball milling.

References

1. Valiev R.Z., Alexandrov I.V., Zhu Y.T., and Lowe T.C., *J. Mater. Res.*, **17** (2002) 5-8.
2. Zhang X., Wang H., Scattergood R.O., Narayan J., Koch C.C., Sergueeva A.V., and Mukherjee A.K., *Appl. Phys. Lett.*, **81** (2002) 823-25.
3. Jia D., Wang Y.M., Ramesh K.T., Ma E., Zhu Y.T., and Valiev R.Z., *Appl. Phys. Lett.*, **79** (2001) 611-13.
4. Schiøtz J., Ditolla F.D., Jacobsen K.W., *Nature*, **391** (1998) 561-63.
5. Kumar K.S., Suresh S., Chisholm M.F., Horton J.A., and Wang P., *Acta Mater.*, **51** (2003) 387-405.
6. Swygenhoven H.Van, *Science*, **296** (2002) 66-67.
7. Yamakov V., Wolf D., Phillpot S.R., Mukherjee A.K., and Gleiter H., *Nature Mater.*, **1** (2002) 45-48.
8. Chen M.W., Ma E., Hemker K.J., Sheng H.W., Wang Y.M., Cheng X.M., *Science*, **300** (2003) 1275-77.
9. Liao X.Z., Zhou F., Lavernia E.J., Srinivasan S.G., Baskes M.I., He D.W. and Zhu Y.T., *Appl. Phys. Lett.*, **83** (2003) 632-34.
10. Liao X.Z., Zhou F., Lavernia E.J., He D.W., and Zhu Y.T., *Appl. Phys. Lett.*, **83** (2003) 5062-64.
11. Liao X.Z., Huang J.Y., Zhu Y.T., Zhou F., and Lavernia E.J. *Phil. Mag.*, **83** (2003) 3065-75.
12. Liao X.Z., Zhao Y.H., Srinivasan S.G., Zhu Y.T., Valiev R.Z. and Gunderov D.V., *Appl. Phys. Lett.*, **84** (2004) 592-94.
13. Rösner H., Markmann J., Weissmüller J., **Phil. Mag. Lett.**, **84** (2004) 321-34.
14. Swygenhoven H.Van, Derlet P.M., Hasnaoui A., *Phys. Rev. B*, **66** (2002) 024101.
15. Yamakov V., Wolf D., Salazar M., Phillpot S.R., and Gleiter H., *Acta Mater.*, **50** (2002) 5005-20.
16. Yamakov V., Wolf D., Phillpot S.R., Mukherjee A.K., and Gleiter H., *Nature Mater.*, **3** (2004) 43-47.
17. Ke M., Hackney S.A., Milligan W.W., and Aifantis E.C., *NanoStructured Mater.*, **6** (1995) 689-97.
18. Shan Z., Stach E.A., Wiezorek J.M.K., Knapp J.A., Follstaedt D.M., Mao S.X., *Science*, **305** (2004) 654-57.
19. Schiøtz J., Jacobsen K.W., *Science* **301** (2003) 1357-59.

20. Liao X.Z., Srinivasan S.G., Zhao Y.H., Baskes M.I., Zhu Y.T., Zhou F., Lavernia E.J., Xu H.F., *Appl. Phys. Lett.*, **84** (2004) 3564-66.
21. Hai S., Tadmor E.B., *Acta Mater.*, **51** (2003) 117-31.
22. Liu C.D., Bassim M.N., You D.X., *Acta Met. Mat.*, **42** (1994) 3695-04.
23. Hansen N., Ralph B., *Acta Met.*, **30** (1982) 411-17.
24. Johari O. and Thomas G., *Acta Metall.*, **2** (1964) 1153-59.
25. Smith C.S., *Trans. Met. Soc. AIME.*, **212** (1958) 574-89.
26. Blewitt T.H., Coltman R., and Redman J.K., *J. Appl. Phys.*, **28** (1957) 651-60.
27. Meyers M.A., Vöhringer O., Lubarda V.A., *Acta Mater.*, **49** (2001)4025-39.
28. El-Danaf E., Kalidindi S.R., Doherty R.D., *Metall. Mater. Trans.*, **30A** (1999) 1223-33.
29. Liao X.Z., Zhao Y.H., Zhu Y.T., Valiev R.Z., Gunderov D.V., *J. Appl. Phys.*, **96** (2004) 636-40.
30. Asaro R.J., Krysl P., Kad B., *Phil. Mag. Lett.*, **83** (2003) 733-43.
31. Zhu Y.T., Liao X.Z., Zhao Y.H., Srinivasan S.G., Zhou F., Lavernia E.J., *Appl. Phys. Lett.*, **85** (2004) 5049-51
32. Kovács I., Zsoldos L., *Dislocation and Plastic Deformation* (Pergamon Press, Oxford, 1973), pp. 40-186.
33. Hirth J. P., Lothe J., *Theory of Dislocations* (John Wiley \& Sons, New York, 1982), pp. 315-39.
34. Yamakov V., Wolf D., Salazar M., Phillpot S.R., and Gleiter H., *Acta Mater.*, **49** (2001) 2713-22.
35. Swygenhoven H.Van, Derlet P.M., and Frøseth A.G., *Nature Mat.*, **3** (2004) 399-403.

High Pressure in Large Plastic Deformation: Effects and Techniques

V.N. Varyukhin* and B.M. Efros
Donetsk Physics & Technology Institute of NASU, 72 R. Luxemburg St., 83114 Donetsk, Ukraine

Abstract

We analyze the effect and physical mechanisms of the influence of pressure on metal materials that are in the state of plastic flow, as well as on characteristics of materials subjected to severe plastic deformation under pressure.

Influence of pressure on deformation behavior of materials

Increasing the plasticity of solids under pressure has been the subject of numerous investigations starting from fundamental works [1,2]. It is currently shown that the effect is due to two interrelated processes: suppression of fracture and defect structure homogenization (DSH) during active deformation under high pressure [3,4].

For a wide class of solids there exist two levels of critical pressure: $P_c \sim \sigma_s \sim 10^{-3}K$ and $P_0 \sim 10^{-2}K$ (σ_s – yield stress, K – compression modulus). When $P = P_c$ the process of non-uniform pore and crack development is retarded as a result of deformation, i.e. the solid plasticizes and the ultimate strain ε_p increases (ε_p - value of strain prior to fracture). For brittle bodies the value of P_c coincides with that of pressure of brittle – plastic transition. When P is increased further, in the $P_c < P < P_0$ range, the process of suppression of nonuniformity origination becomes more intensive and for $P > P_0$ the solid can be ideally plastic (it deforms as a high-viscous liquid). Table 1 list values of P_0 calculated for various metals (see [4]).

Let us consider the process of DSH for a solid subjected to active deformation under high pressure at relatively low deformation velocities. The nucleation and the mobility of dislocations in solids are limited by thermally activated processes. The application of pressure results in decreasing thermal activation and, thus, in the growth of number of dislocation sources and, finally, in the increase of DSH.

It is known that during the deformation the formation of concentrators results in origination of micro – and meso-non-uniformities favoring fracture. For $P > P_c$, in the deformed solid the relaxation processes tend to minimize the probability of nonuniformity origination, otherwise the work $\Delta A_p = P \cdot \Delta V$ (ΔV change in volume due to nonuniformity origination) should be done, which increases with P. This results in deconcentration of stresses in volume macrosamples and, thus, in DSH intensification.

In heterophase and noncubic polycrystalline systems, the process of DSH becomes more intense due to relaxation processes taking place at the expense of pressure – induced shearing stresses at inner interfaces or at elastic nonuniformities. It should be noted that processes of twinning and martensitic transformations during the deformation of metastable system under high pressure also favor the DSH.

---

* Corresponding Author. var@hpress.dipt.donetsk.ua

Y.T. Zhu and V. Varyukhin (eds.), Nanostructured Materials by High-Pressure Severe Plastic Deformation, 13-20.

Table 1. Critical pressure for various metals [4]

Material	Lattice type	P_c, GPa	P_0, GPa
Cr	BCC	0.07 – 0.72	7
Mo		0.11 – 1.13	11
W		0.18 – 1.69	18
Fe		0.08 – 0.84	8.5
Steel 1045		0.09 – 0.91	9.2
Be	HCP	0.17 – 1.72	17.1
Mg		0.16 – 1.67	16.2
Zn		0.05 – 0.54	5.1
Ti		0.04 – 0.43	4.2

In such a way, the main physical reasons of DSH and, thus, of plasticization of crystal solids under high pressure are the decreasing of thermal fluctuations, the suppression of generation and retardation of micro- and mesononuniformity development, as well as the intensification of effects of compression anisotropy in elastically nonuniform and noncubic polycrystals. For $P \geq P_c$, activation of the processes result in the intensification of plasticization effects in solids under pressure that are observed at different scale levels depending on nature of objects observer (mono- and polycrystals, metastable and heterophase systems, noncubic and porous solids) and on value of P realized in the experiment.

When DSH is high, the value of maximal internal stresses decreases and defects of crystal structure tend to distribute more uniformly in volume of macrosamples, witch as a rule results in the improvement of physic-mechanical properties of different materials.

Experiment results on the influence of pressure on grain refinement and plasticity of metals.

The above assumptions concerning the role of hydrostatic pressure in deformation process are supported by the following experimental facts.

When molybdenum is deformed under pressure, on the level of structure the polyhedral grain structure of the annealed material ($d \approx 22$ μm) starts showing local instability to the origination of rotational "vortex" type structure during hydrostatic extrusion of ratio $e > 0.3$. With e increase, the growing instabilities result in a cardinal rearrangement of the initial structure ($e \approx 1.2$-1.4) to a rotational structure of the "vortex" type. With further increase in the extrusion ratio up to $e \approx 1.9$, in sample cross-section the "vortex" character of the structure becomes more pronounced, while in the longitudinal one the fibrous structure is observed (Fig. 1).

At the level of substructure, the low-misoriented cellular structure existing with $e \approx 0.15$-0.3 changes for a fragmented one, which is a population of fragments (subgrains) with a 5-10° (and more) misorientation. For $e \approx 1.2$-1.9, the process of a more uniform grain refinement is genetically related to the process of evolution and branching of nonuniform banded substructure typical of $e \approx 0.4$-1.0 (Fig. 2).

During the hydrostatic extrusion of molybdenum, the increase in pressure level at the expense of backpressure P_{bp} results in shifting the range of deformation ratios necessary for origination of rotational structures of the "vortex" type to lower values of e.

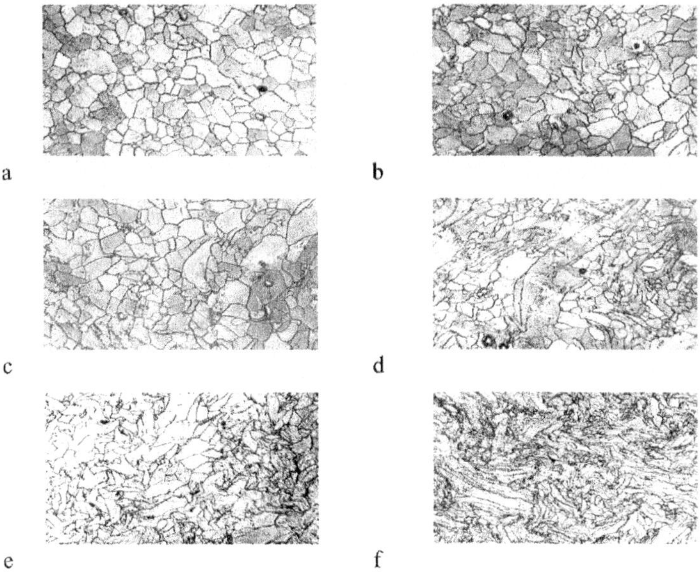

Figure. 1. Microstructure of molybdenum after hydroextrusion (×300): a – $e = 0.36$, $P_{bp} \approx 0.1$ MPa; b – $e = 0.36$, $P_{bp} = 900$ MPa; c – $e = 0.6$, $P_{bp} \approx 0.1$ MPa; d – $e = 0.6$, $P_{bp} = 900$ MPa; e – $e = 0.92$, $P_{bp} \approx 0.1$ MPa; f – $e = 0.92$, $P_{bp} = 900$ MPa.

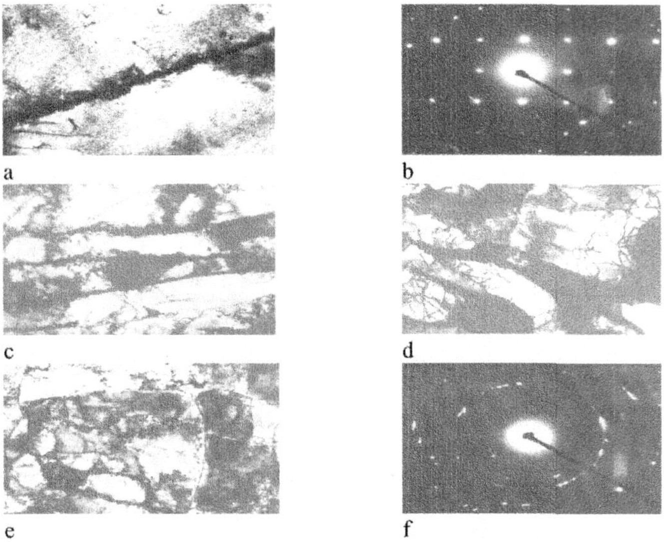

Figure. 2. Substructure of molybdenum as a function of hydrostatic extrusion parameters (cross-section, ×27000); a-e=0; b – microdiffraction of section "a"; c-e=1.2; P_{bp}≈0,1 MPa; d-e≈1.2; P_{bp}≈0.8 GPa *a*; e-e=1.4; P_{bp}≈0.8 GPa; f – microdiffraction of section "e".

The development of plasticization effect in molybdenum under pressure conditions the shifting of the internal of deformation ratios e, that provide the origination of rotational structures of the vortex type, to lower values of e, in the case of hydrostatic extrusion with backpressure P_{bp}~0.8 GPa. In particular, it has been determined that an increase in the level of P_{bp} (e.g. for e=0.6) results in the formation of more dense and narrow walls of subgrains (fragments) as well as in grain shattering. The quantitative analysis off histograms for distribution of grains in sire (in directions of short axis l_1 and long axis l_2) has shown that during the hydrostatic extrusion of molybdenum, with the increase in backpressure and thus in extrusion pressure, mean dimensions of molybdenum, with the increase in backpressure and thus in extrusion pressure, mean dimensions of fragments l_1 and l_2 in cross section of samples are decreasing (Fig. 3).

That was an illustration of the influence of pressure on hardening. The move known effect is its influence on plasticity of materials. The fact that the growth of deformation porosity is retarded or suppressed by the application of high pressure is illustrated in Figure 4 [5]. It is evident that the higher pressure value the later the start of porosity growth, and the higher deformation the material can sustain.

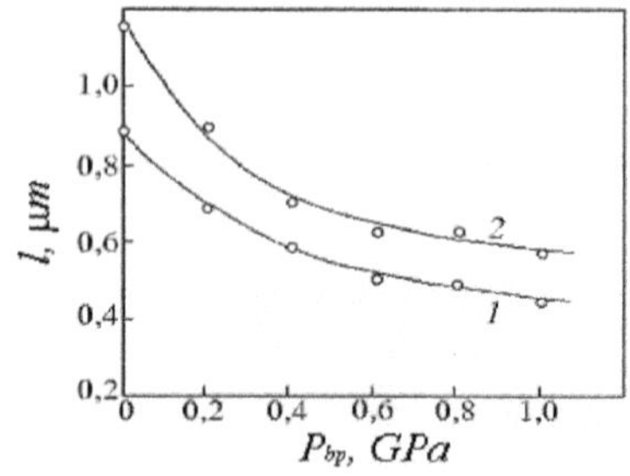

Figure. 3. Change in mean size of grains in the cross-section of molybdenum depending on pressure P_{bp} at hydrostatic extrusion (e=0.6); 1 and 2 are short and long axes of subgrains (fragments), respectively

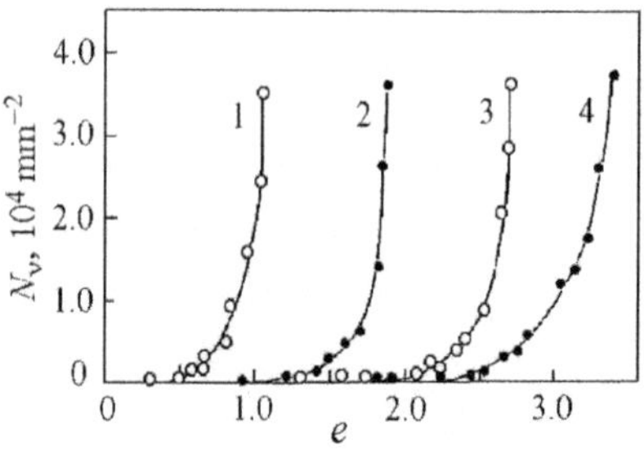

Figure. 4. Density of pores N_v depending on deformation e of steel 1045 samples under pressure [5], MPa: 1 – 0.1; 2 – 420; 3 – 840; 4 – 1120. On the curves, the upper end points correspond to the moment of macroscopic crack origination.

Let us follow the value of hardening for a gasket made of stainless steel T301 upon compression in diamond anvils [5] by measuring its microhardness (Fig. 5a). In the centre of the gasket where pressure is maximum (curve 1), the microhardness is minimum (curve 2). By varying the degree of the diamond anvil compression and thus P_0 one can follow the changes in basic hardening (Fig. 5b). Such a behavior follows from dispersion of the material at the substructural level owing to the formation of misoriented subgrain structure. The intensity of the behavior grows with anvils pressure.

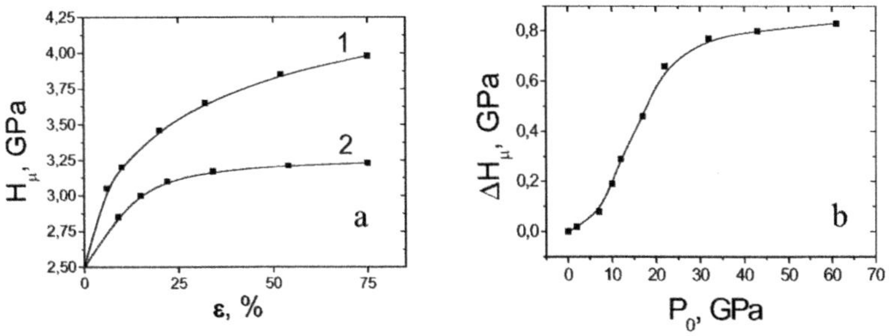

Figure. 5. Changes in characteristics of gasket material hardening during deformation in diamond anvils: a – dependence of microhardness H_μ on degree of plastic deformation ε: 1 – in the centre of gasket ($r \sim 0$, $P \sim P_0$), 2 – at periphery of gasket contact with diamond anvils ($r \sim b$, $P \sim P_{min}$); b – dependence of the degree of gasket hardening ΔH_μ on pressure P_0

Mathematical modeling of the effects resulting from the influence of pressure on plastic deformation of solids.

The development of the SPD technologies should be based on the mathematical models describing fracture and grain refinement of solids and taking into account the pressure effect on this process.

We have proposed a continual model of the deformation of solids describing the fracture of the material under deformation as well as the effect of pressure on the process [6]. The model [6] did not take the interrelation between the formation of microononuniformities and the grain refinement in solids under severe plastic deformation. Another model has been developed to describe the above – mentions effects [7].

In order to incorporate the structure of polycrystals into a continuum model of plasticity, in [7] has been introduced two scalar parameters: the total volume of micro-inhomogeneities θ and the total area of high angular bound-arise per unit volume of the material *S*. Using the assumption of self-similarity of the structure changes during quasi monotone loading and the complementarily of fragmentation and fracture, we obtain a system of kinetic equations for these parameters. We show that the proposed model not only explains a number of known effects, but also suggests new ones. In particular, it turned out that the loading processes that most intensely fragment the metal should be the processes which lead to the highest decrease of plasticity of the given metal (among all processes with the same level of hydrostatic pressure in the center of deformation). To obtain submicro- and nanostructures, these processes need to be carried out under high hydrostatic pressure in the center of deformation. In this case, the relaxation of internal strain will follow the path of crystal fragmentation, not the emergence of micro-inhomogeneities.

To illustrate conclusions following from the model, it was used to study the evolution of metal structure in two processes differing in deformation scheme – forward and twist extrusion [8]. Different levels of backpressure were preset to show how the hydrostatic pressure influences the processes.

The characteristics of carbon steel (0.45% C) were taken for calculation. The deformation scheme was accounted for by value of materials ductility when the hydrostatic component of the stress tensor is zero [7].

The results of calculation are represented in Fig. 6.

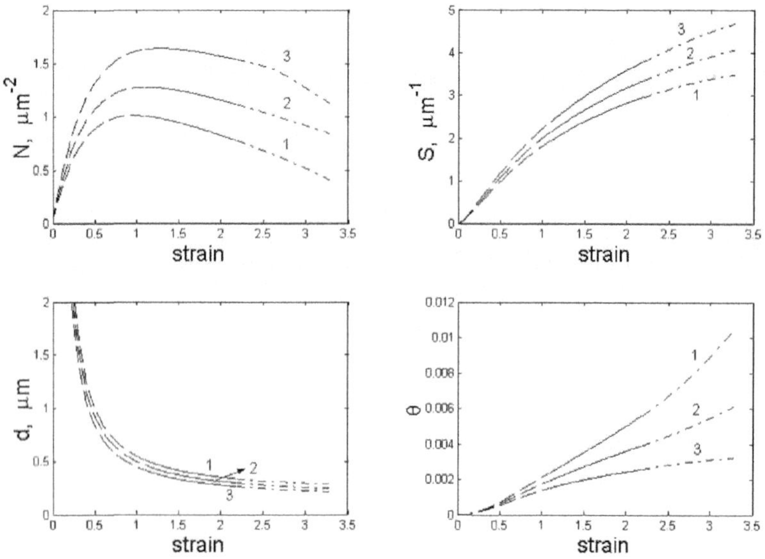

Figure. 6. Characteristics of carbon steel structure as a function of strain during hydrostatic extrusion (each curve corresponds to 3 passes with drawing 3) with different backpressure levels: 1 – P_{bp}~0.1 MPa; 2 – P_{bp} = 400 MPa; 3 – P_{bp}=600 MPa (notations see in the text).

The graphs let us to conclude that the increase in backpressure results in the formation of a great number of the accumulative zones (*N*) that, in its turn, intensify the process of grain (*d)* refinement and suppress the loosening of the material. This is illustrated by the curves *S*(*e*) and *d*(*e*): the more intensive grain refinement and, thus, the smallest size of fragments for the material correspond to the maximum backpressure level. The run of Θ(*e*) curve evidences the positive influence of backpressure on healing the micro-voids of the material that are the basic cause of its fracture. In such a way, the backpressure makes it possible to increase plasticity of the metal, the preferences from the grain refinement viewpoint being preserved.

Influence of the formed nanocrystalline state in metals on phase stability under pressure.

It is known that the formation of nanocrystalline (NC) structure by SPD techniques can essentially influence the kinetics and completeness of phase transformations in metastable materials.

The objects of investigation were Fe-Mn alloys with manganese concentration C_{Mn}=0-55wt% and of different phase composition in the initial state.

For Fe, the NC state was reached by high pressure torsion (HPT). It has been shown that the formation of NC state in Fe (d≈80 nm) in the process of preliminary HPT with e=6-7 and P=10 GPa increases critical points $P^{\alpha\rightarrow\varepsilon}$ of direct transformation and decreases $P^{\alpha\leftarrow\varepsilon}$ reverse transformation by ~4 GPa as compared to coarse-grained (CG) state (d_3~400 μm), as determined by "in situ" investigations in high-pressure chamber (see Fig. 7a). This fact is explained by the stabilizing action of the NC state in Fe, which is the restraining force for the basic γ→ε transformation (the ratio of the volume fraction of the microcrystallite boundaries to the "defect-free" portion in NC Fe is 2-3 orders of magnitude larger than the same ratio in CG Fe). HPT with parameters e≈6.4 and P=10 GPa does not result in α-Fe-Mn alloys (C_{Mn}<10%).

Under SPD by HPT of (γ+ε)– and γ-alloys based of Fe-Mn solid solution a banded type structure originates consisting of twins and stacking-fault twins. When degree of deformation increases (e≥4), the NC state is forded with the average size of microcrystallites ~80 nm (e~7-9) (it should be stated that to NC state is characterized by a 3-4 fold hardening as compared to the initial one).

HPT of (γ+ε)-Fe-Mn alloys results in the increase of high pressure HCP ε-phase as compared to and her types of pressure treatment (hydrostatic extrusion, shock waves). This is confirmed by investigation of changes in phase composition of $Fe_{80}Mn_{20}$ alloy, placed into high-pressure chamber, depending on value of stressed-state indices. HPT with parameters e=6.4 and P=10 GPa induces γ→ε-transformation in γ-Fe-Mn alloys.

In this case, 70 ($Fe_{60}Mn_{40}$ alloy) to 10% ($Fe_{55}Mn_{45}$ alloy) the treatment (the quantity of ε-phase decreases with the increase of stacking-fault energy in γ-phase) HPT with parameters e≈4-5 and P=19-20 GPa has induced γ→ε transformation. After the treatment the high-pressure HCP ε-phase has been conserved completely (the "in situ" investigation in Fe-Mn alloys, C_{Mn}=40-55%, is reversible). Analysis of the results has shown that the increase in hysteresis of the reverse ε→γ transformation due to stabilization of high pressure phase in NC state of γ-Fe-Mn alloys subjected to preliminary HPT is the mechanism of HCP high-pressing ε-phase stabilization under decompression (Fig. 7 b).

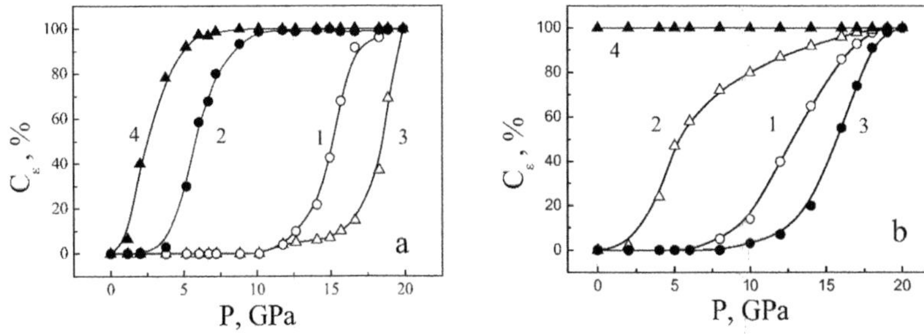

Figure. 7. Pressure dependence of HCP high-pressure ε-phase concentration C_ε in α-Fe (a) and γ-$Fe_{55}Mn_{45}$ alloys (b) in CG and NC states. 1,2-CG structure (d≈400 μm); 3,4 – NC structure (d≈80 nm); 1,3 – loading; 2,4-unloading

Investigation of the temperature stability of HCP high-pressure ε-phase in NC γ-Fe-Mn alloys has shown that the heating to T≥250°C initiates the reverse

$\varepsilon \to \gamma$-transformation (single-phase γ-state needs a 5 min holding at $T \approx 340$-390°C, Fig.8.

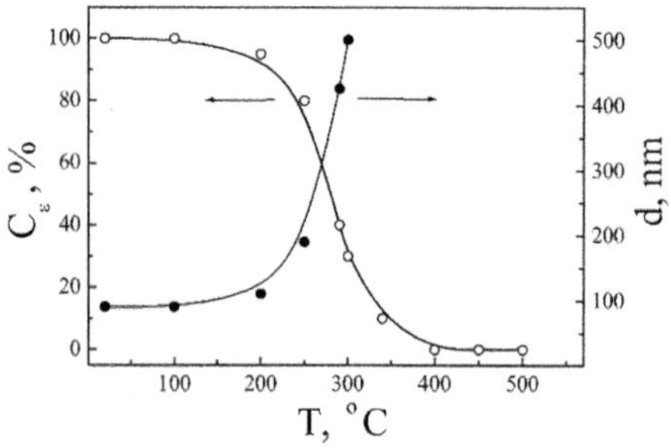

Figure. 8. Influence of tempering temperature on quantity of HCP high-pressure ε-phase for the case of γ-$Fe_{55}Mn_{45}$ alloy. 1 – HCP high-pressure ε-phase concentration $C\varepsilon$; 2 – average size of crystallites (grains) *d*

Conclusion

The obtained result of complex investigation of alloys based of alloys based of Fe-Mn solid solution have shown that their structural-phase state depends on parameters of SPD under pressure, as well as on the initial phase and concentration composition which, in the end, determines the level of mechanical and service properties of this class of materials.

Acknowledgments
Author much appreciate for collaboration: Y.Y. Beygelzimer, O.V. Prokof'yeva, V.P. Pilyugin.

References

1. Bridgman P.V. *Investigation of high plastic deformations and rupture.* Moscow: Foreign Liter. Publ. House), 1955, p. 444.
2. Beresnev B.I., Vereshchagin L.F., Ryabinin Yu.I., and Livshits L.D., *Some problems relating to severe plastic deformation of metals under high pressures.* Moscow: Izd. AN USSR. 1960, p. 58.
3. Beresnev B., Efros B., Streltsov V., *High Pressure Research.* 13 (1995), 281-286.
4. Beygelzimer Y.E., Varyukhin V.N., and Efros B.M., *Physical mechanics of the hydrostatic treatment materials.* Donetsk: Don PTI NAS of Ukraine, 2000, p. 192.
5. Kao A.S., Kuhn H.A., Richmond O., *J. Mater. Res.* 5 (1990), 83-92.
6. Beresnev B.I., Efros B.M., *Physica* 139&140B (1986), 910-915.
7. Beygelzimer Y., Efros B., Varyukhin V., Khokhlov A., *Engin. Fract. Mech.* 48 (1994), 629-640.
8. Beygelzimer Y., *Mechanics of Materials,* (2004) in press
9. Beygelzimer Y., Orlov D., and Varyukhin V., "A new severe plastic deformation method: Twist Extrusion", *Ultrafine Grained Materials II,* ed. By Y.T. Zhu, T.G.Langdon, R.S.Mishra, S.L.Semiatin, M.J.Saran, T.C.Lowe. TMS (The Minerals, Metals & Materials Society), 2002, 297-304.

Towards Comparison of Alternative Methods for Producing Bulk Nanostructured Metals by Severe Plastic Deformation

T.C. Lowe*
Los Alamos National Laboratory, Los Alamos, New Mexico, USA

Abstract

Considerations in comparing alternative methods for producing bulk nanostructured metals are developed. When these are applied to examine current severe plastic deformation techniques, critical needs for future research to aid commercialization become apparent. First, Severe Plastic Deformation (SPD) processing to create ultrafine grain materials must be explored more completely before we can rigorously compare alternative processing methods. Mechanistic and microstructure modeling are particularly critical to the evolution of severe plastic deformation technology and the ability to contrast alternative SPD techniques. Finally, SPD is a single step in a multi-step process to produce nanostructured metals commercially. It is imperative that greater attention be given to processing of nanostructured metals in order to compare alternative techniques from a commercial perspective.

1. Introduction

As the field of severe plastic deformation processing of metals continues to mature, new questions about the readiness of the technology for commercialization emerge. We are presently at a stage in which multiple alternative SPD techniques are available. In principle, the commercial potentials of each technique can be discussed and compared. However, in attempting to do so, one discovers not just the differences between techniques, but more importantly, the limitations of our current state of knowledge of SPD and issues that are common to all SPD processing.

In one of the earliest assessments of the commercial potential of SPD [1], the focus of the review was on equal channel angular pressing, the predominant SPD processing method at the time. Prior to 1999 researchers were exploring the applicability of SPD processing to different classes of metals, though most research on SPD was then, and still is conducted using aluminum and its alloys. By 1999, the collective results of research published in approximately 200 journal papers had established that the method was generally applicable to a wide range of metals. However, difficult to deform alloys and metals substantially strengthened by low-ductility second phases showed relatively less improvement in strength as compared to pure metals or solid solution strengthened alloys.

The body of knowledge on SPD prior to 1999 was dominated by exploratory experimental studies. Much of the literature focused largely on characterization and analysis of the microstructures produced by SPD. There was relatively little

* Corresponding Author. tlowe@lanl.gov

Y.T. Zhu and V. Varyukhin (eds.), Nanostructured Materials by High-Pressure Severe Plastic Deformation, 21-28.

fundamental theoretical work or computational modeling published. This was a significant limitation to the development of the field, since conceptual and computational studies often help frame the key challenges for future studies. Also, there was no commercial enterprise that had attempted to evaluate SPD technology in the context of commercial markets and applications. Whereas the potential number of applications of SPD-processed metals had been noted as being large, the commercial viability of any of these was unexplored.

Today the amount of fundamental theory and computational modeling is substantially greater. There are also now several commercial enterprises that manufacture and market SPD-processed metals. The number of alternative methods for imposing severe deformations has grown so that there are more than a half dozen techniques that are being systematically explored. Examining these alternative methods leads to the three key conclusions of this paper:

1) SPD processing to create ultrafine grain materials must be explored more completely before we can rigorously compare alternative processing methods.
2) Mechanistic microstructure modeling and prediction are critical to the evolution of severe plastic deformation technology and the ability to compare alternative SPD techniques
3) SPD is a single step in a multi-step process to produce nanostructured metals commercially. The SPD research community needs to address post-SPD processing of nanostructured metals.

Each of these conclusions will be developed in the following sections, but only after first establishing a manufacturing context for SPD processing in the next section.

2. Context for manufacturing ultrafine grain metals by SPD processing

The marketing, sales, and distribution of all commercial products depends on an ability to reliably and economically manufacture their constituent materials. Of the many prospective applications of metals whose properties have been enhanced by SPD, the cost of the enhancements must be economically justified for each product. One means to characterize this challenge for SPD processing is to examine the economics of comparable metals processing enterprises. Table 1 below summarizes the financial performance of the steel wire drawing industry in the United States. Steel wire drawing is a reasonable comparison industry to the SPD processing industry since it involves deformation processing of common alloys in a form comparable to SPD products, and for a similarly diverse range of markets and applications.

TABLE 1. Cumulative financial performance measures for the U.S. steel wire drawing industry for the year 1997 [2]

Cost of materials	Value added by	Value of	Capital Investment
$2,923,800,000	$1,996,900,000	$4,897,800,000	$195,400,000

If one assumes that the average financial performance of any single wire drawing facility is similar to the cumulative industry performance, then the value added by wire drawing is 41% of the total value of shipped product. Likewise, 59% of the cost of shipped product is the cost of starting material. Capital investment required is almost 4% of the annual volume of almost $5 billion of steel wire. To the extent that the SPD processing industry would over several years mature to a similar steady state, one can use these ratios as a guide to how SPD processing industry must be capitalized and operated in the long term.

Further insight comes from examining the wire drawing industry in greater detail. Table 2 shows an additional set of financial performance measures, averaged across the U.S. wire drawing industry. In particular, Table 2 considers the costs of labor and the characteristic size of wire drawing establishments.

TABLE 2. Average financial performance measures for the U.S. steel wire drawing industry for the year 1997, as compared with the averages for all U.S. manufacturing [2]

Financial Measure	Average Performance of all U.S. Manufacturing ($)	Average Performance of Steel Wire Drawing Companies ($)
Value added per production worker	$148,769	$447,646
Cost per establishment	$5,437,904	$17,797,753
Cost per employee	$111,874	$340,832
Cost per production worker	$163,846	$540,387
Shipments per establishment	$10,358,230	$32,254,895
Shipments per employee	$224,528	$617,689
Shipments per production worker	$312,098	$979,343
Investment per establishment	$408,343	$676,927
Investment per employee	$8,851	$12,963
Investment per production worker	$12,304	$20,553

The principal point here is that to the extent that SPD manufacturing practice is similar to steel wire drawing, then Table 2 shows the levels of financial performance and reveals the structure of the wire drawing establishments that are typical, and therefore, presumably competitive for the industry. First, note that the average annual sales of a wire drawing establishment is $32 million. This corresponds to an average 52 employees per establishment, 33 of which work directly on production. Thus, the average steel wire drawing establishment is a categorized as a small business (less than 500 employees) in the United States. Also note that the value added per production worker is about three times that of the average for all U.S. manufacturing industry. This implies a significant degree of automation of production. The prospective SPD processing industry will not have the benefit of high degrees of automation in its early stages of development. Also, the number of non-production workers in the SPD processing industry will be significantly higher because of early stage need for significant effort devoted to additional technology development and marketing of the new materials technology.

Within this broad context, one can ask which SPD methods are likely to be most attractive, enabling the emergence of the SPD metals processing industry. The key issues that impact this topic are reviewed in the following sections.

3. Rigorous comparison of alternative SPD processing methods requires greater understanding of SPD processing variables

Well established methods of SPD processing include equal channel angular pressing (ECAP), high pressure torsion (HPT), accumulative roll bonding (ARB), twist extrusion (TE), repetitive corrugation and straightening (RCS), and others. There is no need here to provide an exhaustive list of techniques, since the point of this section is that there is not yet a sufficient conceptual framework underlying the SPD processed materials to rigorously compare any two of these methods.

This point is easily understood by examining a single technique and considering our current knowledge of the processing parameters and impacts of their variation. Examining the most mature SPD technique, ECAP, one can list some of the processing parameters that are already known to be significant. These include: channel angle, channel corner radii, exit channel length/diameter ratio, entrance channel/exit channel diameter ratio, channel cross-section shape, die material, die channel surface finish, die lubricant, sliding die components, punch/die clearance, die preheat temperature, billet temperature, deformation rate/strain rate, homologous temperature ($T/T_{melting}$), stacking fault energy, and initial grain size. The effects of each of these parameters on the microstructures and properties of ECAP-processed material may be known from one or more studies of one or more of these variables. However, these studies provide only a catalog of results that in the aggregate do not provide a comprehensive understanding of the effects of the complete set of processing variables.

Consider the simple example of just one of these variables on processing: the homologous temperature. The microstructures that form during deformation are influenced by the thermally activated mechanisms of deformation and other processes that occur during preheating, deformation processing, and subsequent cooling. SDP processing itself causes microstructural changes that alter the relative importance of sub-processes such as grain boundary diffusion, grain boundary sliding, recovery, and recrystallization. Each of these processes has a distinct temperature dependence that in turn allows variations in the temperature of ECAP to produce dramatically different microstructures. Until the range of variations in microstructures are thoroughly understood for a single SPD technique, it is unreasonable to identify a given set of microstructures or properties with the technique itself. Comparison of two alternative techniques would require comparable levels of insight into the effects of processing variables for both techniques.

One approach for comparing alternative SPD techniques is to process the same starting material by two different methods, keeping as many processing parameters as possible identical. This approach is viable to the extent that we understand the most important parameters governing the results of each technique. For example, Zhilyaev et al.[3] have compared ECAP processing with high pressure torsion (HPT). They generally find HPT can produce significantly finer grain sizes than ECAP for a comparable amount of deformation. Explanations for this difference have been proposed in terms of the differing mechanics of shear deformation during HPT versus ECAP. Yet, only recently has the role of hydrostatic pressure emerged as a variable [4] that perhaps more strongly influences microstructural refinement during SPD. The gigapascal and higher pressures that occur during HPT can diminish the separation of partial dislocations, allowing dislocation motion and dislocation interactions to occur differently than can be achieved via ECAP. If hydrostatic pressure were among the most significant processing variables during

SPD, then one would want to control and make this variable identical in order to compare methods. Thus, for ECAP, one might expect that the effects of the ratio of entrance channel diameter to exit channel diameter during ECAP would be well characterized since this is one of the parameters that impacts the pressure present in the region of high shear at the intersection of channels. Interestingly, this parameter is generally not even reported in most of the literature on ECAP. One can argue that comparisons of alternative techniques of SPD are premature outside of the context of well developed understanding of the mechanisms of substructure formation, grain size refinement, and deformation micromechanics.

When one considers that the combinations of the most significant processing parameters differ for each SPD technique, and that these combinations have not been explored extensively for most techniques, the challenge of comparing alternative techniques becomes clearer. Figure 1 illustrates two-dimensional parameter domains that could be explored for two alternative SPD techniques, ARB and ECAP. Note first that the ranges of processing and material characteristics that have been studied for each technique is limited. For example, in the case of ECAP, most research has been reported for one of only a few channel angles (e.g. 90 and 135 degrees). Yet this parameter likely influences microstructural evolution by altering the processing conditions in a manner the goes beyond its obvious impact on the shear strain intensity. Stacking fault energy is indicated in the figure as a significant companion variable to channel angle since differences in shear strain intensity, in combination with the associated variation in hydrostatic pressure, may correlate with how microstructural evolution may occur differently in materials with different stacking fault energy. Further note, that the processing parameters for alternative methods don't necessarily have self-similar analogs. For instance, the reduction-per-pass variable ARB has no analog for ECAP

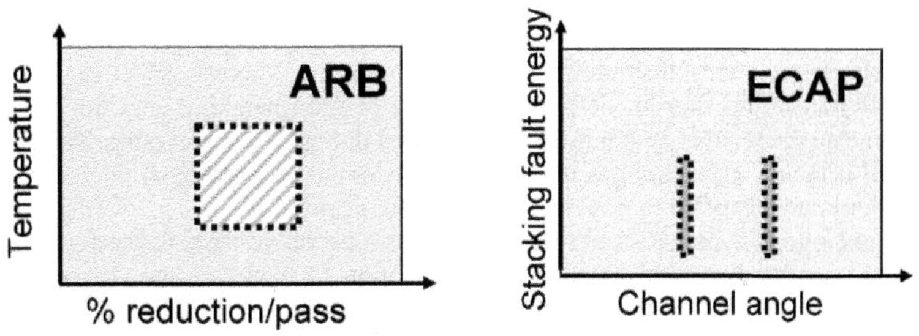

Figure 1. Illustration of two-parameter processing domains for ARB and ECAP

Referencing again the manufacturing context for this work, it is indeed possible to compare the economics of alternative SPD methods. One can calculate and contrast such factors as costs of machinery, tooling, consumable supplies, labor, starting materials and then consider product yield and other factors that determine manufacturing efficiency. For the most part, these factors have not been established, much less optimized for a single SPD technique. Also, the various implementations of SPD processing methods have yet to be optimized for production environments. Until this is done, it is premature to compare the economics of different methods. Thus, we are still capable of offering only the most general conclusions. First and foremost of these is that the degree to which any method can be automated for continuous production will be critical. As we saw in the example of the wire drawing, this industry appears to benefit from a significant degree of automation. Also, since starting material costs are likely to be a large fraction of total production cost, just as in wire drawing, material utilization efficiency will likely be an important factor to optimize. Finally, the stability and repeatability of SPD processing will need to be established. In general, SPD processing tends to place high demands on tooling and the performance of lubricants. The ability to establish and maintain windows of stable operating parameters may be among the more critical factors differentiating alternative SPD processing approaches.

4. Microstructure modeling and prediction are critical to the evolution of severe plastic deformation technology

As in most endeavors in physical sciences, theories and models provide the framework for communicating and exploring our understanding of physical processes. The need for this framework is critical for evaluating SPD processing methods. In the previous section the large number of processing variables was noted. The notion that one or more variables be held constant when comparing alternative SPD techniques was introduced. In this section several variables are proposed that may be of greatest value for being held constant in such comparisons. These are 1) total principal shear, 2) homologous temperature, 3) stacking fault energy, and 4) activation energies (e.g. for self diffusion and creep). Ultimately, the need to hold any of these variables constant can be eliminated by the development and utilization of models that appropriately account for the effects of their variation. Such models are indeed the topic of other contributions in this volume, including Zhu [5], Zehetbauer *et al.* [6], Kim [7].

Briefly, the reasoning underlying the choice of the four variables proposed for consideration is as follows. Principal shear strain is superior to effective strain as a correlating variable since shear deformation is particularly important for introducing the material rotations that promote grain refinement. Whereas effective strain is generally a useful scalar for defining magnitude of deformation, it does not allow us to delineate the relative amounts of shear accrued during SPD processing. The notion of total principal shear implies that it is the predominant shearing strain component that should be identified to correlate deformation magnitude.

Homologous temperature is well accepted as a useful variable for correlating the different temperature dependencies of deformation for different metals, especially when it is clear that thermally activated mechanism play a significant role. Since virtually all SPD processing occurs at elevated temperatures, thermally activated mechanisms will influence modes of dislocation motion, diffusional mass transport, grain boundary sliding, recovery, and recrystallization processes.

Stacking fault energy appears as a significant variable in the models of Zhu [5] and others. It may be one of the most important variables for understanding how the high pressures that are characteristic of SPD processes allows for dislocation mechanisms to operate that would not otherwise be found in conventional deformation processing (e.g rolling, drawing, extrusion, forging).

Finally, the activation energies for diffusional processes help differentiate the relative contributions of temperature and strain rate sensitive mechanisms of deformation. When considered in combination with homologous temperature, the ability to analyze thermally activated processes is more comprehensive. Note that the largest amount of research is still conducted using aluminum and its alloys, often subject to SPD at or near room temperature. Room temperature is approximately 1/3 the melting point for aluminum. Deformation at the same homologous temperature for other commonly studied metals, such as copper or titanium would occur at 162 C and 348 C, respectively.

5. Processing of nanostructured metals

Most research to on SPD materials and processing methods focusing on the SPD process and resulting microstructures and properties. However, any comparison of alternative SPD techniques from a commercial perspective must address the additional deformation processing that will invariably be needed following SPD processing. While SPD produces the ultrafine grain structures that impart the high strength and other desirable properties, other deformation processing such as rolling, drawing, or extrusion will generally be more cost effective for imparting required product dimension. In many cases annealing treatments may also be employed. Thus, thermomechanical deformation processing of SPD-processed metals becomes as important as the SPD processing itself. Unfortunately, there is scarcely any research that addresses post-SPD processing. The differences in post-SPD processing required for different SPD techniques may ultimately be the most significant factor that differentiates their commercial viability. This area clearly demands more attention.

6. Conclusion

SPD processing of metals is still in such an early stage of development that comparison of alternative SPD techniques from a commercial perspective is difficult. The directions for research that will allow alternative SPD methods to be compared have been summarized. First, the effects of variation of key processing parameters needs to be more thoroughly established for any two methods to be compared. Models that explicitly represent the most important variables for any SPD method are needed to facilitate direct comparisons. Finally, since SPD-processed metals will invariably undergo subsequent thermomechanical processing, it is important that these post-SPD processing steps be thoroughly studied. Since SPD processing is only a single step in a multi stage metals fabrication process, it is necessary to understand the interactions between the SPD stage and the other stages before we can establish the relative virtues of alternative techniques for the SPD stage.

7. References

1. Lowe, T.C., Zhu, Y.T. Semiatin, S.L. and Berg, D.L. (2000) "Overview and outlook for materials processed by severe plastic deformation," Investigations and Applications of Severe Plastic Deformation, Edited by T.C. Lowe and R.Z. Valiev, Kluwer Academia Pub. Dordrecht, 347-356.
2. Darnay, A.J. editor (2000) Manufacturing and distribution USA: industry analyses, statistics, and leading companies, Volume 1, Gale Group, Detroit.
3. Zhilyaev, A.P., Cheng, X., Furukawa, M., Horita, Z., and Langdon, T.G. (2004) "Principles of high-pressure torsion and equal-channel angular pressing: a comparison of microstructural characteristics," Ultrafine Grained Materials III, Edited by Y.T. Zhu et al., TMS, Warrendale, PA, 75-80.
4. Zehetbauer, M.J., Stuwe, H.P., Vorhauer, A. Schaffler, E., and Kohout, J. (2003) The role of hydtrostatic pressure in severe plastic deformation, *Advanced Engineering Materials*, **5**, 330-337.
5. Zhu, Y.T. (2005) "Deformation Mechanisms of Nanostructured Materials," this volume

6. Zehetbauer, M.J., Zeipper, L., and Schaffler, E. (2005) "Modelling mechanical properties of SPD materials during and after severe plastic deformation," this volume
7. Kim, S.K. (2005) "Process modeling of equal channel angular pressing for ultrafine grained materials," this volume

Applications of Severe Plastic Deformations for Materials Nanostructuring Aimed at Advanced Properties

R.Z. Valiev*

Institute of Physics of Advanced Materials, Ufa State Aviation Technical University, Ufa, 450000 Russia

Abstract

Recent studies show that extraordinary mechanical properties, i.e. very high strength and ductility, as well as low-temperature and high-strain-rate superplasticity, can be observed in nanostructured metals produced by severe plastic deformation (SPD). However, the appearance of these advanced properties has originated not only from the presence of fine grains, but also from the structure of grain boundaries and internal stresses, as well as disperse precipitations of second phases. The strategy for advancement of mechanical properties in SPD metals is considered in this work with a focus on operating deformation mechanisms. Examples of development of several nanostructured metals and alloys with enhanced properties for new structural and functional applications also are reported and discussed in the present paper.

Introduction

It has been established that nanostructured materials with a grain size about 100 nm or less typically possess very high hardness and strength but exhibit low ductility during tensile tests [1-3]. This type of behavior is often observed in SPD-produced metals and alloys as well [4]. Enhanced strength of SPD materials originated from both strong grain refinement and high dislocation density, but during straining this leads to strain localization due to shear banding and as a result to limited ductility. The reason for this is of fundamental character: the plastic deformation mechanisms associated with generation and movement of dislocations may not be effective in ultrafine grains [3]. In its turn, reduction of ductility provokes a decrease in toughness and fatigue properties, two essential characteristics for development of advanced structural materials.

In this connection, recent findings of extraordinarily high strength and ductility in a number of nanostructured metals produced by severe plastic deformation [4-6] have a special interest. This is conditioned by the fact that SPD-processed materials have a very complex microstructure and can differ not only in ultrafine grain (UFG) size, but also in atomic structure of grain boundaries, crystallographic texture, morphology and nature of the second phases [7,8], depending though on processing routes and regimes. Hence there opens up a path for control of properties of SPD-produced materials by way of varying their structure at the nanoscale level, e.g. by nanostructuring of materials using severe plastic deformation [8].

The present work has two aims. First, it gives consideration to strategies for reaching high strength and ductility in bulk nanostructured metals. Second, using a new approach it demonstrates the possibility of a considerable enhancement of properties in several SPD-produced metals and alloys.

---

* Corresponding Author. RZValiev@mail.rb.ru

Y.T. Zhu and V. Varyukhin (eds.), Nanostructured Materials by High-Pressure Severe Plastic Deformation, 29-37.

Strategy for Enhancement of Strength and Ductility in SPD Nanomaterials

As it was mentioned above, grain refinement typically promotes mechanical strength, and thus one can expect ultrafine-grained materials to possess very high strength. Moreover, introduction of a high density of dislocations in SPD-processed nanometals may result in even greater hardening. However, all this normally decreases ductility, since strength and ductility are typically opposing characteristics. Meanwhile, recent studies have shown that material nanostructuring may lead to a unique combination of exceptionally high strength and ductility, but this task calls for original approaches to fabrication of UFG metals [8].

One such new approach to the problem was suggested recently by Mughrabi et al. [9] and Wang et al. [5]. They created a bimodal nanostructured copper by heavily rolling the metal at low temperature and then heating it to around 450 K. The result was a structure of micrometer-sized grains (at a volume fraction of about 25%) embedded in a matrix of nanocrystalline grains. The material showed extraordinarily high ductility, but also retained its high strength. The reason for this behavior is that, while the nanocrystalline grains provide strength, the embedded large grains stabilize the tensile deformation of the material. What is more, the investigation [9] has shown that bimodal structures can increase ductility not only during tensile tests but also during cyclic deformation. This observation is important for improving fatigue properties.

Another approach suggested recently [10] is based on formation of second-phase particles in the nanostructured metallic matrix, which are to modify shear-band propagation during straining, thereby increasing the ductility. A systematic study of both hard and soft second-phase particles with varying sizes and distributions has been started here, enabling optimization of mechanical properties.

A third approach to the problem of strength and ductility may be the most universal of the three, because it can be applied both for metals and for alloys. This approach is based on formation of ultrafine-grained structures with high-angle and non-equilibrium grain boundaries capable of grain-boundary sliding (GBS) [4,7]. It is well known that sliding, which increases ductility, normally cannot develop at low-angle boundaries. The importance of high-angle grain boundaries was verified in the work [4] on the mechanical behavior of metals subjected to different degrees of severe plastic deformation resulting in formation of various types of grain boundaries. As it will be discussed below, sliding can be easier when non-equilibrium boundaries are present. An indirect evidence of the important role GBS plays in mechanical behavior of SPD metals is observations of an increased strain-rate sensitivity of flow stress that has been reported in several recent works investigating high strength and ductility in nanometals [4,7,11]. High strain-rate sensitivity indicates viscous flow and plays a key role in superplasticity in materials [12], but on the other hand it is associated with the development of grain-boundary sliding, and therefore depends on grain-boundary structure. This fact is in agreement with the recent results of computer simulation of deformation mechanisms active in nanostructured metals. Such molecular dynamics simulations have provided a valuable insight into the deformation behavior of nanometals [13-15], and the results of simulation show that ultrafine grains may also aid in specific deformation mechanisms such as grain-boundary sliding or nucleation of partial dislocations [14-17].

However, there is a question: why should grain-boundary sliding in nanostructured materials produced by SPD, take place at relatively low temperatures? GBS is a diffusion-controlled process and usually occurs at high temperatures. A possible explanation of this interesting phenomenon given in [8] is that diffusion may be faster in SPD-produced ultrafine-grained materials with highly non-equilibrium grain boundaries. Experiments have shown that in SPD-produced metals the diffusion coefficient grows considerably (by two or three orders), and this is associated with non-equilibrium grain boundaries [18,19]. Therefore grain-boundary sliding can be easier in ultrafine-grained metals and develops during straining even at lower temperatures, producing increased ductility. It is well known that enhanced sliding in nanostructured metals can lead even to superplasticity at relatively low temperatures [20].

Processing of nanomaterials to improve both strength and ductility is of primary importance for fatigue strength and fracture toughness. An extraordinary increase in both low-cycle and high-cycle fatigue strength was observed in several recent publications [21,22]. Below some new results will be presented, reporting enhancement of strength and ductility owing to formation of UFG structures in metals and alloys induced by SPD processing.

Development of Nanostructured SPD Materials with Advanced Properties

In this study the materials under investigation were nanostructured pure Ni, commercially pure (CP) Ti and Al-based alloys. The materials were subjected to severe plastic deformation using equal-channel angular pressing (ECAP) and high pressure torsion (HPT).

Nanostructured Ni

In the experiments we used Ni rods of 99.98 wt.% purity and diameter of 20 mm. The rod was subjected to 8 passes of ECAP at room temperature with a 90° channel-intersection angle by route C [23]. The sample after each pass was turned along a longitudinal axis by 180°. Subsequently the material was further treated by cold rolling (CR) with 90% reduction in order to additionally refine the microstructure and obtain a uniform microstructure.

TEM studies showed that this ECAP-processed material had a homogeneous ultrafine-grained structure with a mean grain size of about 400 nm (Fig. 1a). The results of the TEM investigations of the Ni samples subjected to ECAP and cold rolling (90%) are summarised in Table 1. The cold rolling of the ECAP samples led to slightly elongated grains in the direction of rolling. Dislocations inside the grains were found to form walls and cells. The arrangement of dislocation cells and subgrains (which have low angle misorientations) is combined with grain boundaries having high angle misorientations. The important feature of the structure obtained is the absence of extinction contours at most grains boundaries. It was shown by HREM [7] that such a picture corresponds to grain boundaries with a nonequilibrium structure and high defect density. The SAED pattern of the material in this state displays circles of spots having pronounced azimuthal spreading (Fig. 1b), which is typical for an ultra-fine grained material with high level of internal stresses.

In the state of ECAP + CR + annealing, after annealing at 200°C for 1 hour, an growth of the mean grain size up to 450 nm is observed and the number of spots on the diffraction pattern decreases, as shown in Table 1. We conclude that grain growth starts at 200°C.

Figure 1. TEM micrographs of Ni after ECAP (a) and after ECAP followed by cold rolling (b).

Table 1. The average grain size (D), number of diffraction spots at SAED patterns (N), and X-ray data for SPD Ni as a function of treatment.

States of Ni	D, nm	N	FWHM		Microstrains, $x10^{-3}$
			(111)	(220)	
ECAP + CR	330	119	0.1784	0.2328	5
ECAP + CR + annealing	550	72	0.1467	0.1994	3

Table 2 gives the ultimate tensile strength (UTS), yield stress (YS) and elongation to fracture (EL) related to processing. High ultimate tensile stress was observed for nickel after ECAP and CR: 1050 MPa, i.e., more than 3.5 times the UTS of coarse grained Ni (280 MPa) (Table 2). Nickel in this state has limited elongation - 7%. A higher elongation value of 12% was demonstrated by the material after annealing at 200°C. The UTS value for this sample is 890 MPa, which is still 3.2 times higher than for the initial coarse-grained nickel.

Table 2. Tensile properties of Ni processed by SPD at 20°C and strain rate $\varepsilon = 2x10^{-3}s^{-1}$.

States of Ni	UTS, MPa	YS, MPa	El, %	YS*El
ECAP + CR	1050	990	7	6930
ECAP + CR + annealing	890	835	12	10020
Initial material	280	200	40	8000

UTS – ultimate tensile stress, YS – yield stress, EL – elongation to fracture.

SEM investigations of the deformation relief on the surface of the UFG Ni samples, which were well polished before the mechanical tests, revealed after the tensile test at room temperature a network of crossed shear bands oriented at an angle of 35°-45° to the axis of tensile deformation (Fig. 2a). The shear bands are best visible near the fracture region of a sample. At the same time, no appearing slip lines were observed on the surface of the samples.

The shear bands characterise areas of large local deformation (as shown by arrows on Fig. 2a), and grain boundary steps are observed on the borders of the bands. The grain groups are displaced in respect to each other along grain boundaries, and the shear bands have propagated along boundaries that are parallel to the plane of maximal shear stress. The lengths of the shear bands typically extend for about 10 micrometers. The observed specific 'step' relief with no slip lines looks quite similar to a surface relief originating from grain rotation and grain boundary sliding, e.g. as during superplastic deformation [18]. This indicates that these processes take place in this material even at room temperature. It appears that this phenomenon has a co-operative character, since there are no rotations of individual small grains, instead relative sliding of a whole group of grains takes place. Unfortunately, at present, insufficient resolution in SEM did not allow us to reveal the differences between the as-processed and as-annealed states, but these investigations are planned to be conducted in the future.

The investigations of the fracture surface of SPD-produced nickel reveal the formation of a specific dimple relief with extended edges (Fig. 2b), which is typical for ductile materials. The dimple size is large in comparison with a grain size in UFG Ni and the same order with the sizes of localization area between shear bands. The coarse-grained material shows intergranular fracture with traces of dislocation slip at grain surface (Fig. 2c).

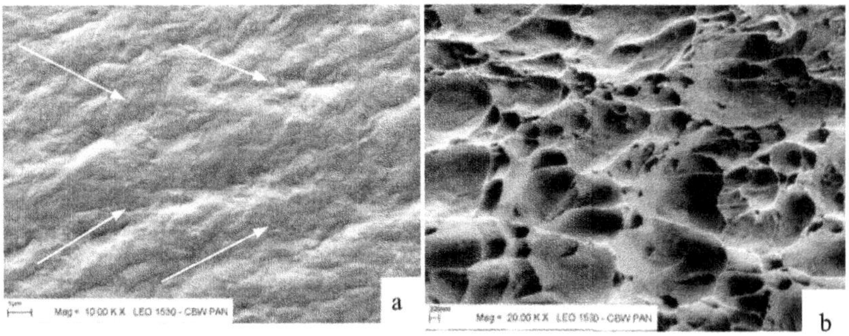

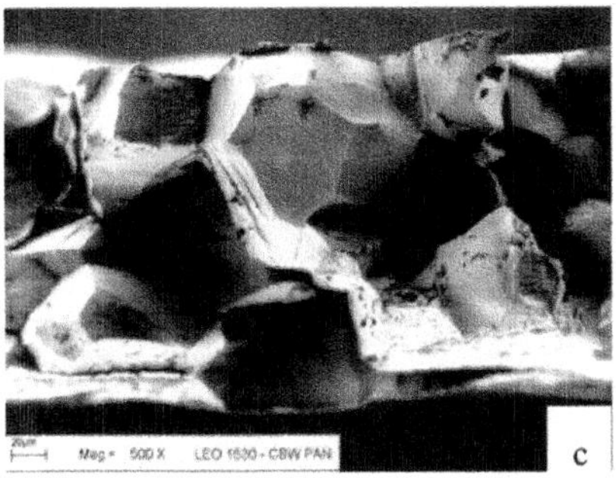

Figure 2. Deformation relief (a) and fracture surface of UFG (b) and coarse-grained nickel (c) after mechanical tests.

Nanostructured CP Ti

Enhanced strength and ductility was observed in CP Ti after HPT and short annealings at temperatures below 300 °C [24]. Tensile tests at room temperature have shown that HPT-processed titanium demonstrates yield stress value higher that 800

MPa, ultimate strength of 980 Mpa, and an elongation to fracture of 12% (Fig. 3). A strain-rate jump test of this material has revealed an increased strain rate sensitivity value of 0.12-0.15, which is visibly higher than that for conventional coarse-grained Ti where this parameter is less than 0.05.

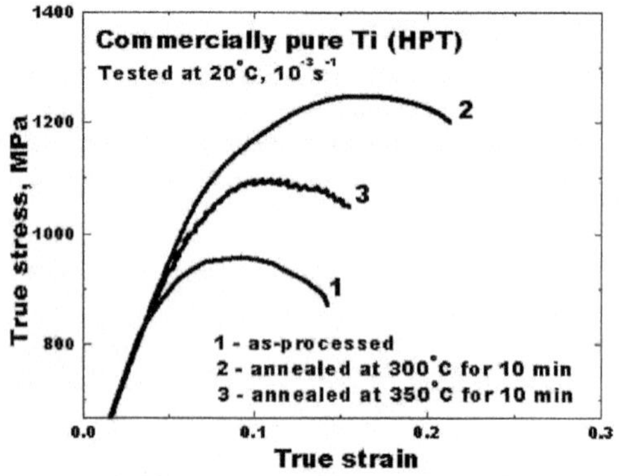

Figure 3. True stress-strain curves of CP Ti after HPT and annealing.

The short annealing at 300 °C results in a 30% increase in strength combined with enhanced ductility (Fig. 3, curve 2) as compared to the as-produced state (Fig. 3, curve 1) or with higher temperatures annealing. The material showed the peak in strength and ductility after annealing at this temperature range.

TEM studies of HPT-processed CP Ti have shown formation of a homogeneous ultrafine-grained structure with a mean grain size of about 120 nm (Fig. 4a and b). It is typical for this structure that the grain boundaries are not well-defined and there is no banded contrast at the boundaries which is characteristic of well-formed grain boundaries as observed by TEM. The HREM images of grain boundaries in the processed material are usually wavy, curved, or corrugated (Fig. 4c) indicating their highly non-equilibrium character [7,25]. Moreover, we observed a variation in the misorientation angle along the same boundary by about 5 °, probably as a result of the presence of disclinations. A high level of defects was also observed in grain interiors. In many regions of grain interior the lattice fringes are poorly defined (area marked by square in Fig. 4c) suggesting local distortions of the crystal lattice.

The short annealing for 10 min at temperatures up to 300 °C resulted in a significant decrease in the lattice distortions without visible grain growth during annealing. This is confirmed by the SAED pattern where the level of asterism or streaking of diffraction spots is decreased. Investigations at higher resolution showed also that annealing at 250-300 °C revealed a rearrangement of defects resulting in their movement from grain interiors to the near grain boundary region [24]. However, as with any ordering phenomenon, this process lowers the internal stresses in the structure as a system that was corroborated by analysis of XRD patterns which exhibited peak sharpening after this heat treatment. The results demonstrate that although during low-temperature annealing the total density of dislocations is decreased, the local density of dislocations at grain boundaries can grow thus increasing their non-equilibrium. These changes in the structure have correlated with the results of SEM studies of deformation mechanisms operating during tension of UFG Ti.

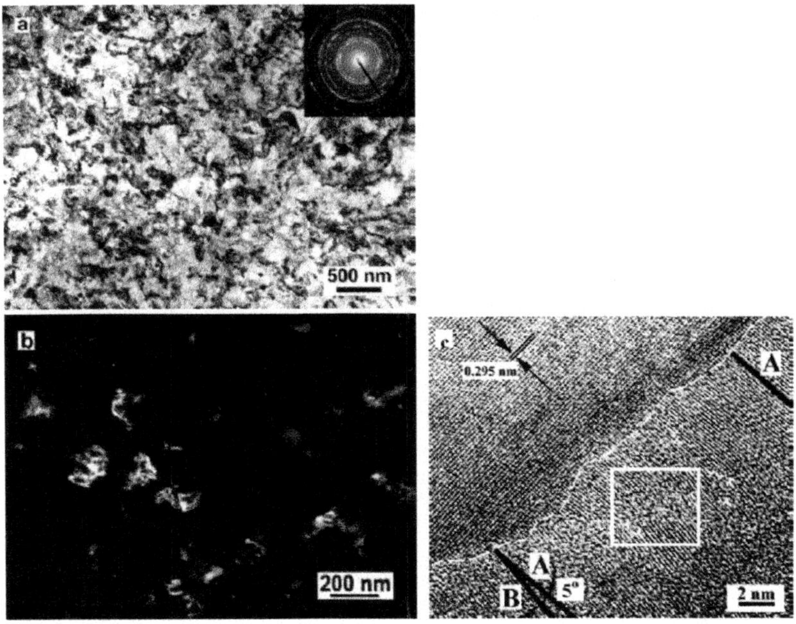

Figure 4. Typical TEM bright field (with SAED as an insert) (a) and dark field (b) micrographs and HREM image (c) of as-processed CP Ti.

These observations showed [24] that fracture surface of the HPT-processed Ti tested at room temperature has the dimpled surface indicating viscous grain boundary processes such as sliding. The SEM analysis of the formation of the surface relief of the samples' gages after testing has also revealed a specific 'step' relief with no slip lines. It was quite similar to a surface relief originating from grain rotation and grain boundary sliding, as is well known from superplasticity studies [18]. This indicates that these processes take place in this material, as in UFG Ni, already at room temperature.

The first semi-commercial products recently developed at our laboratory, i.e. long-sized rods of nanostructured Grade 4 CP Ti (Fig. 5) also demonstrate extraordinary strength and ductility. The values are as follows: UTS = 1150 MPa, YS = 1040 MPa, El = 18%. To produce the rods we used a new combined SPD technology leading to grain refinement in CP Ti down to 100 nm. A thorough characterization of this new product is in progress at present.

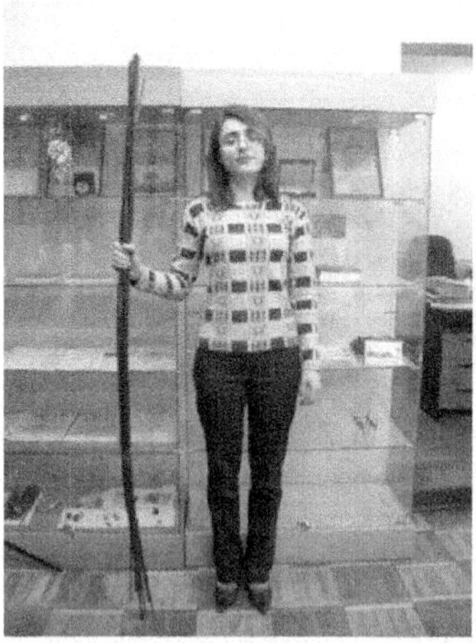

Figure 5. Semi-products of nanostructured Grade 4 CP Ti.

Nanostructured Al alloys

Recent developments in SPD processing under elevated pressure have allowed us to process several Al alloys in the quenched state [26,27], normally characterized by low workability. After subsequent ageing these alloys demonstrated a record high strength with UTS ≈ 750-800 MPa. Here formation of a UFG structure with an ability for grain boundary sliding and the presence of disperse particles make it possible to achieve also very high ductility (with elongation to failure up to 25%) [27]. This paves a way for engineering of nanostructured Al alloys for advanced structural applications.

Summary and Conclusions

To date we see more and more examples demonstrating that the use of SPD enables fabrication of bulk nanostructured materials with exceptionally high strength and ductility. This is also true for commercially pure metals and alloys. One of the approaches bringing us closer to achievement of such unique properties is considered in the present work. This approach is formation of a UFG microstructure with high-angle and non-equilibrium grain boundaries with an ability for grain boundary sliding. The given paper illustrates several new examples of this approach. Development of grain boundary sliding at relatively low temperatures is an extraordinary phenomenon, but probably this is associated with the features of grain boundary structure and diffusion in UFG materials produced by SPD. At the same time, it becomes clear that advanced properties in nanostructured metals and alloys are a realizable goal which can be attained through their nanostructuring using SPD techniques.

Acknowledgements

This work was partly supported by LANL under the NIS-IPP project T-0197. Cooperation with Prof. A.K. Mukherjee at UC Davis, USA, is also gratefully acknowledged.

References

1. Weertman J.R., *Mater. Sci. Eng. A* 166 (1993), 161-167.
2. Shen T.D. and Koch C.C., *Acta Mater.* 44 (1996), 753-761.
3. Koch C.C., Morris D.G., Lu K., Inoue A., *MRS Bull.* 24 (1999), 54-58.
4. Valiev R.Z., Alexandrov I.V., Zhu Y.T., Lowe T.C., *J. Mater. Res.* 17 (2002), 5-8.
5. Wang Y., Chen M., Zhou F., Ma E., *Nature* 419 (2002), 912-915.
6. Valiev R.Z. "New Observations of High Strength and Ductility in SPD-Produced Nanostructured Materials," *Ultrafine-Grained Materials III*, ed. Y.T. Zhu, T.G. Landgdon, R.Z. Valiev, S.L. Semiatin, D.H. Shin, T.C. Lowe. Warrendale, PA: TMS (The Minerals, Metals and Materials Society), 2004, 497-502.
7. Valiev R.Z., Islamgaliev R.K., Alexandrov I.V., *Prog. Mater. Sci.* 45 (2000), 103-189.
8. Valiev R.Z., *Nature Mater.* 3 (2004), 511-516.
9. Mughrabi H., Höppel H. W., Kautz M. and Valiev R. Z., *Z. Metallkunde* 94 (2003), 1079–1083.
10. Zhang X. , Wang H., Scattergood R.O., Narayan J., Koch C.C., Sergueeva A.V., Mukherjee A.K., *Acta Mater.* 50 (2002), 4823–4830.
11. Valiev R Z. *et al., Acta Metall. Mater.* 42 (1994), 2467–2475.
12. Nie T.G., Wadsworth J. and Sherby O.D., *Superplasticity in Metals and Ceramics.* Cambridge: Cambridge Univ. Press, 1997.
13. Van Swygenhoven H., *Science* 296 (2002), 66–67.
14. Yamakov V., Wolf D., Phillpot S.R., Mukherjee A.K. and Gleiter H., *Nature Mater.* 1 (2002), 45–49.
15. Schiøtz J. and Jacobsen K.W., *Science* 301(2003), 1357–1359.
16. Budrovic Z., Van Swygenhoven H., Derlet P.M. ,Van Petegem P. and Schmitt B., *Science* 304 (2004), 273–276.
17. Van Swygenhoven H., Derlet P.M. and Frøseth A.G., *Nature Mater.* 3 (2004), 399–403.
18. Zelin M.G., Krasilnikov N.A., Valiev R.Z., Grabski M.W., Yang H.S., Mukherjee A.K., *Acta Metall. Mater.* 42 (1994), 119-126.
19. Würschum R., Herth S. and Brossmann U., *Adv. Eng. Mater.* 5 (2003), 365–372.
20. McFadden S.X., Mishra R.S., Valiev R.Z., Zhilyaev A.P. and Mukherjee A.K. *Nature* 398 (1999), 684–686.
21. Höppel H.W., Zhou Z.M., Mughrabi H. and Valiev R.Z., *Phil. Mag. A* 82 (2002), 1781–1794.
22. Vinogradov A. and Hashimoto S., *Adv. Eng. Mater.* 5 (2003), 351–358.
23. Krasilnikov N., Lojkowski W., Pakiela Z., Valiev R., *Mater. Sci. Eng. A*, submitted.
24. Valiev R.Z., Sergueeva A.V., Mukherjee A.K., *Scripta Mater.* 49 (2003), 669-674.
25. Horita Z., Smith D.J., Nemoto M., Valiev R.Z., Langdon T.G., *J. Mater. Res.* 13 (1998), 446-450.
26. Zhao Y.H., Liao X.Z., Jin Z., Valiev R.Z., Zhu Y.T., *Acta Mater.* 52 (2004), 4589-4599.
27. Islamgaliev R.K., Yunusova N.F., Sabirov I.N., Sergueeva A.V., Valiev R.Z., *Mater. Sci. Eng. A* 319-321 (2001), 874-878.

New Applications of the SPD Concept: μSPD

Y. Estrin[1*], E. Rabkin[2], R.J. Hellmig[1], M. Kazakevich[2] and A. Zi[1]

[1]Institute of Materials Science and Technology, Clausthal University of Technology, Clausthal-Zellerfeld, Germany

[2]Department of Materials Engineering, Technion, Haifa, Israel

Abstract

Extreme grain refinement by severe plastic deformation (SPD) is an established processing approach broadly applied to bulk materials. The dimensions of specimens or workpieces used are typically in the centimeter range. We propose to adopt an analogue of equal channel angular pressing in the context of microforming. It is suggested that using sub-millimeter sized channels ("angular vias") in a process similar to conventional forcefill technique, fibres or wires with ultrafine grain size can be manufactured. The cross-sectional dimensions of the channels can be in the micrometer range. An alternative technique that may lead to ultrafine grained fibres or wires is pressing of metals through porous "filters" with micrometer or submicrometer scale open porosity. First results demonstrate the viability of this approach which we refer to as μSPD.

Introduction

Equal channel angular pressing (ECAP) is a commonly accepted technique for producing an ultrafine grain structure in bulk materials by severe plastic deformation [1]. The major challenges in the area of severe plastic deformation are to scale-up the process for industrial applications and/or to achieve a continuous process. Typical methods developed for this purpose are accumulated roll bonding (ARB) [2], continuous confined strip shearing (C2S2) [3] and the recently developed ECAP-Conform [4].

To manufacture small wires or fibres having diameters in the submicrometer range, the aforementioned methods are not suitable, however. We suggest a new method based on the forcefill technology [5] in combination with a porous medium used as a severe plastic deformation device. Forcefill is a process used in microelectronics to interconnect layers of a thick film circuit or conductors on different sides of an insulating film through so called "vias". The vias are small channels of micrometer- or submicrometer size, which are filled with a metal, for example aluminium, as conducting material. The metal is forced into a via, while in solid state, by a high pressure. Using this method, very fine high quality connections through straight via channels can be realized.

Our idea is that the concepts of severe plastic deformation can be introduced by providing channels with some angularity, as in the ECAP process, while adopting the forcefill technology of filling the vias. Material passing through angular channels with diameter in the micrometer or submicrometer range will be severely deformed by shear, which may lead to grain refinement in the way it happens under conventional ECAP. For channels with multiple kinks, the process would be similar to multi-pass ECAP. It is admittedly hardly possible to manufacture a multi-pass

* Corresponding Author. juri.estrin@tu-clausthal.de

Y.T. Zhu and V. Varyukhin (eds.), Nanostructured Materials by High-Pressure Severe Plastic Deformation, 39-46.

ECAP die with very small channel dimensions. An alternative solution we suggest is to use a porous material with open porosity as a tool through which a metal is pressed by forcefill. Tortuous paths that will be taken by the material plastically flowing through a system of pores would be akin to what it would experience under multi-pass ECAP. The randomness of the paths and the angles encountered introducing natural averaging would provide a high degree of reproducibility of properties of the fibers/wires leaving such a "multi-channel die".

Experimental Procedures

To realize this idea, we used porous steel disks (Mott Corp., Farmington, Conn., USA) of various grades as die material. These steel disks produced by sintering have a steel skeleton with open porosity. Therefore, coherent pores can form a random channel through the whole disk serving as a "die". Obviously, there can be many channels leading through a porous steel disk. A schematic picture of the experimental set-up and the process proposed are presented in Fig. 1.

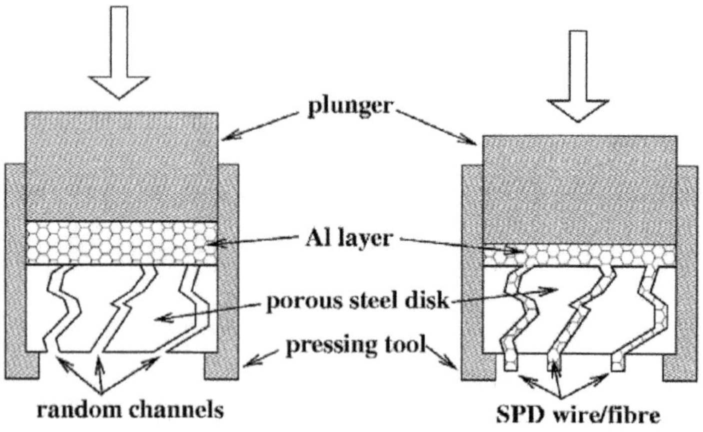

Figure 1. Schematics of the process: a plunger exerts a pressure on an Al layer forcing the material to flow through a porous steel disk (left); material flowing through random channels to exit the disk on the opposite side (right).

The porous steel dies used had a cylinder geometry with one inch diameter and a height of about 3 mm. Dies with various average nominal pore sizes from 0.3 μm to 100 μm were used. The cylindrical tool set used, Fig. 2, was manufactured from stainless steel.

Figure 2. Experimental set-up: tool set

The tool set with an aluminium layer sitting on the top of a porous steel disk was placed in a universal testing machine fitted with a furnace. A constant load of 80 kN was applied through the plunger for 24 hours at a processing temperature of 450°C. These conditions were suggested by simulations of the forcefill of Al [5] and were not optimised otherwise.

After removing the specimens from the fixtures they were broken at liquid nitrogen temperature to investigate the fracture surface aligned with the load direction. Light microscopy and scanning electron microscopy (SEM) were used. Energy dispersive X-ray analysis (EDX) was employed to identify aluminium penetrated into the porous steel disks. Investigating the fracture surfaces obtained in this way has the advantage of not having to deal with the problem of material being smeared across the surface due to cutting and polishing.

For comparison, specimens were also prepared by sawing (3000 rpm) and light polishing using a 6 μm polishing paper for 5 to 15 s. These specimens were investigated using high resolution SEM and energy dispersive spectroscopy (EDS) for detection of aluminium which penetrated larger pores at a distance from the Al-steel interface.

Results and Discussion

To assess the results of this process, a few steel disks "infiltrated" with aluminium by forcefill were fractured at liquid nitrogen temperature and analysed using SEM. Figure 3 shows the structure of the specimens with a nominal pore size of 5 μm (left) and 100 μm (right).

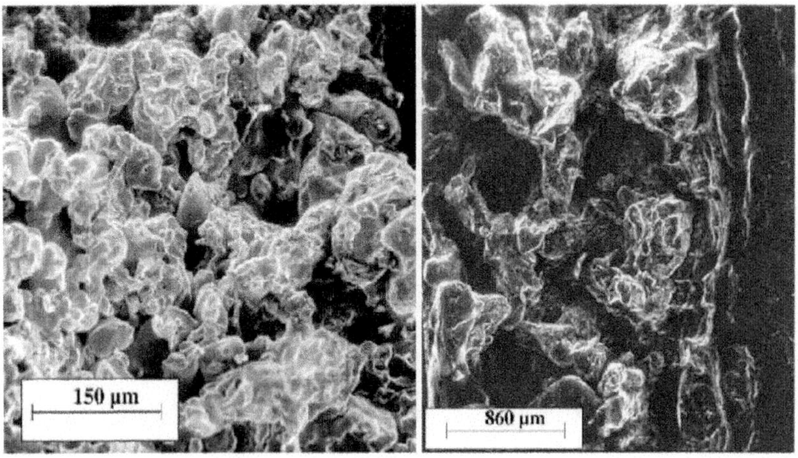

Figure 3. SEM micrographs of the fracture surfaces of the 5 µm grade (left) and 100 µm grade (right) porous steel disks.

As can be seen on the micrographs, the porous structure inside the steel disks does possess a system of really random channels that may allow percolation of aluminium by severe plastic deformation. The larger pores seem to exceed significantly the nominal pore size as given by the porous steel specification (grade). Also, it appears rather unlikely for smaller grades that there exist many channels leading through the steel disks to allow percolation of aluminium.

Figure 4 shows a typical picture of a broken specimen containing the Al layer (bottom part) as well as the porous steel disk (top part).

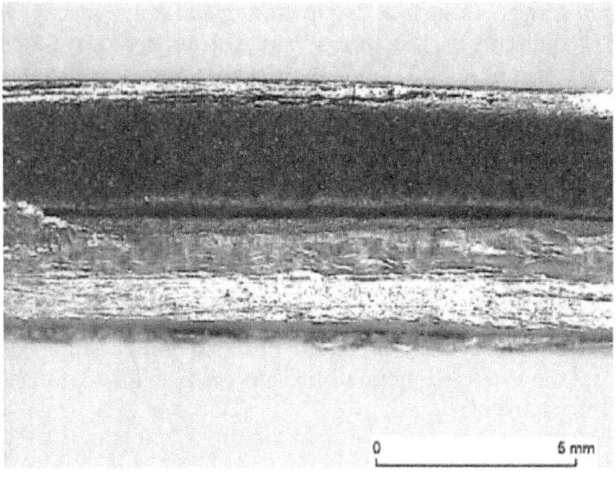

Figure 4. Overview picture of a fracture surface across a 2 µm grade steel disk and an aluminium layer.

The top of the steel disk is deformed because a notch for easier breaking was sawn into the specimen. Near the Al-steel interface a layer of aluminium that has penetrated into the steel disk is visible.

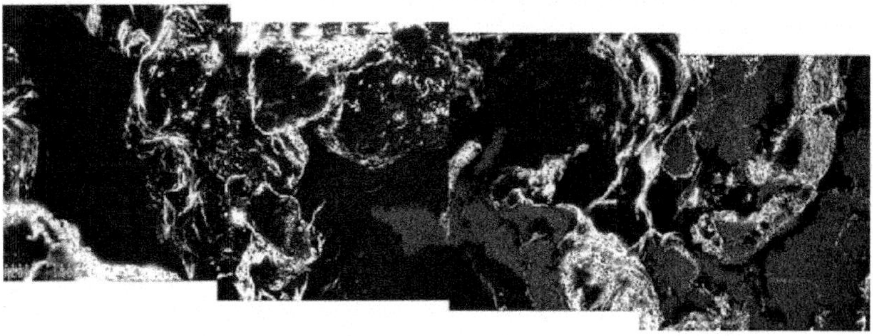

Figure 5. SEM micrograph of a fracture surface showing penetration of Al (red) into a porous steel disk (100 μm grade).

In Figure 5, aluminium that has penetrated into the 100 μm grade steel disk is visualised. The SEM micrograph demonstrates that aluminium does find its way through the disk using a random path. Some pores get filled as material flows through random channels. In this micrograph the path taken by the aluminium ends before it can reach the other side of the porous steel die. To demonstrate that aluminium does pass through the dies, high resolution SEM micrographs were taken on the polished surfaces.

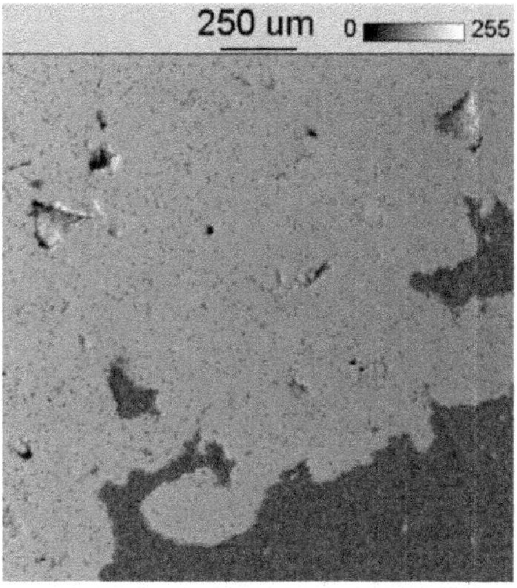

Figure 6. High resolution SEM micrograph of a polished cross-sectional cut through the specimen (100 μm grade). Aluminium detected by EDS appears in red.

Figure 6 shows a high resolution SEM micrograph on a polished surface of the porous steel disk. Many small pores vanished due to polishing, but the larger pores are still clearly visible. The aluminium/steel interface is seen in the bottom right corner. The picture is an unambiguous proof that aluminium has intruded into the porous steel die. As there are isolated pores visibly filled with aluminium, material must have flown along a kinked, angular path to reach these pores. Furthermore, near the bottom left corner an almost 90° angle of material flow can be observed. This evidence strongly suggests that severe shear deformation seen as a pre-requisite for grain refinement by SPD was involved in the process. It should be mentioned that as the EDS detector was located under an acute angle relative to the specimen, signals from the bottom of cavities were absorbed by the cavity walls. Therefore, aluminium present at the bottom of some pores may have not been necessarily detected.

Figure 7 (left) shows another high resolution SEM micrograph, this time from a position near the rear side of the specimen, far away from the aluminium-steel interface. Also here filled pores can be clearly detected indicating that there may be a channel through the 100 μm grade specimen. In the top left part of this micrograph the rear specimen surface can be seen. To investigate if material did pass through the specimen, the rear surface was investigated as well, which is shown in Figure 7 (right). It is rcognised that aluminium did pass through the whole porous disk exiting the "die" on the rear side. Several exit channels are visible in the small section presented in this high resolution SEM micrograph. It is remarkable that it was not necessarily the large pores at the rear side that provided the outlets of the deformed aluminium.

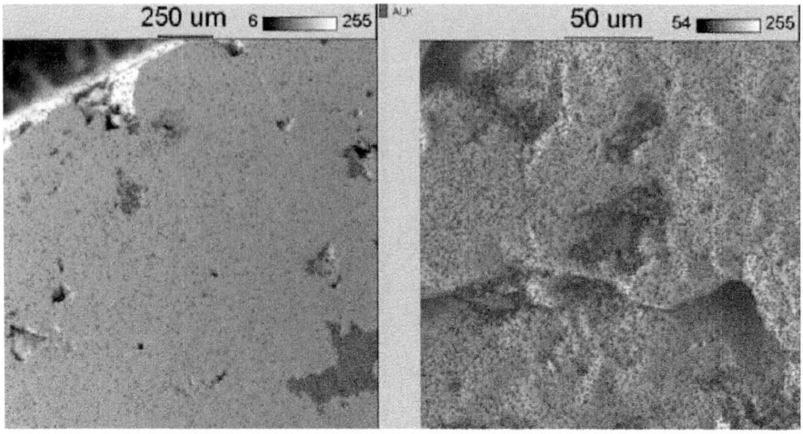

Figure 7. High resolution SEM micrograph: a cut made near the rear surface (left) and the rear surface itself (right) showing aluminium detected by EDS (appearing in red).

Table 1 summarizes the results obtained with porous steel disks of various grades. The average penetration depth is seen to increase with increasing pore size.

Table 1. Penetration depth versus grade of porous steel disks

grade [μm]	penetration depth [μm]
0.2	≈ 30
0.5	≈ 50
2.0	≈ 200
5.0	≈ 200
10.0	≈ 350
100	passing through

It may be conjectured that in the specimens with the smaller grades no channels leading across the entire disk are available, which strongly limits the penetration depth. In addition, a very high pressure is required to fill smaller pores.

So far, no systematic studies of the mechanical properties of the deformed material filling the channels or exiting from the die have been carried out. This is a subject of further investigation. As the temperatures for this first, rather preliminary, experiment were quite high, no optimum processing conditions are expected to have prevailed in this case. The high process temperature was chosen to be in the range of typical forcefill temperatures, just to see whether the process is possible at all. As the feasibility of pressing of aluminium through a 100 μm grade porous steel disk has now been established, a reduction in process temperature will be a next step to make. It is expected that processing in the temperature range of 250°C to 300°C, with appropriately chosen pressures, may lead to a pronounced grain refinement and a change in mechanical properties, as obtained by conventional ECAP technique.

An alternative to porous steel disks will be a die with a system of equal angular channels proper, yet with channel diameter in the sub millimeter range. A schematic view of this kind of set-up, with only two channels to illustrate the principle, can be seen in Fig. 8. This kind of set-up that will be tried out next ensures more controlled processing conditions than porous steel disks with their random channel geometry. However, due to the fact that there are many channels present in the porous steel disks, reproducibility in properties can be assumed from the statistical point of view.

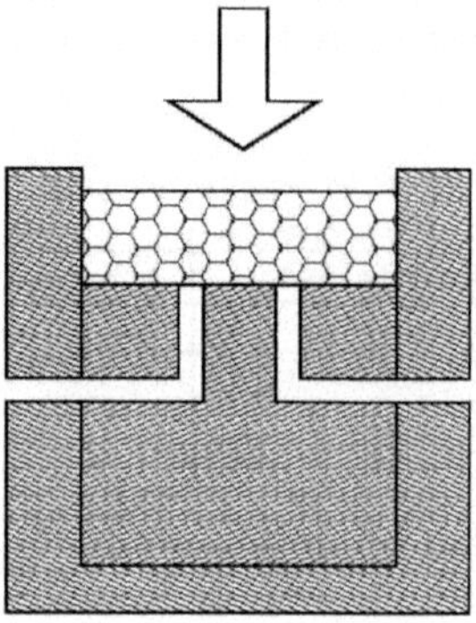

Figure 8. Schematic view of a μSPD tool using a ceramic die with several 90° angular channels for producing wires in a process similar to forcefill.

Conclusion

A new method to produce ultrafine grained wires and fibres has been proposed. It was experimentally demonstrated that this approach combining the forcefill technique with pressing through angular channels is possible in principle. Further experiments are necessary to optimise processing conditions. Two different kinds of possible "dies" are suggested, *viz.* porous steel or ceramic disks for a random-channel pressing and a multiple channel ECAP arrangement with a small channel diameter to be used in conjunction with the forcefill process.

Acknowledgements

The authors gratefully acknowledge the donation of porous steel disks by Mr. W. Krause of Mott Corp., Farmington, Conn., USA.

References

1. Valiev R.Z., Islamgaliev R.K., Alexandrov I.V., *Progress Mater. Sci.* 45 (2000) 103-189.
2. Tsuji, N., Saito Y., Utsunomiya H., Tanigawa S., *Scr. Mater.* 40 (1999) 785-800.
3. Lee J.-C., Seok H.-K., Suh J.-Y., *Acta Mater.* 50 (2002) 4005-4019.
4. Raab G.J., Valiev R.Z., Lowe T.C., Zhu Y.T., *Mat. Sci. Eng.* A382 (2004) 30-34.
5. Estrin Y., Kim H.S., Kovler M., Berler G., Shaviv R., Rabkin E., *J. Appl. Phys.* 93 (2003) 5812-5815.

Mechanisms of Submicron Grain Formation in Titanium and Two-Phase Titanium Alloy during Large Strain Warm Working

G.A. Salishchev*, S.Yu. Mironov, S.V. Zherebtsov
Institute for Metals Superplasticity Problems of Russian Academy of Sciences, Ufa, 450001Russia

Abstract

The mechanisms of grain refinement to submicron scale during warm "abc" deformation were studied in commercial pure titanium and two-phase Ti-64 titanium alloy. Large strains were obtained via successive compression of sample along three orthogonal directions. Grain refinement in titanium occurred due to grain subdivision mechanism because of formation of deformation induced high angle boundaries created by slip and twinning that led to strengthening. The steady state flow, which was observed at the last stages of deformation, was due to accommodation of microstructure arisen from the change in the strain path. In two-phase alloy during straining the initial lamellar structure transformed into the submicron globular one. Unlike titanium, there occurred flow stress softening and superplasticity was observed.

Introduction

The mechanisms of formation of the SMC structure during severe plastic deformation (SPD) were extensively studied [1-6]. In the single-phase materials new grains are being formed as a result of development of deformation-induced boundaries (DIBs) [1,2]. They can be divided into two groups depending on their origin, namely dislocation and twinned origin [1,2,4]. The latter is typical for hcp metals such as titanium. For dislocation boundaries the misorientation angle increases gradually while during twinning the high-angle boundary forms at once and then develops into a random boundary in the course of deformation [4]. It is supposed that considerable increase in the fraction of high-angle boundaries in spite of hardening on the final stage of deformation results in initiation of grain boundary sliding and thus steady state flow. In the two-phase materials with the fraction of the second phase more than 20%, such as titanium alloys, the mechanism of the SMC structure formation is not clear yet. Microstructure development in these alloys takes place in each phase independently [7]. In the two-phase titanium alloys with initial lamellar structure the grain refinement at high temperatures occurs due to globularization [8]. But for formation of the SMC structure subsequent deformation at lower temperatures is required. However it is not clear what mechanisms are responsible for formation of new grains and whether the mechanical behavior of these alloys is similar to the single-phase materials.
The goal of the present study was to perform comparative analysis of the microstructure evolution, the mechanisms of formation of new grains and the features of mechanical behavior of titanium and the two-phase titanium alloy Ti-6Al-4V during SPD. SPD was applied by successive compression of a sample along three orthogonal directions (so-called "abc" deformation).

* Corresponding Author. gensal@imsp.da.ru

Y.T. Zhu and V. Varyukhin (eds.), Nanostructured Materials by High-Pressure Severe Plastic Deformation, 47-54.

Experimental Procedures

The commercial pure titanium (impurities content, wt %: 0.25 Al, 0.15 Fe, 0.07 Si, 0.05 C, 0.12 O, 0.02 N, 0.005 H) with a mean grain size of 35 μm and the alpha/beta titanium alloy Ti-6Al-4V (wt %: 6.3 Al, 4.1 V, 0.18 Fe, 0.03 Si, 0.02 Zr, 0.01 C, 0.18 O, 0.01 N, 0.002 H) after beta-annealing and water-quenching with a mean β-grain size of 250 μm were used to establish methods for producing SMC structure. The Ti-64 alloy had a beta-transus temperature of 995°C.
Successive compression along three orthogonal directions was carried out using samples with a prismatic shape (Fig. 1a). The initial size of samples were 16×18×20 mm^3 in both cases. The deformation temperatures of 400°C for Ti and 550°C for Ti-6Al-4V corresponded to the conditions of SMC structure formation, which were selected from early experiments [9]. Prior to each rotation, a prismatic shape was restored to the sample by machining the curved faces. A total of 12 and 7 increments of deformation for Ti and alloy respectively were imposed by this means. The initial strain rate and the true strain per deformation step were $10^{-3}s^{-1}$ and ~0.4, respectively. Compression tests were performed in air using a SCHENK machine. The surface relief occurring on the surface of samples at different steps of deformation was studied after additional compression of the samples by $\varepsilon = 15\%$.

The microstructure was studied using a light microscope, transmission electron microscope (TEM) JEOL JEM-2000EX and scanning electron microscope (SEM) JEOL JSM-840. Crystallographic analysis of structure parameters was made by EBSD-technique on a SEM LEO-440 Stereoscan. The textures were determined by measuring incomplete pole figures (002), (100), (101) and inverse pole figures by means of the X-ray diffraction goniometer "DRON".

Results and Discussion

Microstructure Evolution and Mechanical Behavior of Ti during "abc" Deformation.

Stress versus cumulative strain curve (S-Σe) for Ti was plotted under "abc" deformation at temperature of 400°C and strain rate of $10^{-3}s^{-1}$ (Fig. 1b). One can see a non-monotonic dependence of yield strength on a number of compressions. During the first three steps one could observe a sharp increase in its value, but then yield strength weakly depended on cumulative strain.

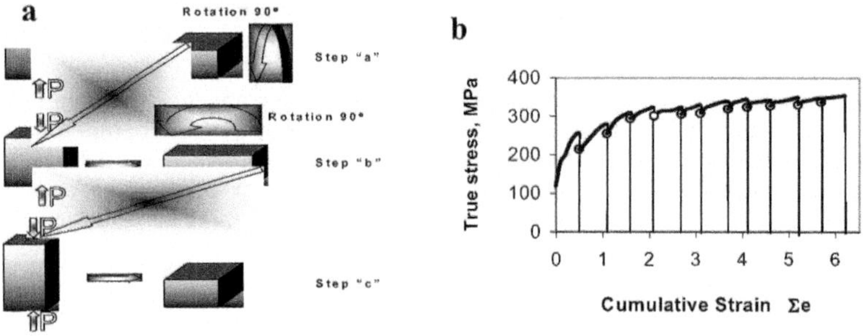

Figure 1. Scheme of "abc" deformation (a) and cumulative S-Σe curve (b) for "abc" deformation of Ti at temperature of 400°C and strain rate of $10^{-3}s^{-1}$.

Investigation of the deformation relief occurring on the surface of samples additionally subjected to 15% strain after different compression steps displayed the localization of strain in macro shear bands (MSB) (Fig. 2). Its appearance was a result of "scattering" of some dislocation boundaries into individual dislocations (Baushinger effect) and formation of "avalanche" of mobile dislocations [9]. In this case the yield stress was controlled by substructure that, probably, led to weak dependence of yield stress on strain. Macrobands, which were formed at the first step of deformation, remain during all steps. So, plastic flow of titanium is localized at the macroscopic scale.

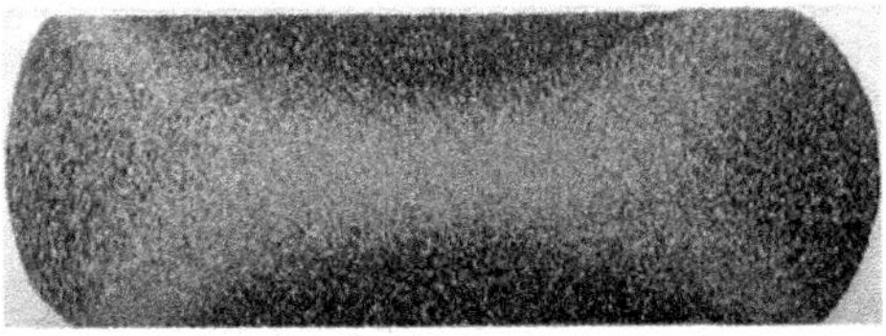

Figure 2. Macro shear localization bands in Ti deformed at temperature of 400°C and strain rate of $10^{-3}s^{-1}$ to height reduction of 40%.

Investigation of texture showed that each change in the strain path resulted in formation of texture being qualitatively similar to the previous one and the prevailing orientation of its texture remained stable. This meant that after changing the strain path the structure rearranged and adapted to new conditions of deformation.

Study of deformation relief in all stages of plastic flow revealed parallel microshear bands. However, if, at small strains, these bands were arranged within the grains and in different grains their direction was different, then, with increasing the number of steps above 3 they became rectilinear (in spite of local bends) and their length was about few hundred micrometers. The initial polyhedral structure, which was quite dominant at small strain, was not revealed after 3 compression steps. At strain above 5 steps two systems of parallel bands intersected at an angle of ~ 90° were revealed in the structure (Fig. 3a). The angle between deformation axis and the bands was approximately 45°. In sites of mutual intersection the local bends, shears and even gaps at boundaries were observed. The final structure after 12 steps was characterized by a significant smoothing of the relief, namely, "smoother" and shorter bands were observed (Fig. 3b).

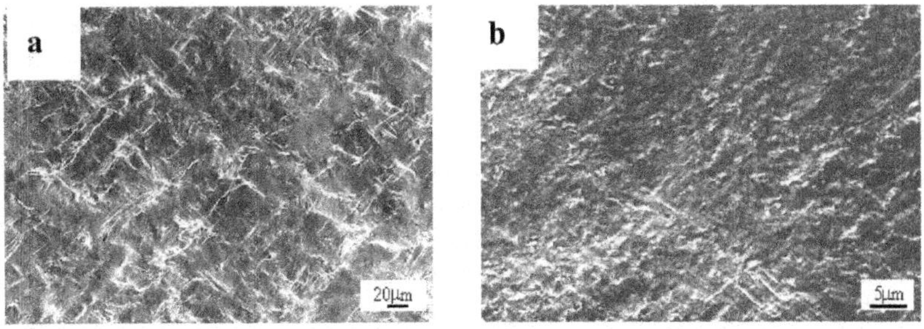

Figure 3. SEM images of surface relief on the titanium sample during different steps of compression strain at 400°C and strain rate of 10^{-3} s^{-1}: a - 5 steps + ε = 15%; b –12 steps + ε = 15%. The deformation axes were vertical in the both cases.

The results of TEM and EBSD analysis showed that at all stages the deformation occurred by slip and twining. The main feature of structure inside the MSB was intense formation of DIB. The evolution of these boundaries transformed the structure, at first, to the lamellar-type structure (Fig. 4a) and, finally, to the equiaxed-type one (Fig. 4b). Simultaneously, DIB development refined the microstructure very strongly, so, after 12 compression steps a mean grain size was about 0.3 μm. The formation of MSB caused significant macroscopic heterogeneity of the structure, which became negligible with increasing the number of compression steps and spreading of new bands.

The TEM analysis indicated that the volume fraction of low angle grain boundaries decreased with strain. The study of conjugated crystallites misorientation via EBSD-scanning (Table 1) confirmed the decrease in the volume fraction of low angle grain boundaries with strain and this decrease was extremely pronounced at the 5th step.

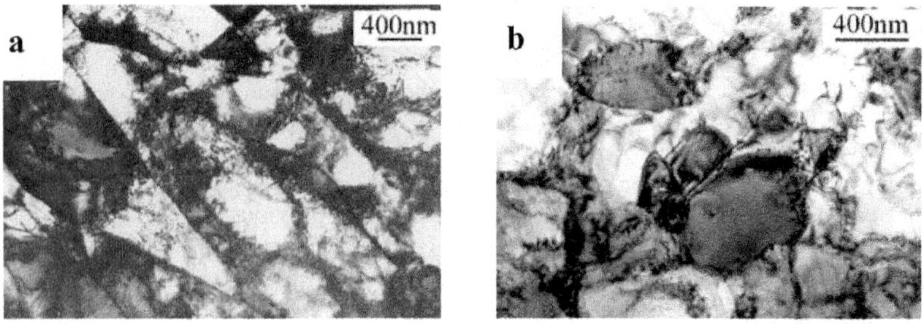

Figure 4. Microstructure evolution of Ti during different steps of “abc” deformation at 400°C and of 10^{-3} s^{-1}: a –3 steps (Σe≈1.5); b –12 steps (Σe≈6).

EBSD-analysis showed that at the initial stage of deformation the twinning on the {101} $\langle\bar{2}\ \bar{1}\ 6\rangle$ system was prevailing. As strain increased the twinning at the {102} $\langle\bar{1}\ \bar{1}\ 2\rangle$ and {113} $\langle\bar{2}\ \bar{1}\ 3\rangle$ systems became more active. The twinning on the last two systems took place over the whole investigated range of strain up to the 12th step. As seen from Table 1, the volume fraction of twins depended rather weakly on the amount of compression steps. However, misorientations axes and angles of boundaries of twins did not correspond presciently to twin-matrix conjugation (Fig. 5a). Their values can be deviated in a very broad range and even leave Brandon’s interval (interval of angles and axes of misorientations, inside that boundary was close to coincidence site lattice boundary).

It should be noted that the inconstancy of crystallographic parameters along the boundary plane was apt to all boundaries and not only to twin ones. In Fig. 5b one can see that the angles and the axes of misorientations were not discrete and constant along the random grain boundary in titanium sample subjected to strain of 40% but they varied within a rather wide range. One of the prominent features of structure evolution was gradual change of a shape of twins and misorientation on its boundaries. This meant that misorientation of high-angle boundaries was changed during deformation.

Investigation of evolution of the misorientation distribution function showed that the almost single consequence of the cumulative strain increase was a gradual reduction of the low-angle boundaries fraction. Other characteristics of the misorientation distribution function, namely the distribution of boundaries by misorientation angles and axes, as well as the content of the most typical coincident side lattice boundaries and their general content did not change.

Table 1. Effect of the cumulative strain on the fraction of different types of boundaries

Type of boundaries	Number of compression steps (ε)				
	1 (30%)	1 (60%)	3	5	12
Low-angle* (from 2° to 15°)	90%	90.3%	81.1%	54.2%	31.1%
High-angle* (from 15° to 93.8°)	10%	9.7%	18.9%	45.8%	68.9%
Twins** (Σ7b, Σ11a, Σ11b, Σ13b, Σ13c, Σ19c)	1.3%	4%	2.1%	2.2%	2.7%

*Step of scanning – 0.1 – 0.2 μm;
**Step of scanning – 0.1 – 5 μm

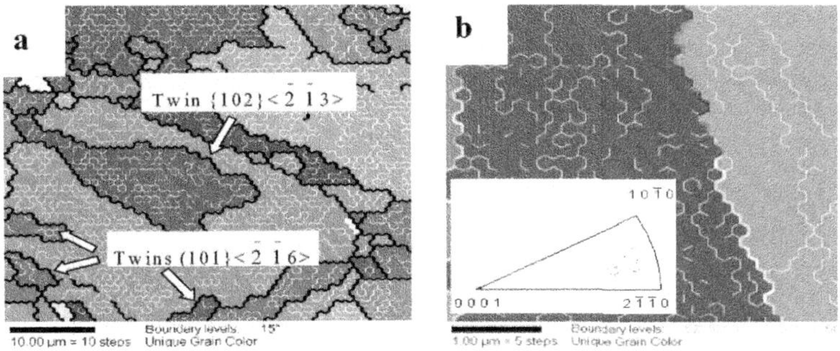

Figure5. Illustration of the changes of crystallographic parameters of boundaries: a – twined (ε = 60%; special boundaries are indicated by color); b – random high angle (ε = 40%; low angle boundaries with the angle of misorientation up to 15° are indicated by white color).

Values of the apparent activation energy and the coefficient of strain rate sensitivity defined at the 12th step of strain were equal to 257 KJ/mol and 0.1, respectively. The value of activation energy coincided with the values obtained earlier by Conrad [10] and indicated that the typical mechanism of warm plastic deformation of conventional titanium as prismatic slip of dislocations controlled plastic flow. It should be noted that the dynamic strain aging appeared during deformation [11]. This is additional evidence that grain boundary sliding was not a mechanism, which might be responsible for steady state flow. Indeed, the superplasticity in SMC titanium was observed at 450°C only [11].

It was noted that misorientation at a separate grain boundary constantly changed during deformation. This meant that the most typical misorientations repeated again and again. Taking into account the reproduction of textures one can conclude that these two processes were connected. In other words, formation of new DIB was caused by crystals reorientations. Since the shape and size of crystallites inside of macrobands is a result of DIB spreading we can conclude, that all these processes are connected.

So, at a macroscopic scale deformation develops by formation and spreading of

MSB. With changing the strain path new and new MSB occur. It promotes uniformity of deformation at the macroscopic scale. Inside MSB the crystals rotate to stable orientations at the given deformation axis. As a result, deformation develops at the mesoscopic scale - new DIB form and microstructure refine. DIB formation is a result of self-organization of dislocation and twinning. Spreading of new DIB and their intersection with each other provides formation of new submicrocrystalline equiaxed grains. The stability of plastic flow at final strain steps is maintained by accommodation of microstructure to new deformation conditions occurring with changing the strain path.

Microstructure Evolution and Mechanical Behavior of Ti-64 Alloy during "abc" Deformation.

The plastic-flow behavior of Ti-64 in terms of stress versus cumulative strain (S-Σe) at 550°C and 10^{-3} s^{-1} using strain increments of ~0.4 (Fig. 6a) exhibited a peak stress and flow softening during the initial increments of deformation and steady-state flow at large strains. Specifically, during the first two increments, a peak stress and then flow softening were observed. During later steps, the value of the peak stress decreased and steady-state flow occurred, which increased from one step to another. The presence of stress peaks and subsequent flow softening on the S-Σe curves is explained by strain hardening during the initial stages of deformation due to dislocation multiplication and subsequent softening associated with the bending of

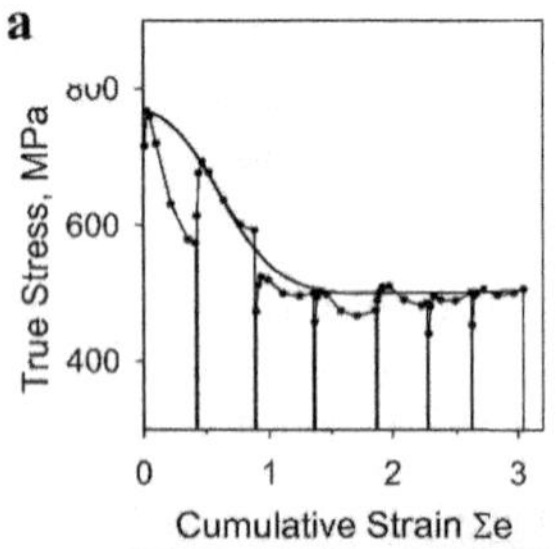

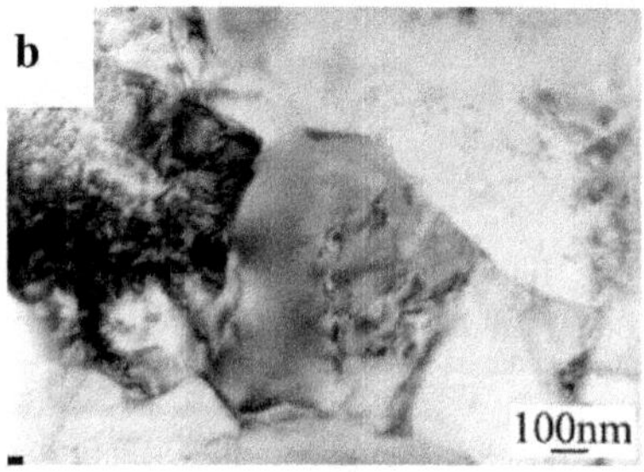

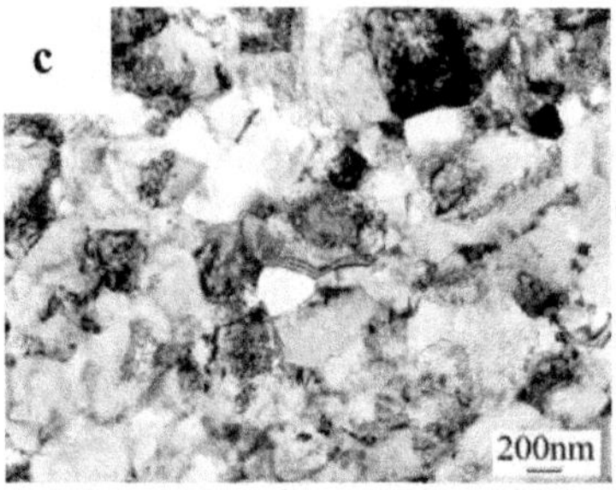

Figure 6. Cumulative S-Σe curve for "abc" deformation of Ti-6Al-4V at 550°C and $10^{-3}s^{-1}$ (a); SEM microstructure (b) and TEM image (c) of Ti-6Al-4V alloy deformed to height reduction of 50%.

non-favorably orientated lamellae and reorientation of favorably orientated ones in regions of localized deformation [8].

At the finer scale transverse sub-boundaries were formed within in α- and β-phases. As deformation increases further, the misorientation of these sub-boundaries increases since dislocations were absorbed into the sub-boundary walls, thereby leading to high-angle boundaries. Concurrently, the semi-coherent interphase α/β boundaries were transformed to non-coherent ones. Due to the transformation of the interphase and development of intra-lamellar boundaries, mass transfer became more rapid, and grooves were formed on the surface of the α-plates, leading to the segmentation of the alpha plates (Fig. 6b). The fragmented α-plates were thus spheroidized.

At the macroscopic scale macro shear bands were formed in the sample. The strain localized into the macro shear bands and therefore microstructure transformation was observed inside them. However, the development of globularization in the central area of the sample led to superplastic flow because of this formation of the new shear bands was retarded at subsequent deformation. This was accompanied by the occurrence of the steady flow stage on the S-Σe curve and the increase in the strain-rate-sensitivity coefficient m from 0.17 for e = 0.4 to 0.35 for Σe = 0.9. Consequently, successive rotations led to accumulation of strain primarily in the center of the sample, and formation of a homogeneous microstructure with globular grains of α- and β-phases about ~0.3 μm in size (Fig. 6c).

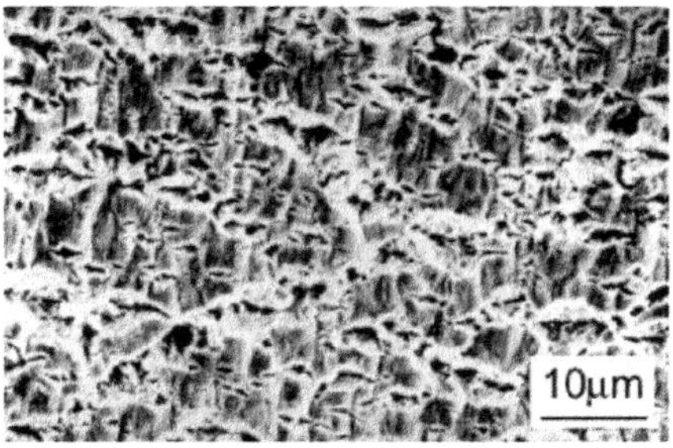

Figure 7. Surface relief of SMC Ti-6Al-4V sample after deformation at 550°C and strain rate of 10^{-3}s-^{1}.

Study of surface relief at the steady state flow stage revealed the bands typical for cooperative grain boundary sliding. The obtained value of apparent energy activation equal to 186 kJ/mol fitted well the date of 182 kJ/mol made by Sergueeva and Mukherejee [12] while studying superplastic flow of SMC Ti-6Al-4V. All these results are evidence of grain boundary sliding occurrence at the steady state flow of alloy at 550°C.

So, at the macroscopic scale the shear localization flow develops in the alloy during initial increments of deformation. Softening and globularization of microstructure in the macro shear bands lead to realization of deformation at the mesoscopic scale. In this case the mesoscopic scale deformation is determined by cooperative grain boundary sliding leading to superplastic flow. Superplastic flow results in deformation accumulation in the central area of the sample and impedes microstructure transformation in periphery regions.

Conclusions

1. The "abc" deformation of Ti at 400°C and 10^{-3} s^{-1} to a cumulative strain of Σe = 6 leads to the formation of a submicrocrystalline structure with a grain size of ~0.4 μm and its hardening. Under "abc" deformation the submicrocrystalline grains are

formed as a result of formation and crossing of deformation induced high angles boundaries created by slip and twinning. At the final steps of deformation the plastic flow close to the steady state one is observed. However, apparent energy activation is equal to 257 kJ/mol, which corresponds to energy activation of prismatic slip at warm deformation of titanium. The stability of plastic flow at final steps of deformation is maintained by accommodation of microstructure to new deformation conditions arisen from the change in the strain path.
2. The process of "abc" deformation of alpha/beta Ti-64 titanium alloy at 550°C and 10^{-3} s^{-1} to a cumulative strain of $\Sigma e = 3$ leads to the formation of a homogenous submicrocrystalline structure with a grain size of ~0.3 μm. A peak flow stress followed by flow softening and then steady-state flow caused by superplasticity are observed on the cumulative stress-strain curve. There occurred globularization controlled by the formation of high-angle boundaries across the lamellas and transformation of semicoherent interface boundaries into incoherent ones. The apparent energy activation is equal to 186 kJ/mol, which corresponds to the energy activation of superplastic flow under deformation condition taken in this experiment.

References

1. Rybin V.V., *Large plastic strains and fracture of metals.* Metallurgy: Moscow, 1986, p. 224.
2. Hughes D.A. and Hansen N., *Acta Met.* 45 (1997), 3871-3886.
3. Hurley P.J. and Humphreys F.J., *Acta Mater.* 51 (2003), 1087-1096.
4. E.V. Nesterova and V.V. Rybin, *Voprosi Materialovedenija*, 33 (2003), no. 1:151- 163.
5. Barnett M.R., Montheillet F., *Acta Mater.* 50 (2002), 2285-2296.
6. Salishchev G.A., Zherebtsov S.V. and Mironov S.Yu., *Voprosi Materialovedenija*, 33 (2003), no. 1:175-184.
7. Humphreys F..J. and Hatherly M., *Recrystallization and related annealing phenomena.* Oxford: Pergamon, 2004.
8. Semiatin S.L., Seetharaman V. and Weiss I., *Advances in the Science and Technology of Titanium Alloy Processing*, ed. , Warrendale, PA: The Minerals, Metals and Materials Society, 1997, .3.
9. Salishchev G.A., Zherebtsov S.V., Mironov S.Yu., Myshlyaev M..M. and Pippan R. "Formation of a submicrocrystalline structure in titanium during successive axial compression in three orthogonal directions", *Nanomaterials by Severe Plastic Deformation – NANOSPD2*, ed. Vienna, Austria, December 9-13, 2002, pp. 523-529.
10. Conrad H, *Prog. Mater. Sci.*, 26 (1981), 123-403.
11. Malysheva S.P., Salishchev G.A., Galeyev R.M., Danilenko V.N., Myshlyaev M.M. and Popov A.A., *The Physics of Metals and Metallography,* 95 (2003), no. 4:98-105.
12. Sergueeva A.V., Stolyarov V.V., Valiev R.Z. and Mukherjee A.K. *Scripta Mater.* 43 (2000), 819-825.

Nanostructured and Polycrystalline Ti Anomalies of Low Temperature Plasticity

V. Bengus[1*], S. Smirnov[1], E. Tabachnikova[1], V. Romanchenko[1], S. Khomenko[1], D. Gunderov[2], V. Stolyarov[2], R. Valiev[2]
[1]B. Verkin Institute for Low Temperature Physics & Engineering
Kharkov, Ukraine
[2]Ufa State Aviation Technical University, Ufa, Russia

Abstract

Systematic experimental study have been carried-out of the commercially pure Ti plasticity under uniaxial tension at 300, 77 and 4,2 K, in the nanostructured states processed by equal channel angular pressing (as well as in the coarse-grained state) with the aim to find explanation of pure Ti increased plasticity at 77 K. Regularities of changing of the ultimate uniform plastic deformation of Ti under cooling below ambient temperature in dependence of the average grain size and on the oxygen concentration in the initial material have been established. The hypothesis is proposed and approved that physical nature of the increased plasticity of nanostructured and polycrystalline Ti at 77 K is caused by a stimulating influence of a thermal anisotropy internal stresses (TAIS) upon mobile dislocation multiplication by the double cross-slip in the large grains and upon nucleation of mechanical twins.

Introduction

Physical nature of the increased yield stress of nanocrystalline metals is well understood in contemporary physics of plasticity, but explanation of their unusual plasticity is still absent [1]. Increased plasticity of polycrystalline iodide Ti at 77 K (in comparison with 300 K) had been discovered almost 40 years ago [2-5]. Such plasticity of Ti at 77 K practically retains at even lower temperatures [4] but the reason of Ti increased low temperature plasticity still has been unknown. Temperature dependence of Ti yield stress has no singularities in the 300-77 K temperature interval [4]. Ti has no any phase transitions in this interval, which can be the reason of its plasticity anomaly [6]. Recent observations [7] had shown that nanostructured Ti processed by equal channel angular pressing (ECAP) to the average grain size of 0,26 μ also showed at 77 K higher plasticity than at 300 K.

This report is devoted to explaining the Ti plasticity anomaly in the nanostructured and polycrystalline states through the increasing of mobile dislocation production intensified by the thermal anisotropy internal stresses

Experimental Procedures

Starting material for producing nanostructured state of Ti was polycrystalline Ti of the different commercial purity: Grade 2 (O_2 - 0.16 % weight), VT1-0 (O_2 - 0.17 % weight), and Grade 4 (O_2 - 0.34 % weight).

* Corresponding Author: bengus@ilt.kharkov.ua

Y.T. Zhu and V. Varyukhin (eds.), Nanostructured Materials by High-Pressure Severe Plastic Deformation, 55-60.

Plastic deformation of these different commercial purities of Ti was studied in the several structural states: Initial state 1 (processed by hot forging) has an average grain size d of 20 μ for the Grade 2 Ti, d of 15 μ for the Grade 4 Ti, and d of 15 μ for the VT1-0 Ti. The state 2 for all studied purities was processed by intensive plastic deformation of initial Ti by 8 runs of ECAP at 450 C. It has an average grain size d of 0.8 μ along ECAP direction for Grade 2, 0.85 μ along ECAP direction for Grade 4, and 0.3 μ along ECAP direction for VT1-0. The state 3 for VT1-0 with an average grain size d of 0.1 μ along ECAP direction was processed by intensive plastic deformation of initial Ti by 8 runs of ECAP at 450 C; and subsequent 75% cold rolling, which followed by 300 C annealing during 1 hour.

All specimens were cut from rods in parallel to ECAP axis. Mechanical characteristics were investigated under uniaxial tension, with the approximately $4\cdot10^{-4}$ s^{-1} rate of a relative deformation with a straining machine of the 3 кN/mm stiffness at 300 K, 77 K and 4,2 K. The micro specimens for tension had the 6 mm gage length and a square cross section of 0,7 x 1,5 mm.

Experimental Results

For the Grade 2 Ti, temperature dependences of the ultimate plasticity ε_{unif} are shown on the Fig. 1a. It can be seen that ε_{unif} for the nanostructured Ti increases from 18 to 25% under cooling from 300 to 4.2 K (with the 30% maximum near 77 K). Values ε_{unif} for the Grade 2 polycrystalline Ti are almost two times larger than for the nanocrystalline one.

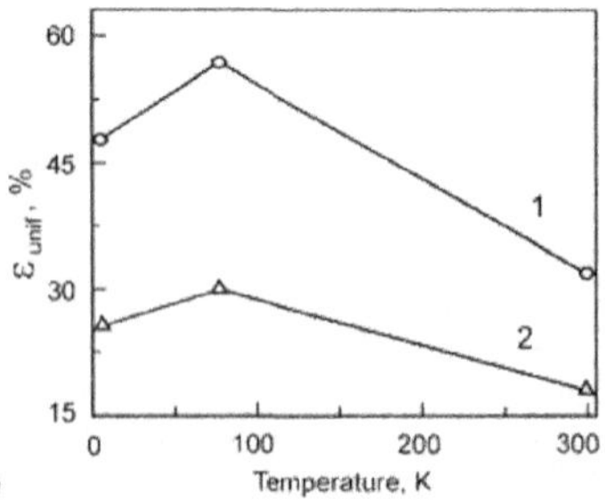

a)

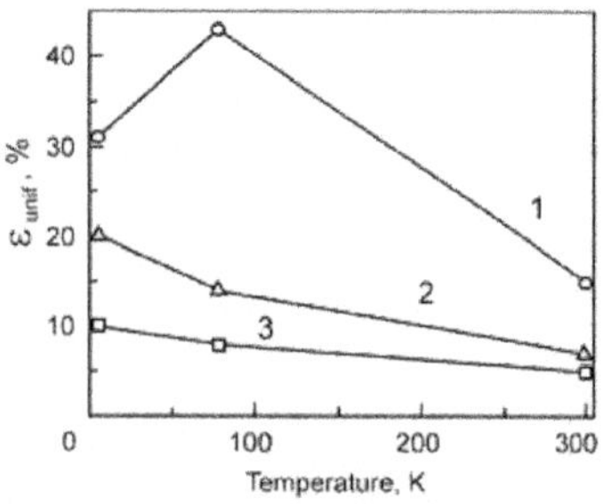

b)

Figure 1. a) Temperature dependence of the ultimate plasticity ε_{unif} for the Grade 2 Ti, the initial average grain size dimension 20 μ – O, and ECAP processed average grain size dimension 0.8 μ – Δ, along ECAP axis.
b) Temperature dependence of the ultimate plasticity ε_{unif} for the VT1-0 Ti, the initial – O (1), and ECAP processed to the 0.3 μ average grain dimension – Δ, along ECAP axis (2), and the 0.1 μ average grain dimension – □ (3), obtained after the additional thermo mechanical treatment.

For the VT1-0 Ti, temperature dependences of the ultimate plasticity ε_{unif} are shown in the Fig. 1b. It is seen that ε_{unif} for the nanostructured Ti increases monotonously: from 7 to 20% for the state 2 and from 5 to 10% for the state 3. Values ε_{unif} for the VT1-0 polycrystalline Ti, increase from 15 to 31% under cooling from 300 to 4.2 K (with the 43% maximum near 77 K). These values are smaller than in more pure Grade 2 Ti.

For the Grade 4 Ti, temperature dependences of the ultimate plasticity ε_{unif} are shown in the Fig. 2. It is seen that ε_{unif} for the nanostructured Ti Grade 4 decreases monotonously from 12% at 300 K down to 7% at 4.2 K. Values ε_{unif} for the Grade 4 polycrystalline Ti increases from 16 to 18% under cooling from 300 to 77 K and then decreases to 7.5% at 4.2 K. Values ε_{unif} for the Grade 4 polycrystalline Ti are more than two times smaller than for the more pure VT1-0 one. ε_{unif} for the nano structured Ti Grade 4 at 4.2 K is practically the same as for the polycrystalline one.

From data presented it is obvious that plasticity of the nanostructured Ti (as well as of the polycrystalline one) depends substantially on the oxygen concentration in Ti. Increasing of the oxygen concentration in the nanostructured Ti decreases magnitudes of the ε_{unif} and can drastically change its temperature dependence.

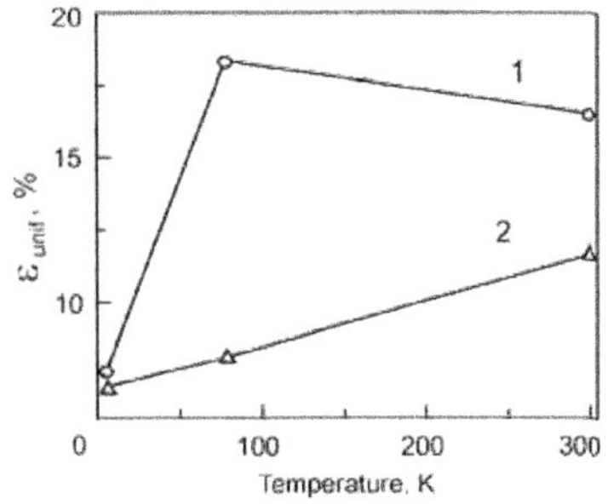

Figure 2. Temperature dependence of the ultimate plasticity ε_{unif} for the Grade 4 Ti, the initial average grain size dimension 15 μ – O (1), and ECAP processed average grain size dimension 0.85 μ – Δ (2), along ECAP axis.

Discussion

Stimulating of Dislocation Multiplication by the Double Cross-Slip in Grains by Thermal Anisotropy Internal Stresses in Titanium

Stimulating influence of internal stresses upon nucleation and multiplication of mobile dislocations is an important factor, which determines plasticity resource of a material. Thermal anisotropy internal stresses (TAIS) tensor $\hat{\sigma}^{th}$ [8, 9] is one of the additive components of internal stresses, which have random character, and for their consistent description need use of statistical methods. But estimation of the TAIS values and finding of their temperature dependence can be done by more simple method. In [8] the model has been proposed, which allows to estimate the order of value of the $\hat{\sigma}^{th}$ tensor diagonal components in polycrystals of all non-cubic symmetry classes. According to [8] the TAIS tensor $\hat{\sigma}^{th}$ has a diagonal form in the Cartesian coordinates: the $x^{(c)}$ and $y^{(c)}$ axes are along $[2\bar{1}\bar{1}0]$ and $[01\bar{1}0]$, respectively; the $z^{(c)}$ axis is oriented along [0001]. Here $\sigma_{11}^{th(c)} = \sigma_{22}^{th(c)}$ and TAIS action upon shear processes in grains are determined by $\sigma_{th} \equiv \sigma_{33}^{th(c)} - \sigma_{22}^{th(c)}$, as shown in [10]. Estimated numerically from elastic constants and the temperature dependences thermal expansion coefficients TAIS temperature dependences are shown in Fig.3 (T_0 is temperature of the 'initial" state of the sample).

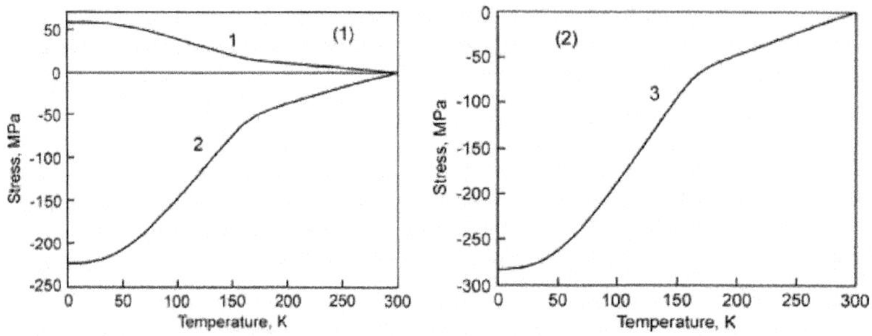

Figure 3: Temperature dependences of TAIS of the iodide Ti for T_0 =300 K:
1 - $\sigma_{22}^{th(c)}(T)-\sigma_{22}^{th(c)}(T_0)$; 2 - $\sigma_{33}^{th(c)}(T)-\sigma_{33}^{th(c)}(T_0)$; 3 - $\sigma_{th}(T)-\sigma_{th}(T_0)$.

It is seen that the highest changes in these values occur in the interval 300-40 K. Since $\sigma_{33}^{th(c)} < 0$, while $\sigma_{11}^{th(c)}$ and $\sigma_{22}^{th(c)} > 0$, below 300K each grain in iodide Ti experiences compressive stress along the six-fold axis and tensile stress in the perpendicular direction [10]. It is evident that tensor $\hat{\sigma}^{th}$ shear components will have the same order of value and similar temperature dependences. The total stress initiating plastic deformation in the sample is described by the tensor $\hat{\sigma}$:

$$\hat{\sigma} = \hat{\sigma}^e + \hat{\sigma}^{th} + \hat{\sigma}^i ,$$

where $\hat{\sigma}^e$ is the tensor of elastic stresses induced by the applied force; $\hat{\sigma}^i$ is the tensor of internal stresses induced by the lattice defects. The $\hat{\sigma}^i$ tensor is temperature dependent in accordance with the temperature influence on defect structure evolution. But reliable data on $\hat{\sigma}^i$ in studied by us Ti is absent today that makes difficult synonymous interpretation of experimental data. Therefore, it can only be supposed that observed in Fig. 1-2 strong temperature dependences of ε_{unif} are mainly caused by TAIS contribution to local values of $\hat{\sigma}$, which initiate micro mechanisms of nucleation and multiplication of mobile dislocations and mechanical twins that control a plastic deformation value.

Considerable influence of TAIS can be expected upon dislocations multiplication by the double cross-slip (DCS) process, which must be developed intensively in the polycrystalline Ti and in large grains of the nanostructured Ti. According to [11] the dislocation multiplication coefficient through the DCS k_{DSC} is expressed by:

$$k_{DSC} = k_1 \exp(-w_2 h_1 / b) , \quad (1)$$

where k_1 is number of cross-slip events, which a screw dislocation undergoes while slipping over one unit area; w_2 is the probability of a second cross-slip event; h_0 is the critical dimension of the edge dipole; b being the Burgers vector magnitude.

Increasing of the TAIS amplitude in some region must lead to stresses increasing in cross-planes, stimulate cross-slips and lead to increasing of the k_1 value. Simultaneously in the region, which is favorable for the DSC, decreasing of w_2 and of h_0 can be expected. Therefore according to (1) it can be expected that in the commercially pure polycrystalline Ti growth of TAIS under cooling from 300 to 77 K must lead to increasing of k_{DSC} and of a number of mobile dislocations and consequently stimulate increasing of the plasticity resource and the ε_{unif} value.

With decreasing of grain dimensions to smaller than dislocation free path and transition to a nanostructured state processes of DCS will proceed in a smaller number of grains and exercise smaller influence upon proceeding of plastic deformation. This must be displayed in decreasing of a ε_{unif} value and reducing its temperature dependence. These regularities have been really observed, as it is demonstrated by the experimental dependences of Fig. 1 and 2.

Stimulating of Dislocations Heterogeneous Nucleation at Grain Boundaries by Thermal Anisotropy Internal Stresses in Nanostructured Titanium

Structural investigations of nanostructured Ti had shown that there are no dislocations in the grains, which are smaller than some characteristic dimension d_0, which in the commercially pure Ti is $d_0 \approx 100$ HM [12]. It seems that d_0 magnitude is a characteristic of the specific structural state and depends on way of its obtaining and on the material purity. It is natural to consider that plastic deformation of such grains realized by dislocations, which are nucleated at grain boundaries, and the complete area of the slip plane swept by the each mobile dislocation is equal approximately to the average cross-section area of the grain. The value of the uniform plastic deformation ε_d in such grains (at some point of the stress-strain curve) is determined by the number of mobile dislocation loops N_d emitted by sources at grain boundaries $\varepsilon_d = b\,N_d./d$. A macroscopic specimen deforms uniformly in average before necking with continuity conservation, and ε_d magnitude also must achieve values of the order of ε_{unif}. Hence, it follows that observed increasing of ε_{unif} under cooling of nanostructured Ti from 300 to 77 K testifies increasing of N_d value since b and d values are constant. It is natural to suppose that the reason of N_d increasing consists in increasing of the total stress in the specimen under strain, which initiate action of mobile dislocation sources located at grain boundaries.

Under transition from the polycrystalline to the nanostructured state of Ti the grains volume fraction with dimensions smaller than d_0 increases. Therefore, under such transition one can expect changing of the relative contribution of different mechanisms controlling plastic deformation development: weakening of the DSC role and intensification of contribution of mobile dislocations nucleation at grain boundaries. Decreasing of the mobile dislocation multiplication processes must lead to decreasing of the specimens macroscopic plasticity resource that has been observed experimentally (as Figures 1-2 shows): average ε_{unif} value decreases with decreasing of grains dimensions.

Twinning contribution to the enhanced plasticity of Ti at 77 K

Essential influence of TAIS upon twins' nucleation in iodide Ti had been shown yet in [10]. Twinning contribution to the enhanced plasticity of Ti at 77 K can be evidenced indirectly by decreasing of VT1-0 Ti plasticity owing to transition from the state 2 to the state 3 due to rolling at room temperature. It is known that due to ECAP (the state 2) rather perfect axial crystallographic texture arises with the orientation of the 6-order axis perpendicularly to the ECAP axis [13]. Also it is known that at cryogenic temperatures twinning of Ti realizes intensively mainly along the $<\bar{2}\bar{2}43>\{11\bar{2}4\}$ system [6]. Therefore the texture in the state 2 is rather favorable for twinning development along $\{11\bar{2}4\}$ and $\{1\bar{1}23\}$ planes. Rolling decreases perfect texture and blurs distributions of <0001> directions [13]. Therefore one can wait decreasing of twinning intensity of samples in the state 3. As Figure 1b evidences, nanostructured Ti plasticity in the state 3 smaller than in the state 2, and the ε_{unif} value temperature dependence is weaker. It can be concluded that observed differences are caused not only by grains dimensions decreasing from 0.3 to 0.1 mkm, but partially are due to decreasing of twinning contribution to nanostructured Ti plasticity.

Conclusion

Increased plasticity at the uniaxial tension of pure Ti at 77 K can be explained by the stimulating influence of the thermal anisotropy internal stresses upon mobile dislocation production through: 1) the double cross-slip mechanism in the large grains; 2) the heterogeneous nucleation of mobile dislocations and mechanical twins at grain boundaries in the small grains.

Acknowledgements

This work have being carried out with a partial financial support by the project № 3 –026/2004 of the NASU Complex Program "Nanosystems, Nanomaterials and Nanotechnologies" and by the INTAS-01-0320 Project.

References

1. Zhu Yuntian T. & Liao Xiaozhou, *Nature Materials*, 3 (2004), 351-352.
2. Ul'yanov R.A. and Moskalenko V. A., *Metallovedenie i Termicheskaya obrabotka metallov*, 1966, no. 10:48-51.
3. Garde A. M. and Read-Hill R. E., *Metallurgical Trans.*, 2 (1977), 2885-2888.
4. Moskalenko V., Startsev V., Kovaleva V., *Cryogenics*, 20, (1980), 503-508.
5. Collings E. W., *The Physical Metallurgy of Titanium Alloys*, American Society for Metals, Metals Park, 1984, 223 p.
6. Zwicker U., *Titan und Titanlegierungen*, Springer-Verlag Berlin Heidelberg New York, 1974, 511 p.
7. Wang Y.M., Ma E., Valiev R.Z., and Zhu Y.T., *Advanced Materials* (in press).
8. Likhachev V. A., Solid State Physics, 3, (1961) 1827 (in Russian).
9. Bengus V. Z., Tabachnikova E. D., Natsik V. D., Miskuf J., Csach K., Stolyarov V. V., Valiev R. Z., Low Temperature Physics, v. 28, № 11 (2002) p. 864-874.
10. Bengus V., Smirnov S., Influence of the thermal anisotropy internal stresses on low temperature mechanical behavior of polycrystalline and nanostructured Ti. *Nanomaterials by Severe Plastic Deformation, Proceedings of the Conference "Nanomaterials by Severe Plastic Deformation - NANOSPD2" (December 9 – 13, 2002, Vienna, Austria)*, ed. Michael Zehetbauer and Ruslan Z. Valiev, J. Wiley VCH, Weinheim, 2004, 151-157.
11. Wiedersich H., Journal of Applied Physics, 33 (1962), 854-858.
12. Zhu Y.T., Huang J.Y., Gubicza J., Ungár T., Wang Y.M., Ma E., Valiev R.Z., *J. of Materials Research* 18 (2003), 1908-1917.
13. Kilmametov A., Stolyarov V., Shestakova L., Aleksandrov I. "Structural peculiarities of high strength nanostructured titanium processed by severe plastic deformation" *Structure and Properties of Nanocrystalline Materials*, ed. Taluts G. G, Noskova N. I., Ekaterinburg: UDRAS, 1999, 185-195 (in Russian).

Modeling of Grain Subdivision during Severe Plastic Deformation by VPSC Method Combined with Disclination Analysis

A.A. Nazarov[1,2], N.A. Enikeev[1*], A.E. Romanov[3], T.S. Orlova[3] ,
I.V. Alexandrov[1], I.J. Beyerlein[4]
[1]Ufa State Aviation Technical University, Ufa, 450000
[2]Inst. for Metals Superplasticity Problems, Russian Academy of Sciences, Ufa, 450001
[3]Ioffe Physico-Technical Inst, Russian Academy of Sciences, St. Petersburg, 194021
[4]Los Alamos National Laboratory, Los Alamos, NM 87545, USA

Abstract

Microstructure development during severe plastic deformation by simple shear is modeled using a combination of the visco-plastic self consistent (VPSC) method and a disclination model. Strain incompatibilities between a homogeneous effective medium and a grain are calculated by VPSC. These are assumed to result in an accumulation of disclinations in the junctions of a grain that are relaxed by a growth of low-angle dislocation boundaries from the junctions. Predicted misorientation distributions between subgrains and their parent grains agree semi-quantitatively with experimental misorientation distributions for geometrically necessary boundaries. The texture after 100% simple shear was found to be insensitive to the presence of subgrains with misorientations less than 15°.

Introduction

Grain fragmentation and refinement is one of the most important consequences of dislocation substructure evolution during large plastic deformation common for crystals of different types [1,2]. Mechanisms and characteristics of microstructural refinement is a matter of great theoretical and practical importance, particularly in producing ultrafine grained materials using equal-channel angular pressing (ECAP) and other methods of severe plastic deformation [3]. In principle, the refinement process can be rigorously described in terms of dislocation dynamics, though full 3-D modeling is a challenging task due to the complex nature of dislocation multiplication, interaction and annihilation in real crystals. Recently, the disclination approach has been shown to be very helpful for the description of elementary processes of grain subdivision [1,4]. It has been proposed that grains subdivide by the growth of low-angle boundaries due to the motion of partial disclinations generated at grain boundaries and grain boundary junctions. In turn, this formalism alone cannot be used to analyze the whole process of grain subdivision, since it neglects the distribution of plastic deformation on slip systems and does not consider important crystallographic and geometric factors influencing grain refinement.

On the other hand, polycrystal models are successfully used to study the plastic deformation of polycrystals and texture formation. These models account for the distribution of strain among crystallographic slip systems and grains, albeit as an approximate continuum approach. The visco-plastic self-consistent (VPSC) model is

* Corresponding Author: carabus@mail.rb.ru

Y.T. Zhu and V. Varyukhin (eds.), Nanostructured Materials by High-Pressure Severe Plastic Deformation, 61-66.

one such approach [5]. This model considers each grain as an ellipsoidal inclusion deforming in a homogeneous equivalent medium (HEM) representing the whole polycrystal. Very recently, this model has been applied to simulate the texture evolution during ECAP [6]. However, the one-site VPSC model used in the cited paper neglects the non-uniformities of the stress and strain within grains that lead to subdivison. Instead, a very simple geometric criterion of grain subdivision was used and thus internal grain microstructural development was neglected.

The joint use of disclination models with polycrystal modeling combines the advantages of both methods and makes a powerful tool for modeling microstructure development during large plastic deformation. In this combination, VPSC provides the data on the incompatibilities of strain between the HEM and a grain, while the disclination model predicts subgrain formation and characteristics. The aim of the present paper is to develop such a combined simulation scheme for severe plastic deformation. By modeling simple shear we will demonstrate that this scheme allows one to describe the main microstructural characteristics of grain subdivision, such as the evolution of misorientations and texture.

Description of the model

Following the basic ideas of the VPSC method, consider an ellipsoidal grain, Fig. 1b, deforming in a HEM, representing the average behavior of the aggregate. Let the strain rate of the HEM be given by a tensor $\dot{\mathbf{E}}$. For the case of simple shear, Fig. 1a, with a velocity gradient tensor in the 1,2,3 coordinate system

$$\mathbf{D} = \begin{pmatrix} 0 & D & 0 \\ 0 & 0 & 0 \\ 0 & 0 & 0 \end{pmatrix}, \tag{1}$$

the strain rate of the HEM is $\dot{\mathrm{E}}_{12} = \dot{\mathrm{E}}_{21} = D/2$, equal to the applied strain rate. The strain rate tensor of a grain is denoted as $\dot{\boldsymbol{\varepsilon}}$ and depends on its orientation. If the grain deforms differently than the HEM, strain incompatibilities will result in an accumulation of disclinations in the junctions of this grain that will force its splitting. Splitting occurs by redistribution of slip in different domains of the grain and formation of low-angle dislocation boundaries. To be able to introduce junction disclinations and obtain a criterion for the splitting, we approximate the grain by a parallelepiped circumscribing the corresponding ellipsoid (Fig.1b). Edges of this parallelepiped coincide with the axes of ellipsoid. The difference between the shear strain rates of the HEM and a grain, $\dot{\mathrm{E}}_{12} - \dot{\varepsilon}_{12}$, are then transformed to the coordinate frame made by the axes of the ellipsoid (1′,2′,3′) (See Fig. 1b). Let ς_{ij} denote the components of this transformed tensor. Using the formulae in [7] one can calculate the rates of disclination accumulation on the edges of the parallelepiped as follows:

$$\dot{\vec{\Omega}}_1 = [2\varsigma_{23}, \varsigma_{13}, \varsigma_{12}], \quad \dot{\vec{\Omega}}_2 = [\varsigma_{23}, 2\varsigma_{13}, \varsigma_{12}], \quad \dot{\vec{\Omega}}_3 = [\varsigma_{23}, \varsigma_{13}, 2\varsigma_{12}]. \tag{2}$$

If $\dot{E}_{12} - \dot{\varepsilon}_{12} > 0$ and when the maximum of $\Omega_1, \Omega_2, \Omega_3$ exceeds the critical value of $\Omega_c = 1^\circ$, it is assumed that a pair of dislocation boundaries with a misorientation vector $\vec{\theta} = \vec{\Omega}_1 + \vec{\Omega}_2 + \vec{\Omega}_3$ is formed along the diagonals of the parallelepiped as shown in Fig.1b. Full relaxation of junction disclinations and a rotation of two of the four subgrains formed with this misorientation vector result (See Fig. 1c). The sense of this rotation is chosen such that the subgrains rotate in a direction nearly opposite of that of the parent grain. Nucleation of subgrain boundaries as described above will be referred to as the *splitting* of a grain.

Further straining accumulates new disclinations on the junctions unless the strength of one of them exceeds Ω_c. These subsequent disclinations will relax by evolving along the initially formed low-angle boundaries causing a step-by-step increase in their misorientation angle. The rotation matrix for these step-wise subgrain rotations is accumulated unless the angle of rotation becomes larger than a value delimiting low- and high- angle grain boundaries ($\theta_0 = 10^\circ$ to 15°). When this

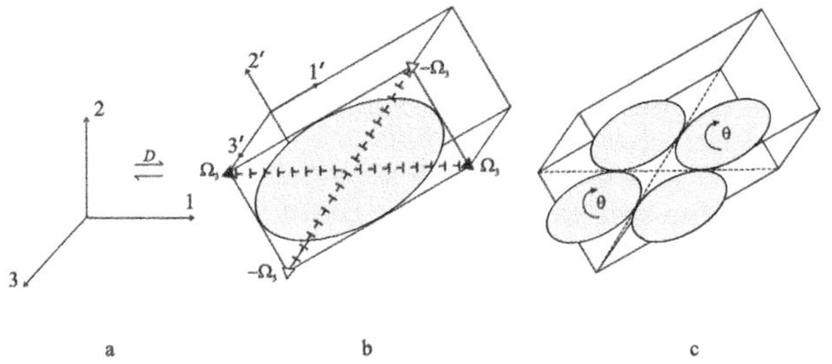

Fig.1. A model of grain subdivision by a growth of a pair of subboundaries along the grain diagonals. (a) Coordinate frame, 1-shear direction, 2-normal direction, 3- transverse direction; b) A junction disclination quadrupole and a mode of its relaxation; (c) New grains (subgrains) formed due to subdivision

occurs, the grain is considered to *subdivide* into four grains, which are assigned corresponding orientations and further included into the aggregate simulated by VPSC. In turn, these new grains can split and subsequently subdivide during further straining thus leading to the formation of next-generation subgrains and new grains. Thus, the number of grains will progressively increase during deformation. Each time a grain can contain only one generation of subgrains; new generations will form when subgrains become independent grains and accumulate their own internal misorientation.

Results of simulations

To test the proposed simulation scheme simple shear of a polycrystalline aggregate containing initially 200 randomly oriented grains is modeled (See Fig. 1a). The evolution of the microstructure is described in terms of the calculated number of splitting/subdivison events per original grain, total number of split/subdivisions, misorientation distributions and texture for different values of applied strain. The low/high angle boundary limit is assumed to be 15°.

For the aggregate a simulated shear strain of $E_{12} = 1.0$ (equivalent von Mises strain $E_{VM} = 1.12$) resulted in 30 subdivisions; one grain subdivided 6 times, some grains 2 and 3 times, and others only once. The lower the strain rate of a grain, the more rapidly it divides into subgrains. Accumulation of a 15° misorientation sufficient for the first subdivision occurred after $E_{12} = 0.36$. Among 230 grains in the final aggregate, 152 grains are split into subgrains with misorientations less than 15°.

For each new grain and subgrain, the identity of their parent grain is recorded, which allowed a conditional misorientation distribution function $f(\theta)$ to be calculated. In calculating this function, misorientations θ of all new grains and subgrains with respect to the current orientation of their parent grain is taken into account. This function is calculated instead of the true distribution function, since in the 1-site VPSC neighborship of grains is not defined. Nevertheless, $f(\theta)$ can give semi-quantitative information on the distribution of misorientations of subboundaries. Histograms representing the distribution for three values of the shear strain are plotted in Fig. 2a-c. In accordance with trends observed in experimental data [8-10], the distribution becomes wider and the average misorientation increases with increasing strain. The dependence of the average misorientation angle on the strain plotted in Fig.2d agrees quantitatively well with experimental data obtained on the substructure of cold-rolled metals [8-10].

Presented in Fig.3 are the pole figures calculated for the polycrystal subjected to $E_{12} = 1.0$. Fig 3(a) has been calculated using output from the standard VPSC code, Fig. 3(b) for the aggregate of 230 grains resulting from combined VPSC/disclination simulations, i.e. taking account the orientations of only grains including those newly formed during deformation, while Fig. 3(c) additionally includes all subgrains rotated to less than 15° with respect to their parent grains. Generally, during the simulation with grain subdivision texture develops in the same way as in the standard VPSC code. Maximums are located along the same directions and have nearly the same value for all three cases (9.54, 9.64, and 9.56, respectively). It would be expected that the presence of low-angle misorientations of up to 15° inside grains would result in a significant spread of the texture. Surprisingly, however, only slight differences in the pole density in the vicinity of some maximums are observed in Fig. 3c as compared to Fig.3a. This result may not be too surprising as standard VPSC has proven to predict well first pass textures in many materials (See [13] and [14] and references therein). However, the coupling between misorientation evolution and texture is expected to become stronger under larger strains and strain path changes, such as in ECAP.

Discussion

We have presented a simple model, which combines the advantages of both the VPSC polycrystal model and a disclination model for subgrain boundary nucleation and evolution and provides a possibility to study the substructure evolution during large plastic deformation. The key feature of this model is a splitting/subdivision criterion that is based on the disclination concept.

The relaxation of junction disclinations occurs by a complicated process of redistribution of slip near the junctions under the combined action of applied and internal stresses. This process has to be analyzed with a model that accounts for the non-uniformity of stress and strain inside grains. This can be done in terms of the n-site VPSC model [11] or a finite-element simulation. However for multiscale modeling, these methods would be computationally too expensive. Instead,

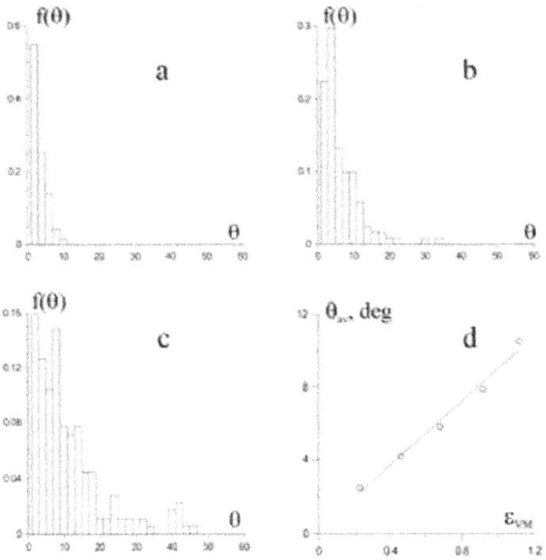

Fig.2. Evolution of the conditional misorientation distribution function during simple shear strain up to $E_{VM} = 0.23$ (a), 0.68 (b), and 1.12 (c) and a plot of the average misorientation angle of subgrains with strain (d).

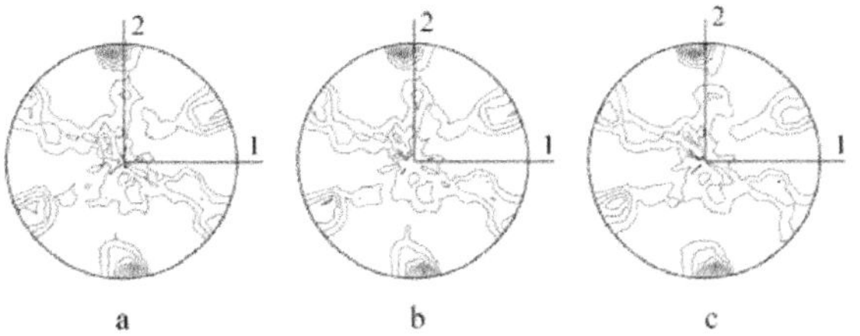

Fig.3. Pole figures for a polycrystal with 200 initially random grains deformed by simple shear up to $E_{VM} = 1.12$. (a) After a standard VPSC simulation (number of grains: 200); (b) After a VPSC/disclination simulation neglecting subgrains with misorientations $\theta < \theta_0 = 15^\circ$ (final number of grains: 230); (c) As in (b), additionally including subgrains (final number of grains and subgrains: 382). Intensity level increment is equal to 0.95. Direction of shearing and axes are defined in Fig.1a.

physically-based models can be used to understand the modes of relaxation of disclinations. The model presented here is the simplest one of these. More sophisticated models can be put forward on the basis of an energetic analysis of disclination systems. The development of such models is under way [12]. Nevertheless, even with the simplest disclination subdivision criterion, this combined approach is capable of describing some important structural characteristics of heavily deformed polycrystals and thus can be used to simulate the ECAP process.

Recent electron microscopic studies of cold-rolled polycrystals provided extensive data on the substructure evolution during large deformation. One of the important findings of those works is that geometrically-necessary boundaries (GNBs) form on two different scales [9]. The boundaries formed on a smaller scale, slightly larger than the cell size have misorientations less than 6°, whereas the GNBs formed on a larger scale slightly smaller than the grain size can have misorientations exceeding 15°. It is the latter type whose average misorientation angle increases with strain in good accordance with the predictions presented in Fig. 2d [9]. One can suggest that the first type of GNBs forms due to an inhomogeneity of strain inside grains, while the second type forms due to the incompatibility of strain of neighboring grains, by an accumulation and relaxation of junction disclinations.

Thus, the model presented in this paper describes well the formation and parameters of the largest-scale substructure in deformed metals, which is characterized by truly large-angle misorientations and most important for grain refinement by severe plastic deformation.

Finally, only slight differences of the final texture are found between the standard and modified VPSC codes for the case of simple shear up to 100%. However, for deformation with strain path changes, such as in multi-pass ECAP, a more significant influence of grain subdivision on texture can be expected.

Acknowledgements

This research is funded by a Los Alamos National Laboratory – Initiatives for Proliferation Prevention Project #T2-0197.

References

1. Rybin V.V., *Large Plastic Deformations and Fracture of Metals*. Moscow: Metallurgia Publ., 1986, 224 p.
2. Hansen N., *Scripta Metall. Mater.* 27 (1992), 1447-1452.
3. Valiev R.Z., Islamgaliev R.K., and Alexandrov I.V., *Prog. Mater. Sci.* 45 (2000), 103-189.
4. Romanov A.E. and Vladimirov V.I. "Disclinations in Crystalline Solids", *Dislocations in Solids*, vol.9, ed.: F.R.N. Nabarro, Amsterdam: North Holland, 1992, p. 191.
5. Lebensohn R.A. and Tome C.N., *Acta Matell. Mater.* 41 (1993), 2611-2624.
6. Beyerlein I.J., Lebensohn R.A., and Tome C.N., *Mater. Sci. Eng. A* 345 (2003), 122-138.
7. Rybin V.V., Zisman A.A., and Zolotarevsky N.Yu., *Acta Metall. Mater.* 41 (1983), 2211-2217.
8. Hughes D.A., Liu Q., Chrzan D.C., and Hansen N., *Acta mater.* 45 (1997), 105.
9. Liu Q., Jensen D.J., and Hansen N., *Acta Mater.* 46 (1998) 5819.
10. Hughes D.A. and Hansen N., *Acta Mater.* 48 (2000), 2985.
11. Lebensohn R. A., *Acta Mater.* 49 (2001), 2723.
12. Orlova T.S., Nazarov A.A., Enikeev N.A., Alexandrov I.V., Valiev R.Z., and Romanov A.E., *Phys. Solid State*, submitted.
13. Li S., Beyerlein I.J., Necker C.T., Alexander D.J., and Bourke M.A.M., *Acta Mater.* 52 (2004) 4859.
14. Vogel S. C., Alexander D. J., Beyerlein I. J., Bourke M. A. M., Brown D. W., Clausen B., Tomé C. N., Von Dreele R. B., Xu C., and Langdon T. G., *Materials Science Forum*, 426-432 (2003) 2661.

II. ADVANCES IN SPD TECHNOLOGIES

Ultimate Grain Refinement by ECAP: Experiment and Theory

Vladimir I. Kopylov[1] and Vladimir N. Chuvil'deev[2*]
[1]Physico-Technical Institute, National Academy of Sciences of Belarus, 10 Kuprevicha St., Minsk 220141, Belarus
[2]Research Physical & Technical Institute, Nizhny Novgorod State University, 23/3 Gagarin Ave., Nizhny Novgorod 603950, Russia

Abstract

The paper describes the results of experimental and theoretical investigations of the grain refinement process during severe plastic deformation. Experimental results on deformation induced grain refinement in pure metals and copper, magnesium and aluminum alloys are analyzed. A model is developed for estimating the minimum grain size which can be attained by equal-channel angular pressing method.

1. Introduction

It is known that severe plastic deformation is accompanied by intensive fragmentation: the formation of disorientated microscopic fragments. As deformation proceeds, the disorientation of fragments increases while their size gradually decreases reaching a certain ultimate value d^*. Parameter d^* is conventionally called the grain refinement limit. The d^* value depends on the nature of a material, deformation temperature and the type and method of deformation.

The objective of this work is experimental and theoretical investigation of the grain refinement limit d^* and the development of a model for estimating the d^* value.

2. An approach to the problem of the grain refinement limit during severe plastic deformation

2.1 Classical theory of grain refinement

Fundamentals of the grain refinement theory were described by Rybin [1]. According to the conception developed therein, the grain refinement process at severe plastic deformations can be briefly described as follows. In the process of intragranular plastic deformation which occurs under the action of external strain, accumulation of defects at inner interfaces (grain boundaries) of a polycrystal takes place. In general, the defect layer formed at interfaces constitutes a complex system of dislocation and disclination type defects [2]. In the first approximation, this defect layer can be described as a system of not self-congruent planar dislocation arrays and a system of junction wedge disclinations [3]. The defects formed at grain boundaries and, first of all, junction disclinations generate a strong field of internal stresses σ_i. These stresses give rise to the accommodation intragranular slip. It is important to note that the stress fields generated by the junction disclinations during severe plastic deformation are so strong that the accommodation glide of dislocations in the grains acquires a cooperative nature. For describing this cooperative glide, a conception of nucleation and motion of special-type defects, broken dislocation boundaries, is used [1, 3]. Branching and intersecting with each other, these broken dislocation

* Corresponding Author. chuvildeev@nifti.unn.ru

Y.T. Zhu and V. Varyukhin (eds.), Nanostructured Materials by High-Pressure Severe Plastic Deformation, 69-76.

boundaries (whose propagation velocity through a grain is proportional to the strain rate) progressively fragment the grain.

2.2. Force condition for fragmentation

In connection with the above described conception, the force condition for fragmentation can be presented as a condition imposed on the intensity of the internal stress field σ_i generated by the junction disclinations. These stresses should be sufficient to cause plastic deformation in the grains:

$$\sigma_i \geq \sigma_1 \quad (1)$$

In a rough approximation, stress σ_1 can be assumed proportional to the yield stress σ_Y of the material. In this case, presenting σ_Y in the conventional form $\sigma_Y = \sigma_o + K/\sqrt{d}$ where σ_o is the stress concerned with obstacles to the dislocation glide within a grain, K is the Hall-Petch coefficient, d is the grain size; the expression for σ_1 is written as

$$\sigma_1 = \psi\left(\sigma_o + K/\sqrt{d}\right), \quad (2)$$

where ψ is a numerical parameter.

To a first approximation, the value of internal stresses σ_i is proportional to the strength of junction disclinations ω [4]:

$$\sigma_i = \phi G \omega , \quad (3)$$

where ϕ is a geometrical factor which is of the order of unity, G is the shear modulus. In this case, the force condition for fragmentation (1) can be presented as $\omega \geq (1/\phi)(\sigma_1/G)$. In view of equations (2) and (3), this condition can be rewritten as:

$$\omega \geq \omega^* = \chi\left[(\sigma_o/G) + K/G\sqrt{d}\right], \quad (4)$$

where $\chi = \psi/\phi$ is a geometrical factor, ω^* is a critical strength of junction disclinations at which the junctions start emitting broken dislocation boundaries, i.e. fragmentation begins.

As a rule, in microcrystalline (MC) materials the second summand in equation (4) substantially exceeds the first one, and at small grain size the fragmentation condition is simplified to give

$$\omega \geq \omega^* = \chi K/G\sqrt{d} \sim 1/\sqrt{d} . \quad (5)$$

2.3. Fragmentation as an accommodation process

In the above concept, fragmentation is an accommodation phenomenon, i.e. a process occurring under the action of internal stresses σ_i and resulting in the relaxation of the elastic energy stored during deformation (in this model, the stored energy is associated mainly with the junction disclinations). This conception permits an obvious generalization: in the conditions of severe deformation, fragmentation may be not a single accommodation process. During severe plastic deformation, a different accommodation process or processes can take place which at certain conditions may be more efficient than fragmentation, i.e. proceed with a higher rate. This means that under certain conditions, despite continuing plastic deformation, the fragmentation can be suppressed and the grain refinement is terminated.

2.4. Diffusion accommodation of junction disclinations

In works [2, 5] an alternative mechanism of the accommodation of junction disclinations, viz. diffusion mass transfer, was considered. It has been demonstrated that in the first approximation the kinetics of changes of the disclination dipole strength at intragranular strain rate $\dot{\varepsilon}_v$ can be described by the following equation

$$\dot{\omega} = \xi\dot{\varepsilon}_v - \omega/t_r \tag{6}$$

where ξ is a geometrical factor characterizing the degree of uniformity of deformation, t_r is a characteristic time of accommodation of junction disclinations.

In the dipole approximation (at d=const), the value of t_r is calculated as:

$$t_r = A_1 d^3/\delta D_b^* \cdot kT/G\Omega \tag{7}$$

where A_1 is a numerical factor, δ is the grain boundary width, D_b^* is the coefficient of diffusion in non-equilibrium grain boundaries [5], T is absolute temperature, k is the Boltzmann constant, Ω is the atomic volume.

In the first approximation, from equation (6) taking into account condition (7) and assuming D_b^*=const we obtain an expression for function $\omega(t)$ in a simple form

$$\omega(t) = \omega^{st}(1 - \exp[-t/t_r]); \quad \omega^{st} = \xi\dot{\varepsilon}_v t_r \approx A_1\xi\dot{\varepsilon}_v d^3 kT/\delta D_b^* G\Omega \tag{8}$$

At a small time ($t \ll t_r$) the strength of disclinations increases linearly with deformation $\omega(t) \approx \xi\dot{\varepsilon}_v t \approx \xi\varepsilon_v$. At a large time ($t \geq t_r$) the strength reaches a stationary value which depends on the dipole arm d, i.e. the size of a grain or fragment, as well as on temperature, strain rate and parameters of grain boundary diffusion $D_b^* = D_{bo}^* \exp(-Q_b^*/kT)$, where Q_b^* is the activation energy and D_{b0}^* is the preexponential factor.

2.5. Effect of non-equilibrium structure of grain boundaries

In the case when a grain boundary contains a high density of excess defects (embedded lattice dislocations or non-equilibrium vacancies), its free volume changes and the grain-boundary diffusion coefficient increases substantially [5]. It has been demonstrated that the activation energy of grain boundary diffusion in non-equilibrium boundaries depends on the excess free volume $\Delta\alpha$ connected with defects embedded into the boundary. At a certain defect density the magnitude of $\Delta\alpha$ may be so high that the activation energy of grain boundary diffusion becomes equal to that of diffusion in a melt Q_L. Thus, depending on the density of embedded defects, the magnitude of Q_b^* can vary from the equilibrium value, which is equal to the activation energy of grain boundary self-diffusion Q_b~9 kT_m [6], up to the value of the activation energy for diffusion in the melt Q_L~3-4 kT_m [7]. Accordingly, by virtue of equations (7) and (8), the values of t_r and ω^{st} may change substantially.

2.6. The condition of fragmentation

As noted above, a necessary condition for the fragmentation to occur is when the internal stresses σ_i attain and exceed the value of σ_1 which is required for the plastic deformation of grains (equation (1)). To a first approximation, the σ_i value is assumed proportional to the strength of junction disclinations ω. In the presence of diffusion accommodation, the strength of junction disclinations ω is determined by expression (8) and its maximum value is ω^{st}. In this case, the condition of fragmentation can be presented as

$$\frac{d^3\xi\dot{\varepsilon}_v A_1}{\delta D_b^*(G\Omega/kT)} = \omega^{st} \succ \omega^* \tag{9}$$

From equation (9) it is seen that fragmentation is possible not at any values of parameters d, $\dot{\varepsilon}_v$, D_b^* but only at a certain combination of the latter. Let's rewrite equation (9) as a limitation imposed on the fragment size (see formula (5))

$$\left(d^*\right)^{3.5} = \chi \frac{K}{G} \frac{\delta D_b^*}{A_1 \xi \dot{\varepsilon}_v} \frac{G\Omega}{kT} \tag{10}$$

From condition (10) it follows that at the attainment of a certain minimum size of the fragment d~d* further fragmentation becomes impossible. This is due to the fact that at small fragment sizes the rate of diffusion accommodation of junction disclinations becomes so high that the strength of junction disclinations cannot reach the critical value ω^* necessary for emitting broken dislocation boundaries. Thus, as a result of the development of diffusion accommodation of junction disclinations, a certain limit of the deformation grain refinement emerges, i.e. a minimum grain size which cannot be reduced at the given conditions ($\dot{\varepsilon}$, T) of plastic deformation of the material. The value of this grain refinement limit is determined by expression (10).

3. Discussion and comparison with experiment

3.1. The effect of temperature: analysis

Evaluation of the temperature dependence of the minimum grain size d^* seems to be a complex problem. This is connected with the fact that almost all the parameters appearing in equation (10) depend on temperature. Let's perform a successive analysis of the effect of temperature on each of these parameters.

(a) Temperature dependence of the strain rate. From the description of the ECAP process [8, 9] it is obvious that equation (10) contains the local intragranular strain rate $\dot{\varepsilon}_v$. This value depends on the level of local internal stresses σ^* arising in the deformation zone and, apparently, on the deformation temperature. According to the conception developed in [6], at relatively low temperatures and high stresses the intragranular strain rate $\dot{\varepsilon}_v$ is determined by the dislocation glide which is limited by their interaction with obstacles. In this case, the $\dot{\varepsilon}_v$ value is calculated by formula

$$\dot{\varepsilon}_v^I = \dot{\varepsilon}_0 \exp\left[-\Delta F\left(1 - \sigma^*/\sigma_0\right)/kT\right], \tag{11}$$

where $\dot{\varepsilon}_o = \rho_v b V_o$, ΔF is the free energy necessary for overcoming the obstacles, σ_o is the stress required for overcoming the obstacles at 0 K, ρ_v is the density of lattice dislocations, b is the Burgers vector and V_o is the characteristic velocity of the dislocation glide.

The values of ΔF and σ_0 depend on the nature of obstacles and in pure metals are determined first of all by the type of atomic bonding. It should be outlined that according to data of [6] the ΔF and σ_0 values for fcc metals and transition bcc metals are substantially different. For fcc metals $\Delta F/kT_m<10^{-5}$, $\sigma_0/G<10^{-5}$; for transition bcc metals $\Delta F/kT_m=3.2\pm0.7$, $\sigma_0/G=(1\pm0.4)\cdot10^{-2}$.

At elevated temperatures the rate of intragranular deformation $\dot{\varepsilon}_v$ is controlled by the diffusion accommodation rate. In this case, the expression for $\dot{\varepsilon}_v$ looks as

$$\dot{\varepsilon}_v^{II} = A\left(\frac{G\Omega}{kT}\right)\left(\frac{\sigma^*}{G}\right)^n\left[\frac{D_{vo}}{b^2}\exp\left(-\frac{Q_v}{kT}\right)+\frac{10\rho_v a^2}{b^2}D_{co}\exp\left(-\frac{Q_c}{kT}\right)\right] \quad (12)$$

Here D_{v0} and D_{co}, Q_v and Q_c are the preexponential factors and activation energies for bulk and dislocation-core diffusion, correspondingly. Superscript n is a coefficient describing the sensitivity of the strain rate to variations of stress σ, its value changes noticeably from a material to material and usually falls within the range 3 to 10; A is the Dorn constant [6].

The temperature at which a transition from dependence (11) to relationship (12) occurs is readily calculated for pure metals. In the simplest case this temperature can be found by equating the strain rates $\dot{\varepsilon}_v^{\,I} = \dot{\varepsilon}_v^{\,II}$. At usual parameter values $A=10^6$ (for fcc metals) or 10^8 (bcc metals) [6], $G\Omega/kT_m=40$, $n=5$, $D_{co}=10^{-2}$ cm²/s, $V_o=10^{-2}$ cm/s, $Q_c=10\ kT_m$ and a high ratio $\sigma^*/G=10^{-2}$, which is typical of ECAP, the transition temperature for fcc metals (for which the relationship $\Delta F(1-\sigma^*/\sigma_0) \ll Q_c$ is valid) equals to $0.65T_m$. For bcc metals where $\Delta F(1-\sigma^*/\sigma_0) \sim 3kT_m$, this temperature is noticeably lower and amounts to $0.33T_m$.

(b) Temperature dependence of the Hall-Petch coefficient. The dependence of the Hall-Petch coefficient, K, on temperature is virtually not studied. A number of indirect data evidences that in the range $(0.1–0.5)T_m$ the K value weakly changes with temperature. Henceforward in estimating $d^*(T)$ the temperature dependence K(T) is not accounted for. The K values used in calculation are listed in Table 1.

(c) The effect of strain-induced grain growth. With increasing the temperature of ECAP, the development of strain-induced grain growth process during deformation becomes possible. It can substantially affect the value of d^* measured after the completion of SPD and thus noticeably alter the dependence $d^*(T)$. To eliminate the effect of strain-induced grain growth, the deformation temperature should be below the recrystallization temperature of the MC material. Estimation of this temperature seems to be a rather complex problem. Simple approaches to its solution can be found in [10]. Below, when comparing the theoretical results with experiment, we consider, whenever possible, only such materials and process conditions where the deformation temperature does not exceed the recrystallization temperature.

3.2. The effect of temperature: comparison with experiment

(a) Pure metals. Only a very limited amount of papers report detailed data on the effect of the deformation temperature on the minimum grain size. However, known are the results of research where severe plastic deformation of different pure metals was performed at room temperature (T = 293 K) to attain a grain refinement limit. As the melting temperatures, T_m, of these materials are different, the obtained results of the grain refinement limit versus inverse homologous temperature T_m/T can be used for performing a preliminary estimate of the temperature dependence $d^*(T)$.

The values of the grain refinement limit d^* attained in a variety of pure metals during ECAP at room temperature are presented in Table 2. The data of d^* obtained in the conditions of torsion under quasi-hydrostatic pressure (TQHP) for the deformation of bcc metals are given in Table 2 too.

Table 1. The parameter values used in calculations

Parameter	Value (data from [6, 7])			
	Fe	Cu	Ni	Al
Activation energy for bulk self-diffusion, Q_v/kT_m	16.7	17.5	19.8	18.3
Activation energy for self-diffusion in equilibrium grain boundaries, Q_b/kT_m	11.6	9.2	8.0	10.7
Activation energy for dislocation-core diffusion, Q_c/kT_m	11.6	10.4	11.9	10.6
Activation energy for self-diffusion in a melt, Q_L/kT_m		3.6	3.3	3.8
Preexponential factor for bulk diffusion (cm^2/s), D_{v0}	2.0	0.2	1.9	1.7
Preexponential factor for the coefficient of grain-boundary diffusion ($10^{15}\cdot$m^3/s), δD_{b0}	11	5.0	3.5	50
Preexponential factor for the coefficient of dislocation-core diffusion ($10^{16}\cdot$cm^2/s), a^2D_{c0}	10	1	31	0.7
Preexponential factor for diffusion in melt (cm^2/s), D_{L0}	$8\cdot10^{-4}$			
Burgers vector (m$\cdot10^{10}$), b	2.48	2.56	2.49	2.86
Grain boundary width, δ	2b			
Atomic volume, Ω (cm$^3\cdot10^{23}$)	1.18	1.18	1.09	1.66
Melting temperature, T_m (K)	1808	1356	1726	933
Shear modulus, G (GPA)		42	79	25
Free energy, $\Delta F/kT_m$	$<10^{-5}$			
Stress required for overcoming obstacles at 0 K, σ_0/G	$<10^{-5}$			
Velocity of the dislocation glide, V_0 (cm/s)	10^{-2}			
Dorn constant, A	10^6			
Hall-Petch coefficient, K (MPa$\cdot$mm$^{1/2}$)	23	1.5	18	3.0
Factor of uniformity of plastic deformation, ξ	10^{-4}			
Numerical parameter, A_1	10.0			

Table 2. Effect of temperature on the grain refinement limit at ECAP and TQHP of pure metals

Metal		T_{def}, K	T_m, K	T_m/T_{def}	d^*_{ex}, µm	d^*_{th}, µm	φ	Refs.
Al	8 ECAP passes	293	933	3.18	1.3	0.87	3.1	this work
Cu	12 ECAP passes		1356	4.63	0.3	0.18		this work
Ni	12 ECAP passes		1726	5.89	0.1	0.12		this work
Fe	10 ECAP passes		1808	6.17	0.1	0.10	1.1	this work
Fe	TQHP		1808	6.17	0.1	#		[11]
Fe	TQHP		1808	6.17	0.16			[12]
Fe	TQHP		1808	6.17	0.4			[13]
Fe	TQHP		1808	6.17	0.15			
Cr	TQHP		2163	7.38	0.10			[14]
Mo	TQHP		2883	9.35	0.06			[15]

The value of strain rate is unknown.

If we apply the above described model of the grain refinement limit to analyzing these results and plot a graph for $y=3.5\ln(d^*/b)$ versus inverse homologous temperature $x = T_m/T$, it would be seen that this dependence for each of the group of the studied materials is well described by a straight line $y=A-\varphi x$. But the inclination of the line is substantially different for metals with different lattice types. For fcc metals it constitutes $\varphi_1\approx3.1$ while for bcc metals $\varphi_2\approx1.1$ (Table 2). According to the above described concept, the theoretical inclination φ is determined as:

$$\varphi = \frac{Q_b^*}{kT_m} - \frac{\Delta F}{kT_m}\left(1 - \frac{\sigma^*}{\sigma_o}\right). \tag{13}$$

(It is easy to make sure that the room temperature at which deformation of the given metals was performed is below the critical transition temperature T_1 and for describing the temperature dependence of $\dot{\varepsilon}_v$ by formula (10), expression (12) should be used). According to the concept developed in work [6], for fcc metals the second term in the right-hand side of equation (13) is small, and for these materials approximate formula $\varphi_1\approx Q_b^*/kT_m$ is valid, i.e. $Q_b^*/kT_m\approx3.1$. According to [6], for bcc metals $\Delta F/kT_m\approx3$ and $(1-\sigma^*/\sigma_0)\sim1$, so we obtain $\varphi_2\approx Q_b^*/kT_m-3$ and hence $Q_b^*\sim4.1$ kT_m. Thus, the values of Q_b^* for both groups of metals are $Q_b^*=(3.1-4.1)kT_m$.

The obtained values of Q_b^* are very close to those of the activation energy, Q_L, for diffusion in melts ($Q_L\sim(3-4)kT_m$) [7]. This surprising result can readily be interpreted within the frame of the conception developed in the theory of non-equilibrium grain boundaries [16].

(b) Al-Mg alloys. The results obtained in studying the effect of the deformation temperature on the grain refinement limit in aluminum alloys of the Al-Mg system with composition Al-X% Mg-0.22% Sc-0.15% Zr (X=0, 1.5 and 3%) have been presented in a paper [17]. ECAP (6 passes) was performed at 100, 160 and 200 °C. The experimental values of parameter φ (which is proportional to the activation energy of grain boundary diffusion $\varphi\sim Q_b^*/kT_m$) constitute $(4-5.6)kT_m$ and depend on the magnesium content. It should be noted that the activation energy of grain boundary diffusion Q_b^* in the magnesium-free alloy is close to Q_L in pure aluminum ($3.8kT_m$). For alloys containing 1.5 and 3% Mg, the Q_b^* value is by $1.7kT_m$ higher. The theoretical values of d^* calculated using the parameters listed in Table 1 are in good conformity with experimental data [17].

In estimating the effect of temperature on the grain refinement limit for alloy 5052 we used experimental data reported in [16] where ECAP was performed in the temperature range 50 to 300 °C with an increment of 50 °C. The calculated value of parameter φ in the temperature range 50–250 °C is about 4.9 kT_m. At T>200 °C, a steep rise in the dependence $Q_b^*(T)$ is observed [18]. This is connected with the fact that at high temperatures the grain refinement is impaired by deformation-induced grain growth (this was noted in [16]). As mentioned above, after the onset of grain growth expression (10) is not longer valid. The values of d^* calculated in the temperature range 50–300 °C using the parameters listed in Table 1 [18] appeared to be in good agreement with experimental data [16] except for T=300 °C when grain refinement is strongly affected by intensive grain growth.

(c) Magnesium alloys. The effect of deformation temperature on the grain refinement limit was examined in a recent paper [18] for the following magnesium alloys and experimental conditions: (a) alloy MA2-1, 6 ECAP passes, temperatures 250, 280 and 380 °C, (b) alloy AZ-91, temperatures 150 and 380 °C, and (c) alloy AZ-91 [19], ECAP temperatures 20, 300, 400 and 480 °C. Since experimental values of Q_L for magnesium alloys are not available in literature, we used the value of $4.5kT_m$ as a

preliminary estimate. The theoretical values of d^* calculated by equation (10) using the parameters shown in Table 1 appear to be in a reasonably good agreement with experimental data [18].

Acknowledgments

The authors are grateful to A.V. Nokhrin, I.M. Makarov and Yu.G. Lopatin from the Nizhniy Novgorod State University for assistance in experimental research and data analysis. The work was supported by the Russian Fundamental Research Foundation (grants 02-03-33043 and 03-02-16923), Russian Ministry of Education (grant E02-4.0-131), International Science and Technology Center (grant No.2809), Program "Basic Research in Higher Education (BRHE)" and Scientific-Educational Center "Physics of solid-state nanostructures" at the Nizhniy Novgorod State University.

References

1. Rybin V.V., *Large Plastic Deformations and Fracture of Metals.* Moscow: Metallurgiya, 1986, p. 224 (in Russian).
2. Perevezentsev V.N., Rybin V.V. Chuvil'deev V.N., *Poverkhnost': Fizika, Khimiya, Mekhanika [Surface: Physics, Chemistry, Mechanics]*, 1983, no. 10: 108-115 (in Russian).
3. Rybin V.V., *Izvestiya VUZov, Fizika*, 1991, no. 3: 7-22 (in Russian).
4. Likhachev V.A. and Khayrov R.Yu., *Introduction to the Theory of Disclinations.* Leningrad: Leningrad State University Press, 1975., p. 182 (in Russian).
5. Chuvil'deev V.N., *Nonequilibrium grain boundaries in metals. Theory and Applications.* Moscow: Fizmatlit, 2004, p. 304 (in Russian).
6. Frost H.J. and Ashby M.F. *Deformation-Mechanism Maps.* Oxford: Pergamon Press, 1982, p. 328.
7. Ubbelohde A.R., *The Molten State of Matter.* London: J. Wiley and Sons, 1978, p. 368.
8. Segal V.M., Reznikov V.I., Kopylov V.I., Pavlik D.A. and Malyshev V.F. *Processes of Plastic Structure Formation of Metals.* Minsk: Nauka i Tehnika, 1994, p. 232 (in Russian).
9. Kopylov V.I., "Application of ECAP-technology for producing nano- and microcrystalline materials". *Investigations and Applications of Severe Plastic Deformation 80*, NATO ASI Series 3, High Technology, Kluwer Academic Publisher. 2000, 23-27.
10. Chuvil'deev V.N., Makarov I.M., Kopylov V.I., *Materialovedenie [Materials Science]*, 1999, no.10: 9-13 (in Russian).
11. Valiev R. Z., Ivanisenko Yu. V., Rauch E. F. and Baudelet B., *Acta Mater.*, 44 (1996), 4705-4712.
12. Ivanisenko Yu. V., Sirenko A. A., Korznikov A. V., *Physics of Metals and Metallography*, 1999, no. 87 (4): 78-83.
13. Chashchikhina T. I., Degtyarev M. V., Voronova L. M. *et al.*, *Physics of Metals and Metallography*, 2001, no. 91 (5): 75-83.
14. Krasil'nikov H. A., *Physics of Metals and Metallography*, 2001, no. 91 (3): 81-87.
15. Korznikov A. V., Idrisova S. and Noskova N. I. *Physics of Metals and Metallography*, 1998, no. 85 (3): 113-117.
16. Chen Y.C., Huang Y.Y., Chang C.P., Kao P.W., *Acta Mater.* 51 (2001), 2005-2015.
17. Perevezentsev V.N., Chuvil'deev V.N., Kopylov V.I., Sysoev A.N., Langdon T.G., *Ann. Chim. Sci. Mat.* 27 (2002), 99-109.
18. Chuvil'deev V.N., Nieh T.G., Gryaznov M.Yu., Sysoev A.N., Kopylov V.I., *Scripta Mater.* 50 (2004), 861-865.
19. Kubota K., Mabuchi M., Higashi K., *J. Mater. Sci.* 34 (1999), 2255-2262.

Twist Extrusion as a Tool for Grain Refinement in Al-Mg-Sc-Zr Alloys

D. Orlov[1*], A. Reshetov[1], A. Synkov[1], V. Varyukhin[1],

D. Lotsko[2], O. Sirko[2], N. Zakharova[2], A. Sharovsky[2], V. Voropaiev[2],

Yu. Milman[2] and S. Synkov[1]

[1]Donetsk Phys.&Tech. Institute of the NAS of Ukraine, Donetsk, Ukraine
[2]Frantcevych Institute for Problems of Materials Science, Kiev, Ukraine

Abstract

Al-Mg-Sc-Zr alloys are attractive materials for aircraft and marine applications and have very good potentials for mechanical property enhancement by ultrafine grain structuring. The most developed technique for producing nanostructures has been equal channel angular extrusion. In this paper we describe our attempts on refining grain in these alloys by the twist extrusion technique. It was shown that twist extrusion allows to obtain fairly homogeneous structure with grain sizes d_{av}=0.33, 0.08, and 0.65 μm. This structure is thermally stable up to 500°C for 1 hour.

Introduction

Al-Mg-Sc-Zr alloys are attractive materials for aircraft and marine applications [1, 2, 3]. They are non-heat treatable alloys. which could achieve excellent formability and even superplasticity at elevated temperatures and high specific strength at ambient temperatures. Besides, these alloys show excellent weldability in combination with good corrosion resistance [1, 4, 5].

As it was shown in paper [6], optimum superplasticity was achieved in an alloy containing 3% Mg. On the other hand, in paper [7] it was suggested that total amount of Sc and Zr additions in Al-Mg alloy should be no more than 0.4 weight percents to achieve full solubility of the alloying elements.

The specific strength and formability of these alloys are determined by mutual influence of two factors: size and coherency of disperse secondary $Al_3(Sc,Zr)$ precipitates and grain sizes of aluminum matrix [4, 5, 8]. Size and coherency of the precipitates could be controlled by thermo-mechanical treatments. Whereas the most powerful ways for grain sizes control are severe plastic deformation (SPD) techniques [9, 10]. The later allows to obtain bulk materials with submicrometer and nanometer grain sizes [5, 8, 9]. For example, extremely high superplasticity in Al-Mg-Sc alloys [11] was achieved on billets processed by equal channel angular extrusion (ECAE) [12].

The present SPD processing techniques are proved as excellent tools for scientific purposes: studying of mechanisms of grain refining, influence of strain path, mode of deformation and other processing variables on microstructural evolution and mechanical properties of miscellaneous materials. But nevertheless

* Corresponding Author. orlov@donec.net (Dmitry Orlov)

Y.T. Zhu and V. Varyukhin (eds.), Nanostructured Materials by High-Pressure Severe Plastic Deformation, 77-82.

industrial implementation of the SPD techniques in their nowadays state is still limited due to unprejudiced reasons: volume of processed billets is relatively small and it is very difficult to be scaled up, the SPD processes are labour-intensive and cost of produced materials is quite high. Therefore developing of new SPD techniques and their implementation into really working industrial technologies is a grand question.

Main purpose of the present study was to show principal possibility of ultra-fined grain sizes developing in Al-Mg-Sc-Zr alloys by twist extrusion (TE), the quite young SPD technique recently proposed by professor Beygelzimer. Twist extrusion of the billet is schematically shown on figure 1. Principles of the twist extrusion are described elsewhere [13-15]. Equivalent strain per the TE pass in case of 60° die is about 1.2 [15].

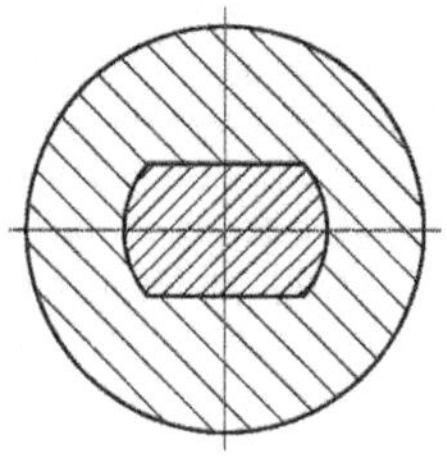

Figure 1. Scheme of the twist extrusion.

Figure 2. Scheme of billets for TE machining from ingots.

Investigation of straining paths during the TE and comparison of this processing with other SPD and conventional metal forming techniques will be reported in following publications.

Experimental Materials and Procedures

The materials used in this study is Al-3Mg-0.3Sc (wt.%) alloy additionally alloyed with 0.05, 0.10 and 0.15 wt.% of Zr. Melting was carried out in an induction furnace in ceramic crucibles in the air with the use of blowing the melt by argon. The melt was poured through a hole in the bottom part of the crucible with using a ceramic filter into a water-cooled copper mold with 55 mm in diameter.

Billets for the twist extrusion with dimensions of 18x28x80 mm^3 were machined from the ingots 55 mm in diameter as shown on figure 2. No thermal treatments or other processing were used before the TE processing. Grain sizes in the billets were about 100 μm.

The TE were performed at elevated (280-300°C) for a billets with 0.15% Zr and room for billets with 0.05% Zr and 0.10% Zr temperatures. The additional forced backpressure during the processing was about 200 MPa. It was performed 5 TE passes through 60° clockwise (CW) die for all the billets (accumulated equivalent strain $e \approx 5.8$). And billets with 0.10% Zr were additionally extruded through 60° counter-clockwise (CCW) die ($e \approx 6.9$).

Annealing of specimens was carried out in a muffle furnace in the air.

Specimens for mechanical testing, structural investigations and measuring hardness were electrolitically polished in the mixture of 70 ml $HClO_4$+930 ml CH_3COOH.

For structural investigations, the optical metallography (microscope MIM-9), transmission (JEM 100CX) and scanning (Superprobe-733) electron microscopes were involved.

Results and Discussion

In ingots containing Zr in the amount of 0.10 wt. % and higher a precipitation of primary intermetallic $Al_3(Sc,Zr)$ particles with size up to 1-2 μm in diameter (figure 3 a) were registered. After the twist extrusion processing all the particles was fragmented and has orientation similar to generatrix of slope of the twist line (figure 3 b). The length of the fibres was up to 30 μm.

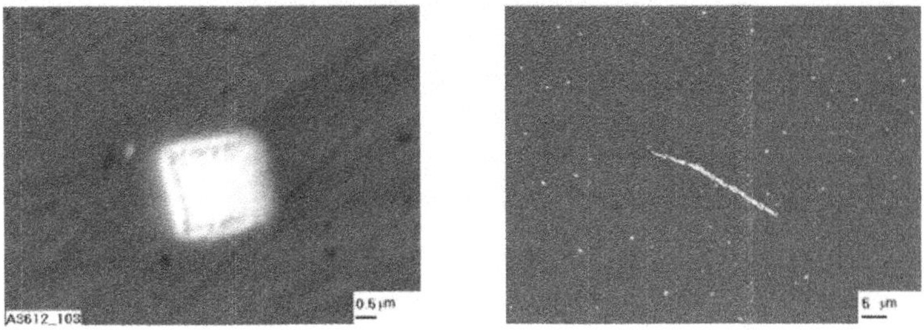

a b

Figure 3. SEM pictures: primary particle of $Al_3(Sc,Zr)$ in cross-section, as-cast structure – a; and the particle in longitudinal section after twist extrusion processing – b.

At the same time in the billet containing 0.05 wt.% Zr no big primary particles in the as-cast condition were observed, but after the TE processing it was observed quite big (up to 6 μm) particles. We attribute this to the fact that during an SPD processing velocity of diffusion increases for few orders [9]. On the other hand, speed of the TE was quite low – about 3 mm/sec. As a consequence, the particles had enough time for coagulation near embryos (small primary precipitates, impurities, etc.) formed during the casting and crystallization.

TEM investigation of the extruded billets has shown formation in them of a very fine wrought structure.

In the alloy contained 0.15 wt.% Zr and processed under 280-300°C grains sizes had the following characteristics: d_{av}=0.97; d_{min}=0.36; d_{max}=2.17 μm. This is higher than after processing by conventional warm direct extrusion under 350°C: d_{av}=0.643; d_{min}=0.26; d_{max}=1.81 μm. Therefore further processing was performed under room temperature conditions.

TE under room temperature of the alloy with 0.05 wt.% Zr showed much more finer structure: d_{av}=0.36; d_{min}=0.14; d_{max}=0.79 μm.

But the finest structure was obtained after the room temperature twist extrusion, 5 passes, through CW die with additional pass through CCW die of the alloy with 0.10 wt.% Zr. As it could be seen from figures 4 a and 5, structure was quite homogeneous in the center and at the edges of the processed billets. In the layer at the billets' edges of several millimeters in thickness a clear cellular structure has been formed with following grains sizes: d_{av}=0.33; d_{min}=0.08; d_{max}=0.67μm. In the center of the billets areas with cellular structure co-existed with areas of structures with continuous bending of the crystalline lattice with a wave length of about 0.5 μm. And a large amount of dislocations in cell body was observed (figure 4 b). The grain sizes are d_{av}=0.33; d_{min}=0.08; d_{max}=0.65 μm. This is comparable with ultra-fine grained structure developed by ECAE [5].

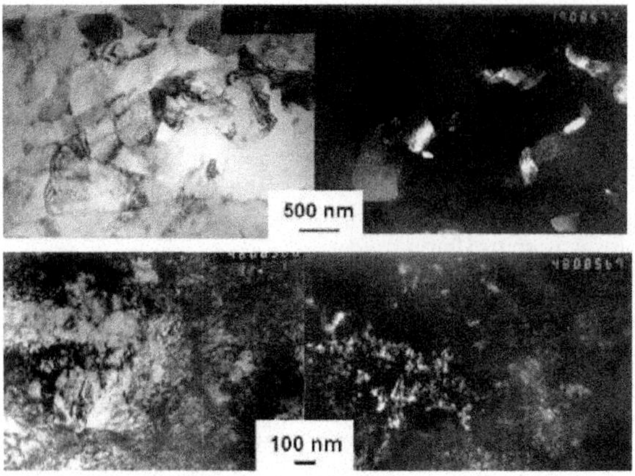

a

b

Figure 4. TEM microstructures: bright and dark field images of the billet's center – a; and enlarged picture to see the high density of dislocations – b.

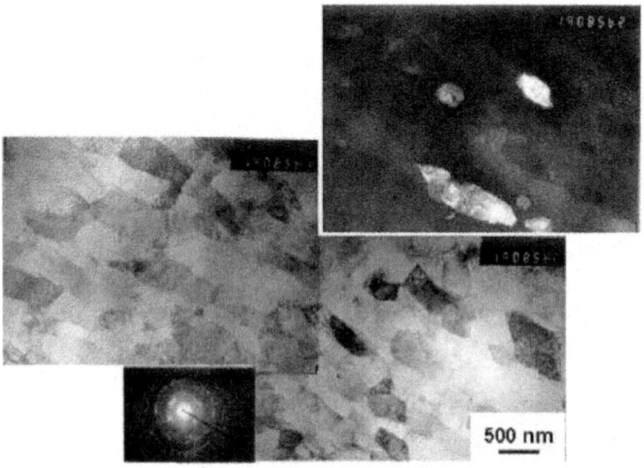

Figure 5. TEM microstructures: bright dark field images and SAD pattern of the billet's periphery.

At the same time, TEM registered no secondary coherent $Al_3(Sc,Zr)$ particles after twist extrusion at temperatures up to 300°C. These particles became visible after annealing extruded samples for 1 h at temperatures of 450°C and higher. Still literary data [3, 4] show that at least after extrusion at elevated temperatures such particles do exist, but particles to 4-5 nm in size are invisible in TEM image due to their coherency. In the billets processed at room temperature such particles are most probably absent, and a high level of their strength is formed by an ultra-fine structure together with strain fields from dislocations inside of grains.

Annealing has shown a rather good thermal stability of the cellular structure. After annealing at 400, 450, and 500 °C the average cross size of grains increased to 0.55 μm, 0.7 μm, and 1 μm, respectively, figure 6. This size is essentially lower than for the same alloy after ECAE treatment (5 μm after annealing at 500^0C in [5])

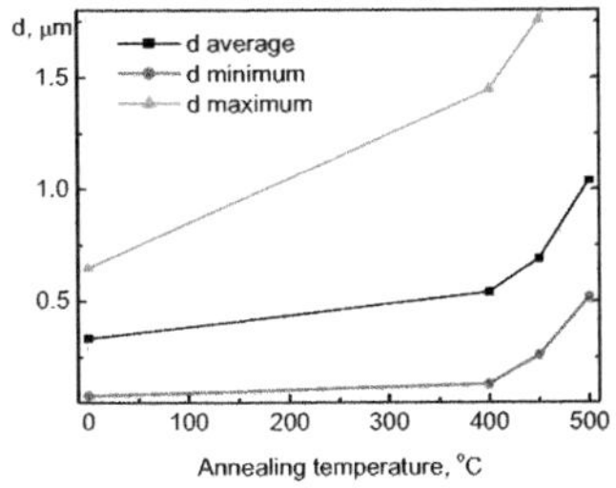

Figure 6. Dependence of the grain sizes on the annealing temperatures during 1 hour.

Conclusions

Twist extrusion of Al-Mg-Sc-Zr alloys under elevated and room temperatures was performed.

Grain structures in the alloys were effectively refined up to sub micrometer sizes by the twist extrusion processing.

The developed microstructure showed quite good thermal stability up to 500°C.

It seems to be good perspectives for enhanced superplastic properties in this materials obtaining by twist extrusion.

Acknowledgements

Authors are gratefully acknowledges National academy of sciences of Ukraine for financial support of the investigation, Grant for young scientists for 2003-2004.

References

1. Yu.A. Filatov, V.I. Yelagin, V.V. Zakharov *Materials Science and Engineering A280* (2000), 97–101.
2. A. Vinogradov, A. Washikita, K. Kitagawa, V.I. Kopylov *Materials Science and Engineering* A349 (2003), 318-326.
3. Yu.V.Milman "Yigh-strrength aluminum alloys" *Metalic Materials with High Structural Efficiency*" ed. Oleg N. Senkov, Sergiy O. Firstov, Daniel B. Miracle Kluwer Academic Publishers. Printed in the Netherlands 2004, 139-150.
4. Yu.V. Mil'man "Influence of Sc on structure, mechanical properties and corrosion resistance in Al alloys" *Progressive materials and technologies*, ed. I.K Pokhodnya, A.G. Kostornov, Yu.V. Koval', L.M. Lobanov, V.L. Naydek, A.G. Naumovets, M.V. Novikov, V.V. Panasyuk, V.V. Skorokhod, A.P. Shpak, B.V. Grinyov, O.V. Kurdyumov and O.V. Paustovs'kiy, Kyiv: National Academy of Sciences of Ukraine, Department of physical and technological problems of materials science, 2003, 335-359.
5. S. Lee, A. Utsunomiya, H. Akamatsu, K. Neishi, M. Furukawa, Z. Horita, T.G. Langdon *Acta Materialia* 50 (2002), 553–564.
6. Furukawa M., Utsunomiya A., Matsubara K., Horita Z., Langdon T.G. *Acta Mater* 49 (2001), 3829–3838.
7. V.N. Perevesentcev, V.N. Chuvil'deev, V.I. Kopylov, A.N. Sysoev, T.G. Langdon *Metally* 1 (2004), 36-43.

8. A. Gholinia, F.J. Humphreys, P.B. Prangnell *Acta Materialia* 50 (2002), 4461–4476.
9. R.Z. Valiev, R.K. Islamgaliev, I.V. Alexandrov *Progress in Materials Science* 45 (2000), 103-189.
10. www.nanospd.org
11. Z. Horita, M. Furukawa, M. Nemoto, A. J. Barnes and T. G. Langdon *Acta Mater.* 48 (2000), 3633–3640.
12. Segal VM. *Mater Sci Eng* A197 (1995), 157.
13. Beygelzimer Y., Orlov D. *Defect and Diffusion Forum*, V. 208-209, (2002), 311-314.
14. Beygelzimer Y., Orlov D. and Varyukhin V. "A new severe plastic deformation method: Twist Extrusion" // *Ultrafine Grained Materials II*; Ed. By Y.T.Zhu, T.G.Langdon, R.S.Mishra, S.L.Semiatin, M.J.Saran, T.C.Lowe. TMS (The Minerals, Metals & Materials Society). – 2002, 297-304.
15. Y. Beygelzimer, V. Varyukhin, D. Orlov, S. Synkov *Twist extrusion – process for strain accumulation* Donetsk: TEAN, 2003, p. 87 [In Russian].

Microstructural Characteristics of Ultrafine-Grained Nickel

Alexander Zhilyaev*

Naval Postgraduate School, Monterrey, CA 93943, U.S.A.

Abstract

A detailed investigation was conducted to evaluate the microstructural characteristics in samples of pure nickel processed using three different procedures of severe plastic deformation (SPD): equal-channel angular pressing (ECAP), high-pressure torsion (HPT) and a combination of ECAP and HPT. Several different experimental techniques were employed to measure the grain size distributions, the textures, the grain boundary character distributions (GBCD) and the boundary surface energies in the as-processed materials. It is shown that a combination of ECAP and HPT leads both to a greater refinement in the microstructure and to a smaller fraction of boundaries having low angles of misorientation. The estimated boundary surface energies were higher than anticipated from data for coarse-grained materials and the difference is attributed to the non-equilibrium character of many of the interfaces after SPD processing.

Introduction

The properties of metallic materials may be advanced through grain refinement by processing that includes severe plastic deformation. Many studies have shown that ultra-fine grain size in the sub-micrometer or even nanometer range can be achieved by imposing extremely large plastic strains during deformation processing (see for example [1, 2, 3]). In addition to equal-channel angular pressing [4], SPD methods include high-pressure torsion [5], accumulative roll bonding (ARB) [6], asymmetric rolling [7], friction stir processing (FSP) [8], redundant forging [9], high-energy ball milling [10] and sliding wear [11]. With appropriate parameters any of these processes may be employed to impart extremely large strains and achieve ultra-fine grains. There are two main issues for ultra-fine grained (UFG) metals and alloys: (i) microstructural characteristics such as a mean grain size, texture and grain boundary statistics achieved during SPD and (ii) thermostability of microstructure. If the ultrafine grain sizes are reasonably stable at elevated temperatures, there is a potential for achieving superplastic ductility at both unusually low testing temperatures and exceptionally rapid strain rates [12]. Very recent results have shown that SPD processing is capable of producing materials that combine both high strength and high ductility and this unusual combination has been attributed specifically to the development of unique nanostructures with non-equilibrium grain boundaries [2, 3].

The present paper is devoted to a thorough characterization of microstructure and grain boundary statistics in pure nickel samples fabricated by means of ECAP, HPT and their combination. The grain boundary (GB) energy of high angle boundaries was also evaluated.

* Corresponding Author: apzhilya@nps.edu. On leave from Institute of Mechanics, Russian Academy of Science, 450000 Ufa, RUSSIA.

Y.T. Zhu and V. Varyukhin (eds.), Nanostructured Materials by High-Pressure Severe Plastic Deformation, 83-88.

Experimental Procedures

High purity (99.99%) nickel was selected as a model material. All details on samples processing can be found elsewhere [13]. Briefly, the nickel was annealed for 6 hours at 973 K to give an initial grain size of ~100 mm before SPD processing.

Cylindrical billets of 16 mm in diameter and of 130 mm in lengths, were pressed at room temperature using a die having an internal angle of 90° between the two channels and relief angle of 20° at the point of intersection of the channels. All of the billets were pressed repetitively for a total of 8 passes under a pressure of ~800 MPa giving an equivalent strain of ~8, with the billets rotated by 90° in the same sense about the longitudinal axis between each separate pass in the processing procedure designated in literature as a route B_C. After pressing, an electric-discharge facility was used to cut small disks from the centers of the cylinders, with the disks lying perpendicular to the longitudinal axes of the billets, and these disks were used for the subsequent measurements. For HPT, samples were prepared in the form of disks having diameters of ~10 mm and thicknesses of ~0.3 mm and they were subjected to a torsion straining of 5 complete revolutions at room temperature under an applied pressure of 6 GPa. To examine the potential for improving the microstructure and the subsequent properties through a combination of ECAP and HPT, an additional sample was prepared using one of the disks cut from a billet after ECAP and then subjecting this disk to HPT for 5 revolutions under an applied pressure of 6 GPa: this specimen is henceforth designated ECAP+HPT.

Experimental measurements of grain size distribution, microtexture, GBs statistics have been performed using a Philips XL-30 scanning electron microscope (SEM) with a TSL orientation imaging system [14]. Internal stress and corresponding elastic energy stored in SPD nickel samples have been evaluated by using a double-crystal diffractometer with negligible instrument-induced broadening [15]. The thermal behavior of the samples was studied by a Perkin-Elmer DSC7 calorimeter within an atmosphere of pure argon [16]. The procedure adopted for estimating the grain boundary energy is outlined in the next section.

Results and Discussion

The grain size distribution. The value of the mean grain size represents the key parameter in defining the nature of the UFG microstructure formed through SPD processing. In general, the mean grain size is usually determined from measurements taken using TEM but these measurements tend to be difficult because it is well-established, after processing by both ECAP and HPT, that many of the

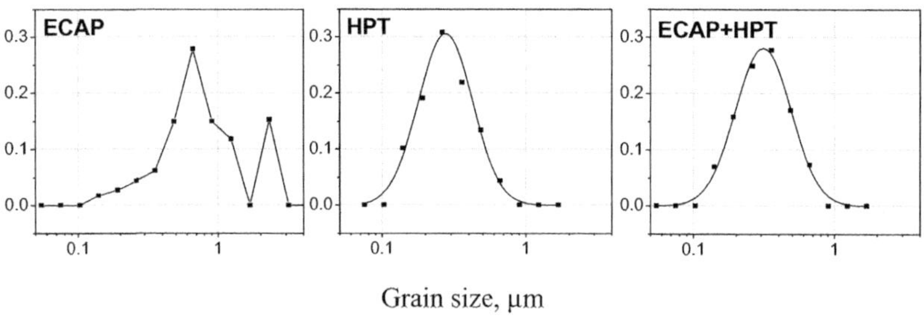

Figure.1. Grain size distribution in UFG nickel by OIM (tolerance angle is 2°).

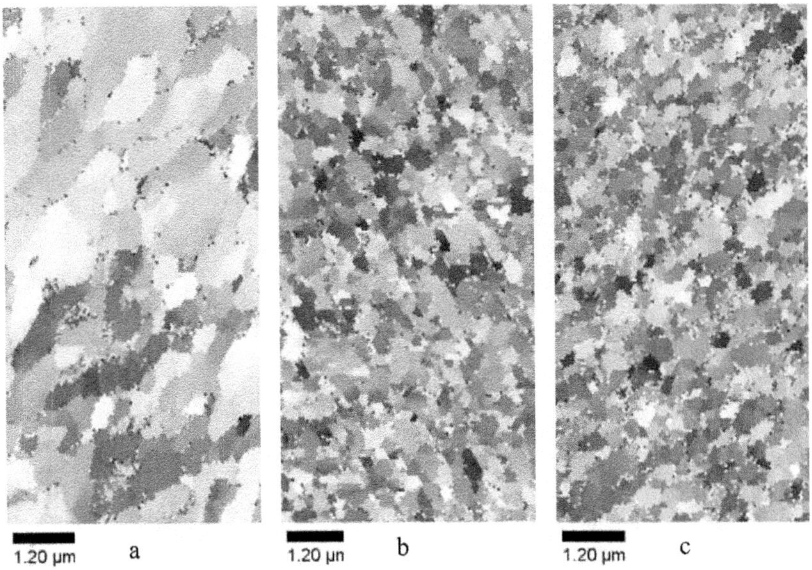

Figure 2. Microstructure of ECAP (a), HPT (b) and ECAP+HPT (c) nickel samples by OIM.

boundaries in the as-processed material are diffuse in nature or represent transition zones between highly deformed grains. In addition, the boundary extinction contours are often very irregular due to the non-equilibrium character of the boundaries and the presence of many extrinsic dislocations. Nevertheless, preliminary observations by TEM gave mean grain sizes in the Ni samples of ~350 nm after ECAP, ~170 nm after HPT and ~140 nm after ECAP+HPT, respectively. The lack of a homogeneous and clearly-defined array of equiaxed grains in many materials after SPD processing makes it difficult to fully characterize the UFG microstructure in terms of the distribution of grain sizes. The problem of microstructural characterization can be significantly overcome through the use of OIM. Figure 1 and 2 illustrate this by showing grain size distributions and images taken by OIM. A complete analysis of the grain size distribution in any UFG structure requires detailed OIM observations and the use of appropriate analytical software. This characterization introduces the ancillary problem of deciding upon the appropriate tolerance angle (TA), where TA is defined as a parameter of the OIM measurements such that two neighboring points are considered to belong to the same grain if the difference in their individual misorientations is less than TA. In general, the grain size distribution after ECAP appears to be multi-peaked whereas the distributions after HPT and ECAP+HPT are smoother and reasonably close to log-normal functions. It appears from Figure 1 that the distributions approximate to a log-normal function for the HPT and ECAP+HPT samples when taking TA = 2° and, despite some perturbations in the curve, the distribution after ECAP may be approximated also to a log-normal function. Thus, it is reasonable to define a tolerance angle of 2° as a suitable cut-off in defining the grain size in UFG microstructures. The grain sizes reported earlier after SPD processing were recorded using TEM and they correspond, to a first approximation, to the use of a tolerance angle of 2°.

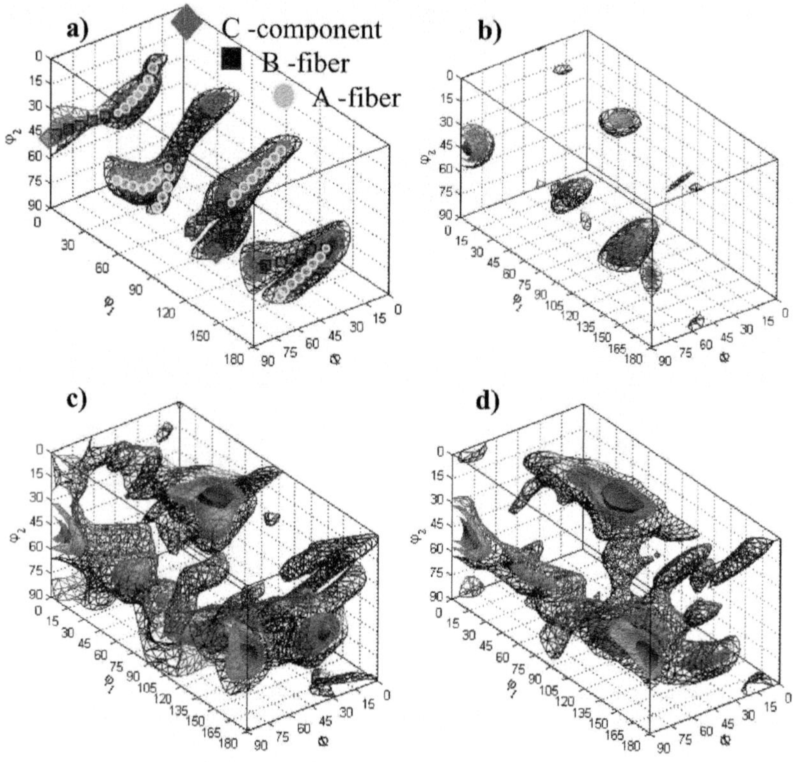

Figure. 3. 3D view of orientation distribution function: (a) – for model polycrystal; (b) – for ECAP nickel; (c) – for HPT nickel; (d) – for ECAP+HPT nickel. Iso-surfaces of ¼, ½ and ¾ of maximum intensity are plotted. Characteristic A-, B- fibers and C-component are presented in right down corner.

Microtexture in SPD nickel. Microtexture data obtained from OIM measurements are shown in Figure 3 for samples prepared through ECAP, HPT and ECAP+HPT as 3D views of the corresponding orientation distribution functions (ODF) calculated using harmonic methods. Following the work by Canova et al. [17], a detailed analysis of texture in pure aluminum after ECAP pressing was accomplished in[18]. In particular, it was established that all three components of torsion texture revealed in [17], namely the C-component ({001}<110>), A- fiber ({111}<hkl>) and B-fiber ({hkl}<110>), are observed in ECAP pure aluminum. Similar features have been discovered in ECAP and HPT nickel samples. Major texture maxima correspond to the C- component schematically depicted on Figure 3a. The three dimensional ODF for ECAP, HPT and ECAP+HPT nickel samples are plotted in the form of iso-surfaces with ¼, ½, and ¾ of absolute maximum for corresponding orientation distributions. It is noted that the ECAP texture (Fig. 3b) consists mostly of C-component and it shows a minor maximum near the end of the A-fiber. The texture of the HPT and ECAP+HPT nickel samples possess more complex structures but nevertheless the C-component dominates

Table1. GBs properties in UFG nickel subjected to severe plastic deformation.

GBs properties		Σ 1 %	Σ30 %	HAB %	ΔH, (J mol^{-1})	W_{EL}, (J mol^{-1})	W_{GB}, (J mol^{-1})
Samples	ECAP	23.2	16.8	60.0	59.3	17.6	41.7
	HPT	15.4	16.5	68.1	102.7	23.3	79.4
	ECAP+ HPT	10.5	18.8	70.7	187.9	42.5	145.4

GB energy. There is a difference in enthalpy released during DSC and elastic energy measured by X-ray and this is related to the liberation of GB energy. It is shown in the last column of Table 1. It is possible to deduce the total GB energy normalized per unit area using the following procedure. The grain boundary energy, W_{GB}, can be represented (in units of J/mol) as [16]:

$$W_{GB} = \alpha \frac{\gamma}{d_0} \Omega \cdot N_A \quad \text{and} \quad W_{GB} = \Delta H - W_{EL}, \tag{1}$$

where γ is the GB energy per unit area (J/m^2), α is a geometrical factor (for an idealized model of spherical grains, α=3), d_0 is the initial grain size, Ω =1.09·10-29 m^3 is the atomic volume and N_A is Avogadro's number. Then, γ can be estimated from the equation

$$\gamma = \frac{d_0 \cdot W_{GB}}{3 \cdot \Omega \cdot N_A} \tag{2}$$

All data required for evaluating of GB energy are shown in Table 1. The GB energy varies from 0.8 J/m^2 for the ECAP nickel to 1.0 J/m^2 for the HPT nickel giving an average value of 0.9 J/m^2 that corresponds to the sample obtained by a combination of ECAP and HPT. Taking into consideration that the real microstructure will decrease the geometrical parameter α to a value of about 1.3 and respectively increase the GB energy, and considering the fact that not all grain boundaries have high angles of misorientation, the evaluated energy will be even higher. From table 1 one can see that fractions of low-angle GBs are 23.2, 15.4 abd 10.5 percents in ECAP, HPT and ECAP+HPT nickel samples, respectively. Considering of the GB energy of low-angle boundaries leads to increase of these values to ~1.1, ~1.2 and ~1.0 J/m^2, correspondingly. These values calculated by this approach correspond now to GB energy of high angle grain boundaries. They are in good agreement with known literature data that gives values for the GB energy between 0.7 and 1.0 J/m^2.

Conclusion

Samples of pure nickel were processed by severe plastic deformation (SPD) using three distinct procedures: ECAP, HPT and ECAP+HPT. The results show the mean grain size is largest after ECAP, intermediate after HPT and the smallest grain size of ~140 nm was achieved after a combination of ECAP and HPT.

The grain size distribution is log-normal after HPT and ECAP+HPT and it approximates to a log-normal function after ECAP. The application of HPT after ECAP appears to significantly suppress the texture developed in ECAP.

Measurements of the grain boundary character distributions reveal the presence of two separate peaks corresponding to low angles ($\Sigma 1$) and high angles (>15°): the fraction of the low-angle peak decreases in the order from ECAP to HPT to ECAP+HPT.

The surface energies were estimated for high-angle random boundaries giving values of ~1.1, ~1.2 and ~1.0 $J \times m^{-2}$ after processing by ECAP, HPT and ECAP+HPT, respectively. These values are higher than anticipated for coarse-grained nickel (~0.7 $J \times m^{-2}$) and the difference is attributed to the non-equilibrium character of the grain boundaries after SPD processing.

Acknowledgements

This work was supported in part by INTAS-03513779 and the National Research Council of the National Academy of Science (U. S. A.).

References

1. Zhu Yu. Th., Langdon T. G., and Valiev R. Z., eds., *Ultrafine Grained Materials III*, Warrendale, PA: TMS, 2004.
2. Valiev R. Z., Islamgaliev R. K., and Alexandrov I. V., *Progr. Mater. Sci.* 45 (2000), 103-189.
3. Valiev R., *Nature Materials*. 3 (2004), 511-516.
4. Segal V. M., Reznikov V. I., Drobyshevskii A. E., and Kopylov V. I., *Izv. Akad. Nauk SSSR, Met.*, 1 (1981), 115-123.
5. Smirnova N. A., Levit V. I., Pilyugin V. P., *Phys. Met. Metall.* 61 (1986) p.127-133.
6. Saito Y., Utsunomiya H., Sakai T., Hong R.G., *Scripta Mater.* 39 (1998), 1221-1228.
7. Choi, C.-H., Kim, K.-H. and Lee, D. N., *Mater. Sci. Forum*, 273-275 (1998), 391-396.
8. Mishra R. S., Mahoney M. W., McFadden S. X., Mara N. A., Mukherjee A. K., *Scripta Mater.* 42 (1999), 163-168.
9. Kaibyshev O. A., Kaibyshev R., and Salishchev G., *Mater. Sci. Forum*, 113-115 (1993), 423-428.
10. Koch C. C. and Cho Y. S., *Nanostruct. Materials*, 1 (1992), 207-212.
11. Rigney D. A., *Ann. Rev. Mater. Sci.*, 18 (1988), 141-211.
12. McFadden S.X., Mishra R.S., Valiev R.Z., Zhilyaev A.P., Mukherjee A.K., *Nature* 398 (1999) 684-686.
13. Zhilyaev A.P., Nurislamova G.V., Kim B-K., Baró M.D., Szpunar J.A., Langdon T.G., *Acta Mater.* 51 (2003) 753-765.
14. Zhilyaev A.P., Kim B-K., Nurislamova G.V., Baró M.D., Szpunar J.A., Langdon T.G., *Scripta Mater.* 48 (2002) 575-580.
15. Zhilyaev A.P., Gubicza J., Nurislamova G., Révész Á., Suriñach S., Baró M.D., Ungár T., *Phys. Stat. Sol.* (a) 198 (2003) 263-271.
16. Zhilyaev A.P., Gubicza J., Suriñach S., Baró M.D., Langdon T.G., *Mater. Sci. Forum*, 426-432 (2003) 4507-4212.
17. Canova G. R., Kocks U. F., Jonas J. J., *Acta Metall.* 32 (1984) 211-226.
18. Swisher D., Oh-Ishi K., Zhilyaev A. P., McNelley T. R., (2004), to be published.

Refinement and Densification of Aluminum Nickelides by Severe Plastic Deformation

Laszlo J. Kecskes,[1*] Danielle M. Gardiner,[1] Robert H. Woodman,[1] Robert E. Barber[2] and K. Ted Hartwig[2]
[1]U. S. Army Research Laboratory, Aberdeen Proving Ground, MD 21005-5069
[2]Department of Mechanical Engineering, Texas Agriculture and Mining University College Station, TX 77843-3123

Abstract

Two nickel (Ni)-coated aluminum (Al) powders were consolidated to full or near-full density by severe plastic deformation using the equal-channel angular extrusion (ECAE) technique. Mixtures of 78Al-22Ni at.% (63Al-37Ni wt.%) or 39Al-61Ni at.% (23Al-77Ni wt.%) were placed in square-shaped copper (Cu) or Ni casings, sealed, preheated to a range of temperatures from ambient to 700 °C, and, once at a uniform temperature, dropped into the ECAE die and single-pass extruded. It was found that the preheating temperature affected the transformation of the initial Al-Ni composition into aluminum nickelide (Al-Ni) intermetallics. It was also found that the use of the nickel can significantly affect the densification of the powders. Scanning electron microscopy, energy dispersive x-ray spectroscopy, x-ray diffraction, and microhardness measurements were used to examine the nature of the resultant intermetallics. The onset and nature of the transformation from the precursors into the products were further studied by differential thermal analysis. These results and the role of ECAE on the transformation are discussed.

Introduction

Aluminum nickelides have been the subject of many studies over the past three decades. As indicated by K. Morsi [1], of the five Al-Ni intermetallic phases in the Al-Ni system, AlNi and $AlNi_3$ have received the most attention. Because of their high strength-to-weight ratio, good high-temperature stability, and good corrosion resistance, Al-Nis may be found in a large variety of structural applications such as aircraft or automotive components. Of the many methods attempted to process aluminides [2-6], combustion synthesis (CS), wherein the products are directly synthesized from their precursors, offers a latitude in process conditions that are unmatched by other, more conventional techniques. CS of metastable Ni + Al mixtures has been studied by several groups [7-12]. It has been shown that the CS reaction initiates once a threshold temperature, T_R, near the melting point of Al, $T_m(Al) = 660$ °C, is exceeded. The stability of the CS reaction strongly depends on conditions such as the initial pressure, temperature, particle size, and heating rate of the reactants.

CS products are often porous; the primary source is the intrinsic specific volume difference between the reactants and the products. Shock compaction techniques have been used to generate Al-Nis directly from Ni + Al powder mixtures [13-17] without altering the metastable structure of the nickelide. During consolidation, interparticle friction can cause CS reaction initiation well below T_R [18]. However,

* Corresponding Author. kecskes@arl.army.mil

Y.T. Zhu and V. Varyukhin (eds.), Nanostructured Materials by High-Pressure Severe Plastic Deformation, 89-94.

product structures remain porous, as CS could take place simultaneously with the passage of the shock wave. Consequently, it was hypothesized that a more benign processing method, wherein heat generation and dissipation is better controlled, while maintaining sufficient consolidation pressure, would more likely result in an improved product structure.

We have shown previously that using ECAE, a metastable Ni + Al precursor could be converted into stable, single- or two-phase Al-Nis [19] and subsequently consolidated to moderate densities. In this follow-up effort, we show new results with improvements on the densification aspects of our previous samples using Ni as an alternate sample containment material. Again, scanning electron microscopy (SEM), energy dispersive x-ray spectroscopy (EDS), and x-ray diffraction (XRD) were used to evaluate the extrudate products.

Experimental Procedures

Precursor Preparation

Two Ni-coated pure Al precursors, Al-22Ni and Al-61Ni (at.%) were prepared from Al powder, Ni sulfate, in an aqueous ammonium hydroxide solution by a hydrometallurgical method. The coated powder was separated, washed several times in water, and dried in hydrogen. Using SEM, the surface morphology and the cross-sectional profiles were observed. The relative amount of each major component was determined by inductively coupled plasma spectroscopy with levels of sulfur and oxygen impurities determined as well. Differential thermal analysis was performed on each powder at a 3 K/min heating rate in air, up to at least 700 °C, beyond T_m(Al), to determine the T_R of the Ni + Al CS reaction.

ECAE Procedures

A multisample billet geometry, designed to simultaneously coextrude both powders, was used. The powders were loaded into either copper (Cu) or Ni billets. Each of powders was placed into 6.4-mm-diameter, 3-to-4-cm-deep hole in the billet. A muffle furnace was heated to 300, 600, or 700 °C, and the tooling was warmed to 300 °C. A thermocouple was placed under each billet in the furnace to verify the temperature, and it was found that the temperature did not vary more than 10 °C from the set point. At the time of extrusion, the temperature varied less than 2 °C. The load and stroke were recorded for each extrusion.

After ECAE, the billets were sectioned. Representative longitudinal and transverse sections were taken, and the remnants of the Cu or Ni casing were removed. These sections were mounted and polished for examination by SEM and EDS. Elemental ratios were obtained using standardless semiquantitative analysis. Other sections were also cut for XRD analysis. Finally, microhardness measurements, using a 100-gf Vickers indenter, were performed on the polished samples. About 15 to 20 hardness measurements were impressed into each sample at arbitrary locations across the cross section.

Results and Discussion

Precursor Characteristics

SEMs of the precursor powders and their respective cross sections are shown in Figure 1. Both precursors have a 20-100 μm, rounded shape. The coating exterior is studded with Ni nodules. The coating of Al-61Ni is about 7-10 μm thick, completely covering the Al particle surface; the thickness, however, is not uniform. The Ni coating of Al-22Ni is thinner, about 1-2 μm thick; it does not completely cover the particle surfaces. Chemical analysis measured negligible amounts of sulfur, 0.42 wt.-% oxygen (O) on Al-61Ni, and 0.28 wt.-% O on Al-22Ni. No other impurities were detected.

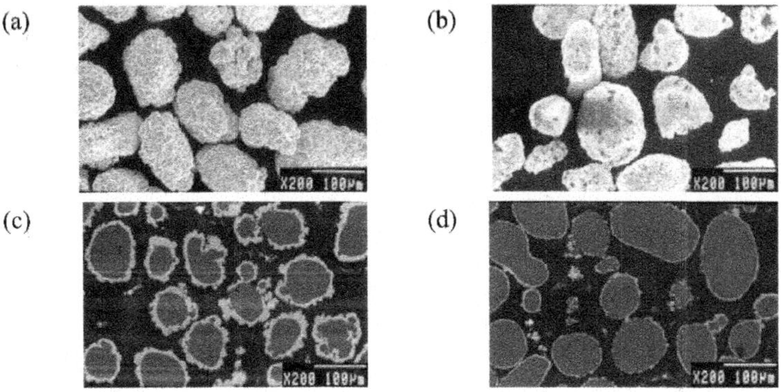

Figure 1. SEMs of the precursors: Al-61Ni is shown in (a), Al-22Ni is shown in (b), with the corresponding cross sections shown in (c) and (d), respectively.

ECAE Results

Table 1 summarizes the XRD and microhardness results for the extrudates. The load versus stroke data for each extrusion indicated that the loads were not excessive. All single-pass extrusions were uneventful. In some cases, slightly higher loads were attributed to poor lubrication and increased flash. The billets did not shorten appreciably, indicating perhaps that compaction or consolidation was not complete. Upon extraction, the powder samples were intact, but remained friable.

Table 1. Summary of the ECAE Al-Ni Results

Billet#	**T [°C]**	**Al-61Ni at.%**		**Al-22Ni at.%**	
		XRD Phases	**100-gf Microhardness [kg/mm^2]**	**XRD Phases**	**100-gf Microhardness [kg/mm^2]**
1-Cu	300	Al, Ni	210	Al, Ni	48
2-Cu	600	Al, Ni	140	Al, Ni	100
3-Cu	700	Al_3Ni_2, AlNi, Ni	226	Al_3Ni_2, AlNi, Al_3Ni	195
4-Ni	300	Al, Ni	126	Al, Ni	74
5-Ni	600	Al, Ni	125	Al, Ni	52
6-Ni	700	Al_3Ni_2, AlNi, Ni	636	Al_3Ni_2, AlNi, Al_3Ni	670

At 300 and 600 °C, for the Cu-cased Al-61Ni, XRD show Al and Ni only. In contrast, at 700-°C, the extrudate shows a partial reaction. Al is absent; in addition to Ni, a small amount of Al_3Ni_2 is detected. XRD of the 300-, 600-, and 700-°C Al-22Ni extrudates revealed similar results. At 300 and 600 °C, only Al and Ni are indicated. In contrast, at 700-°C, XRD reveals no Al or Ni; only Al_3Ni_2 and Al_3Ni are present.

Perpendicular cross sections from the 600- and 700-°C Al-61Ni Cu-cased samples are shown in Figures 2(a) and 2(b). (It may be noted that there was little or no difference between the 300- and 600-°C extrudates for both compositions.) In the 600-°C extrudate, the coated particles are ductile, deform, and flow readily. The Ni coating does not debond during extrusion but adheres well to the Al particles. The sample shows an orientation effect along the extrusion direction, with a grain aspect ratio of 3:1. Transverse to the extrusion direction, grains appear less skewed and are more isotropic. EDS shows Al and Ni only (dark and light gray phases in the images, respectively). The consolidation, however, is incomplete. Isolated voids remain within and at boundary points between particles. At 700 °C, the morphology is markedly different. The structure consists of partially reacted and deformed precursor particles with hollow interiors. A large number of cracks signify that it was brittle during extrusion. It is believed that the interior voids arose from the specific volume change upon conversion from Ni + Al into Al-Nis. Note the interfacial layer of unreacted Ni. EDS verified that the primary gray phase is Al_3Ni_2. In addition, there are regions that correspond to AlNi. Smaller regions with Ni/Al ratios of 4:1 were presumed to be $AlNi_3$.

SEMs of transverse cross sections from the 600- and 700-°C Al-22Ni samples are also shown in Figure 2. Aside from the thinner Ni coating, the morphology in Figures 2(a) and 2(c) are similar. The 600-°C consolidation of Al-22Ni is more complete; little or no voids are observed. In contrast, the morphology of the 700-°C extrudate is different from those at lower temperatures as well as that of the Al-61Ni extrudate. The extrudate is highly porous (see Figure 2(d)) consisting of an isotropic distribution of fine-grained, 3-10 μm particles that are considerably smaller than the initial Al base particles. Most of the particles were determined to be Al_3Ni_2, with a few larger particles being AlNi, Al_3Ni, and Ni.

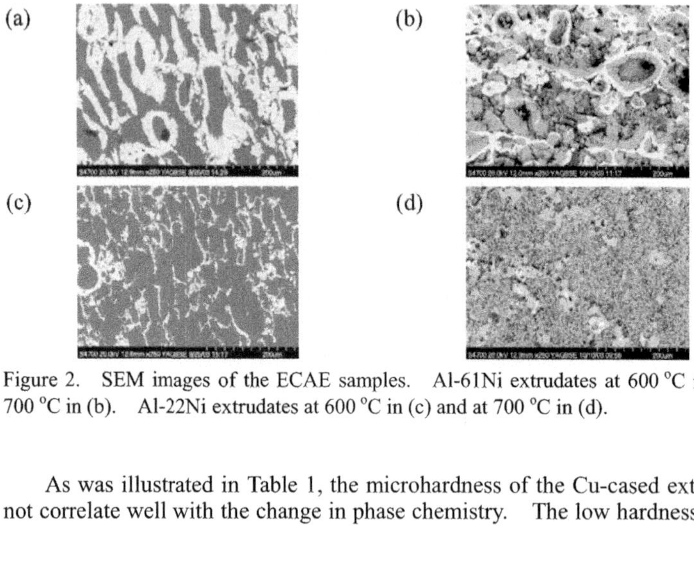

Figure 2. SEM images of the ECAE samples. Al-61Ni extrudates at 600 °C in (a) and at 700 °C in (b). Al-22Ni extrudates at 600 °C in (c) and at 700 °C in (d).

As was illustrated in Table 1, the microhardness of the Cu-cased extrudates did not correlate well with the change in phase chemistry. The low hardness, attributed to lack of consolidation (*i.e.*, the residual porosity was too high), could not be used as a clear diagnostic to confirm that a reaction took place. In Al-61Ni, the samples had only a gradual rise in hardness values, corresponding to a conversion into intermetallics without a clear transition. Data for Al-22Ni show a definite transition from low to high hardness values between 600 °C and 700 °C, suggestive of a reaction. It is also speculated that the thicker Ni coating of Al-61Ni could have biased the hardness to artificially higher values.

Prior data indicated that exotherms initiate in both powders at T_R of 640 ± 6 °C. Al-22Ni was considerably more exothermic than Al-61Ni and, the latter reaction remained incomplete, even when heated to 700 °C. ECAE of the Cu-cased extrudates below T_R was effective in fully consolidating ductile Ni + Al precursors. However, above T_R, full densification by ECAE of fully or partially reacted Al-Nis was unsuccessful, though in Al-22Ni, ECAE sufficiently disturbed the as-formed structure of the Al_3Ni_2, AlNi, and Al_3Ni intermetallics such that the extrudate developed an uniformly dispersed structure.

Additional experiments using Ni casings, showed great improvement in consolidation. As shown in Figure 3 and Table 1, aside from a dramatic rise in sample microhardness at 700 °C, the general appearance of morphology and phase chemistry of the Ni-cased extrudates is unchanged from those of the Cu-cased extrudates. The rise in hardness is most likely caused by the elimination of porosity. It is presumed that the higher flow stress (relative stiffness) of Ni *versus* that of Cu, at the extrusion temperature, effectively resulted an increase in densification.

(a) (b) (c) (d)

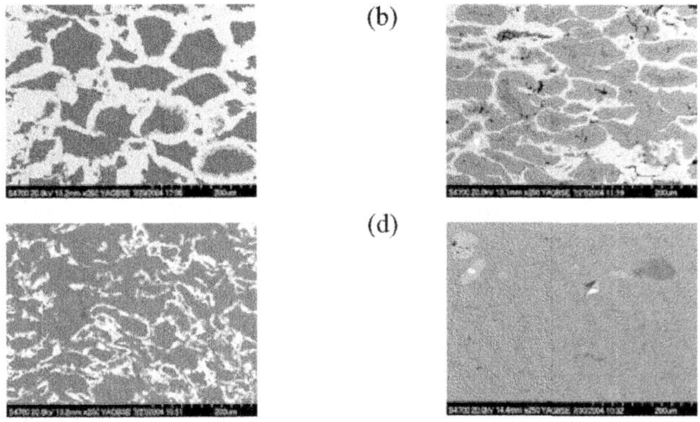

Figure 3. SEM images of the ECAE samples. Al-61Ni extrudates at 600 °C in (a) and at 700 °C in (b). Al-22Ni extrudates at 600 °C in (c) and at 700 °C in (d).

Conclusions

A series of ECAE experiments using Cu or Ni casing materials were conducted to extrude Ni-coated Al to form Al-Nis. Extrudate structures obtained from both powders below T_R reflected the initial coated powder morphology. During preheating, above T_R, Al-Ni phases formed as a result of a reaction between Al and Ni. The resultant Al-22Ni had an isotropic, uniformly dispersed, multiphase structure, consisting of primarily Al_3Ni and Al_3Ni_2, interspersed with AlNi and unreacted Ni. For Al-61Ni, ECAE yielded contrasting results; below T_R, the initial precursor morphology was preserved. Above T_R, the heterogeneous product mainly consisted of more brittle AlNi and $AlNi_3$ intermetallics, and unreacted Ni; lesser amounts of Al_3Ni_2 also formed.

Consolidation quality of the extrudates below T_R was very good. However, above T_R, both Cu-cased extrudates remained highly porous. The uniformly dispersed Al-22Ni product phases were fully consolidated using a Ni casing. The heterogeneous structure of Al-61Ni was also fully consolidated.

Acknowledgements

The authors thank H. Kim for preparing the metallographic samples and performing the microhardness measurements. We would also acknowledge Prof. T. Geleishvili and Dr. A. Peikrishvili for providing us with the coated precursor powders.

References

1. Morsi, K., *Mater Sci Eng A*, A199 (2001), 1-15.
2. Cheng, T., and McLean, M., *Mat Letts*, 29 (1996), 91-99.
3. Deevi, S.C., and Sikka, V.K., *Intermet*, 5 (1997), 17-27.
4. Liu, Z.G., Guo, J.T., and Hu, Z.Q., *Mater Sci Eng A*, A192-193 (1995), 577-582.
5. Cardellini, F., Mazzone, G., and Antisari, M.V., *Acta Mater*, 44 (1996), 1511-1517.
6. Abe, O., and Tsuge, A., *J Mater Res*, 6 (1991), 928-934.
7. Philpot, K.A., Munir, Z.A., and Holt, J.B., *J Mater Sci*, 22 (1987), 159-169.
8. Alman, D.E., *J Mater Sci Lett*, 13 (1994), 483-486.
9. Lebrat, J.P., Varma, A., and McGinn, P.J., *J Mater Res*, (9) (1994), 1184-1192.
10. Li, H.P., and Sekhar, J.A., *J Mater Res*, 10 (1995), 2471-2480.
11. Plazanet, L., and Nardou, F., *J Mater Sci*, 33 (1998), 2129-2136.
12. Fan, Q., Chai, H., and Jin, Z., *Intermet*, 9 (2001), 609-619.
13. Horie, Y., Graham, R.A., and Simonsen, I.K., "Observations on Shock-Synthesis of Intermetallic Compounds," *Metallurgical Applications of Shock-Wave and High-Strain-Rate Phenomena*, (New York, NY: Marcel Dekker Inc, 1986), 1023-1035.
14. Song, I., and Thadhani, N.N., *Metall Trans*, 23A (1992), 41-48.
15. Thadhani, N.N., Work, S., Graham, R.A., and Hammetter, W.F., *J Mater Res*, 7 (1992), 1063-1075.
16. Dunbar, E., Thadhani, N.N., and Graham, R.A., *J Mater Sci*, 28 (1993), 2903-2914.
17. Aizawa, T., *Ceram Eng Sci Proc*, 18 (1997), 573-580.
18. Kecskes, L.J., Szewczyk, S.T., Peikrishvili, A.B., and Chikhradze, N.M., *Met Mat Trans*, 3A, (2004), 1125-1131.
19. Kecskes, L.J., Szewczyk, S.T., Woodman, R.H., Barber, R.E., and Hartwig, K.T., "Processing of Aluminum Nickelides by Equal-Channel Angular Extrusion," *Ultrafine Grained Materials III*, (Warrendale, PA: TMS, 2004), 327-332.

Low-Temperature Sheet Rolling of TiAl Based Alloys

M.R. Shagiev*, G.A. Salishchev

Institute for Metals Superplasticity Problems, 39 Khalturin Str., Ufa, 450001, Russia

Abstract

In the present study a considerable decrease in the sheet rolling temperature of γ-TiAl based alloy was achieved through grain refinement. The use of preforms with homogeneous micro- and submicrocrystalline structure made possible to perform rolling of γ-TiAl alloy at temperatures as low as 800-1000°C. The decrease in the rolling temperature was directly related to a decrease in the brittle-to-ductile transition temperature. A homogeneous microcrystalline structure was found to be an optimal initial condition for pack rolling at strain rates of about 10^{-1} s^{-1}. Along with retained fine-grained microstructure in the γ sheets, which provides superior superplastic elongations, the decrease in rolling temperature leads to a substantial decrease in their cost because of energy savings, applicability of cheaper canning materials than those currently being used, less oxidation during preheating and thus a better surface finish.

Introduction

Sheets of γ-TiAl based alloys that possess improved superplastic (SP) properties are of great interest for subsequent SP forming and production of honeycomb sandwiches for aircrafts [1,2]. At the present time sheet rolling of γ alloys is generally being performed at temperatures as high as 1200-1300°C [1-5]. The reason of such high-temperature processing is rather coarse-grained microstructure (d=10-30 μm) of the rolling preforms which results in brittleness up to 1100-1200°C [6]. Processing at these temperatures is very energy consuming and requires expensive tool and canning materials to protect the intermetallic preforms from severe oxidation. Moreover, the sheets have a coarse-grained structure, and thus high temperatures are required for secondary forming operations, such as forging and SP forming [1,2,7]. On the other hand, it is well known that a finer grain structure results in decreasing SP temperatures and reducing cavitation during superplasticity [8,9]. Besides, the brittle-to-ductile transition (BDT) temperature can be also decreased by grain refinement [10]. The latter will make possible to perform rolling at lower temperatures, at which the microstructure is more stable. Decreasing the rolling temperature will retain the fine-grained microstructure in the sheets thus providing improved SP properties [8,9].

In the present work the effect of grain size on the BDT temperature and thermal stability of the microstructure of the stoichiometric TiAl alloy was studied. Based on these results the temperature ranges for sheet rolling were defined and the optimal microstructural condition in the rolling preforms were determined and used for experimental pack rolling of the TiAl sheet.

Experimental Procedures

Cast Ti-50 at. % Al (hereafter TiAl) alloy was used as a starting material. Micro-

---

* Corresponding Author. marat@imsp.da.ru

Y.T. Zhu and V. Varyukhin (eds.), Nanostructured Materials by High-Pressure Severe Plastic Deformation, 95-100.

(MC condition) and submicrocrystalline (SMC condition) microstructures were produced by multistep isothermal forging. The details of grain refinement method are given in [11].

Flat samples with a gauge size of 10×5×2 mm^3 were tensile tested in a Schenck Trebel RMC 100 machine at temperatures in the range of 700-1100°C with an initial strain rate of 1×10^{-1} s^{-1}. The yield strength and the total elongation both of MC and SMC conditions were determined.

Rolling preforms of the TiAl alloy in both the SMC and MC conditions were sealed in 18/10 stainless steel cans. The packs were soaked in a furnace at temperatures of 750-1000°C for 15 minutes and then hot rolled in a duo rolling mill using a speed of 50 mm/s (a strain rate of about 10^{-1} s^{-1}) and 5-10% nominal reduction per pass with the total reduction up to 50-80%. Intermediate reheats for 5 minutes were conducted between passes.

Microstructure analysis was performed using an optical microscope Neophot-32 and a transmission electron microscope JEM-2000EX operating at an accelerating voltage of 200 kV. The mean grain size was determined from measurements conducted on 70 to 100 grains using the linear intercept method.

Results and Discussion

Microstructure after hot working

Figures 1 and 2 show typical microstructures of the TiAl alloy after isothermal forging. The mean grain size was measured to be 5 μm and 0.4 μm for the MC and SMC conditions, respectively. The volume fraction of the α_2-phase was 3-5% in both conditions. Some variation in the grain sizes in the MC condition (Figure 1a) was caused by a non-homogeneous distribution of the α_2-phase particles. The SMC microstructure was found to be more homogeneous (Figure 2a). The mean size of the α_2-phase particles was about 0.5-1 μm and 0.1 μm in the MC and SMC conditions, respectively.

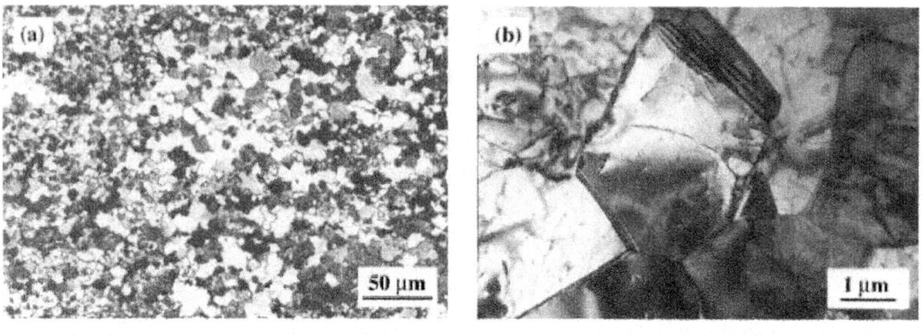

Figure 1. OM (a) and TEM (b) microstructures of the TiAl alloy after isothermal forging at 950°C (MC condition).

A high dislocation density ($\rho\sim5\times10^9$ cm^{-2}) and extinction contours caused by elastic stresses were observed in the SMC specimens (Figure 2b), while a lower dislocation density ($\rho\sim10^8$ cm^{-2}) was detected in the MC specimens (Figure 2a).

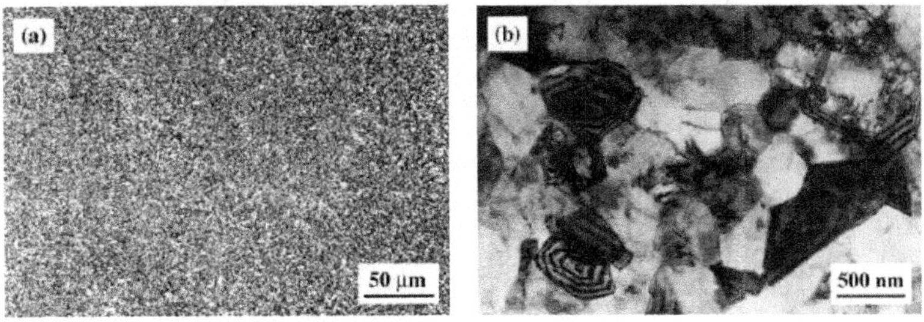

Figure 2. OM (a) and TEM (b) microstructures of the TiAl alloy after isothermal forging at 800°C (SMC condition).

Brittle-to-ductile transition at a strain rate of 10^{-1} s^{-1}

The temperature dependencies of the total elongation and yield strength for the TiAl specimens are shown in Figure 3. The total elongation of TiAl, both in the MC and SMC conditions, increased and the yield strength decreased continuously with increasing temperature from 700°C to 1000°C. The elongation increased rapidly from 0.5-20% at 700°C to 130-140% at 800°C (Figure 3a), indicating the BDT occurred in this temperature interval.

Since sheet rolling should be performed in the ductile region, the BDT temperature determines the lower temperature limit for rolling at strain rates of about 10^{-1} s^{-1}. Thus, the rolling temperature of the TiAl alloy in both the MC and SMC conditions should be higher than 700°C. On the other hand, sheet rolling should be performed at temperatures where the microstructure is thermal stable. If this is not the case, considerable grain growth will lead to increase in the BDT temperature and thus increase the minimum sheet rolling temperature.

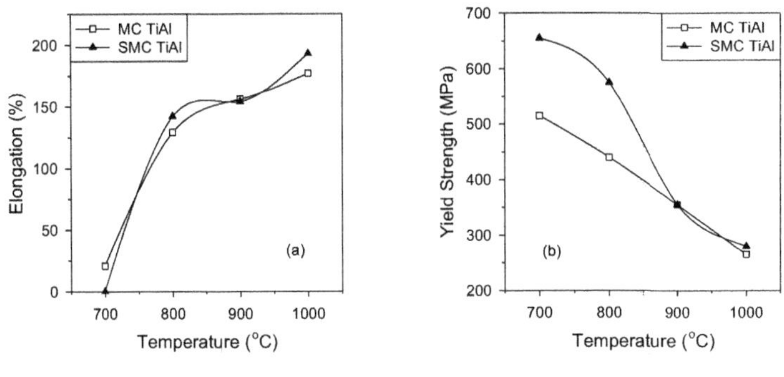

Figure 3. Temperature dependencies of (a) total elongation and (b) yield stress of the TiAl alloy specimens in the MC and SMC conditions. Initial strain rate was 1.0×10^{-1} s^{-1}.

Thermal stability of the MC and SMC microstructures

Figure 4 shows the temperature dependencies of the mean grain size of TiAl in the MC and SMC conditions after annealing for 30 minutes. Only a slight grain growth occurred in the MC condition after annealing up to 1000°C while heating at higher temperatures resulted in rapid grain growth (Figure 4a). In the SMC condition the rapid grain growth started from 850°C (Figure 4b). Grains retained submicron sizes after 30 minutes annealing at temperatures up to 875°C.

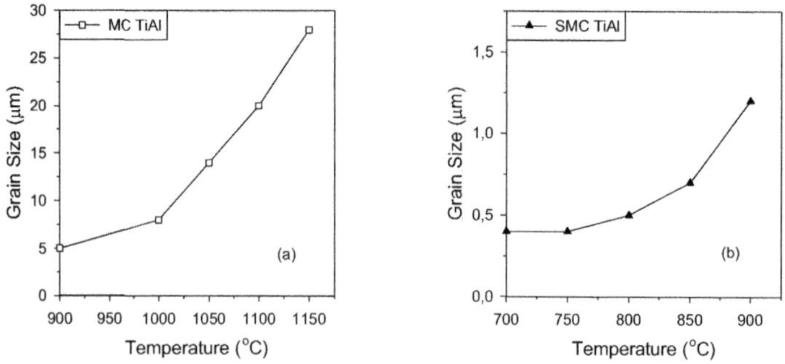

Figure 4. Dependencies of the mean grain size on the annealing temperature for the TiAl alloy in (a) MC and (b) SMC conditions. Annealing time was 30 minutes.

Analysis of the thermal stability of the MC and SMC conditions in the TiAl alloy revealed the upper temperature limit for sheet rolling of the TiAl alloy. For preforms with MC structure it was about 1000°C while for those with SMC structure – less than 850°C. Exceeding these temperatures will lead to considerable grain growth that will shift the BDT temperature and the minimum sheet rolling temperature towards higher temperatures.

Thus, sheet rolling at strain rates of about 10^{-1} s^{-1} should be performed at 750-1000°C in the case of MC TiAl preforms and at 750-850°C in the case of SMC TiAl preforms. According to [4,5], the soaking temperatures during pack rolling should be 100-200°C above the lower rolling temperature since the actual rolling temperatures are lower than the predicted ones, as a result of radiation heat losses and roll chilling. Having defined the processing windows of pack rolling, several sheets of the TiAl alloy were produced (Figure 5).

The results of the present work show that production of the TiAl rolling preforms with a homogeneous fine-grained structure is effective in decreasing the BDT temperature to temperatures as low as 700-800°C. This is substantially lower than the BDT temperature of γ alloys with a coarse-grained structure (above 1100°C) [6]. The refined grain size makes it possible to perform rolling at temperatures between 800 and 1000°C which is 200-400°C below the conventional temperatures for sheet rolling of these types of alloys.

The present results show that TiAl samples in the SMC condition tensile tested at a high strain rate of 10^{-1} s^{-1} were less ductile at 700°C and had higher yield stresses at temperatures below 800°C than the samples in MC condition. These results are somewhat different from results of previous work [10] where properties of the TiAl alloy in similar microstructural conditions were studied at lower strain rates (less than 10^{-2} s^{-1}) and a higher elongation was obtained for the SMC condition. These poorer properties of SMC specimens at the high strain rate can be caused by several reasons. First of all, the strain-rate sensitivity of the flow stress rapidly decreases when the strain rate increases above 10^{-2} s^{-1} [8], promoting strain localization and a decrease in elongation in both the MC and SMC specimens. The effectiveness of grain-boundary sliding to accommodate strain distortions at grain boundaries also decreases rapidly with increasing strain rate [8,12], which leads to stress concentrations at low straining and cracking at grain boundaries, if additional stress relaxation mechanisms are not activated. Such a stress accommodation may be provided by deformation twinning [13]. Development of single-system twinning seems to be the reason of ductility improvement in the MC condition with grain size of 5-10 μm [13]. However, when the grain size decreases to a submicron level, higher stresses, which may exceed the fracture stress, are required for twin development, and rapid fracture may occur before twins start to operate. On the other hand, when the grain size is relatively large (above 10 μm), multi-system twinning develops leading to fast fracturing due to crack formation at twin intersections [13]. That is why decrease in the grain size down to 5-10 μm leads to a considerable decrease in the BDT temperature while grain refinement to submicron level does not further affect the BDT temperature.

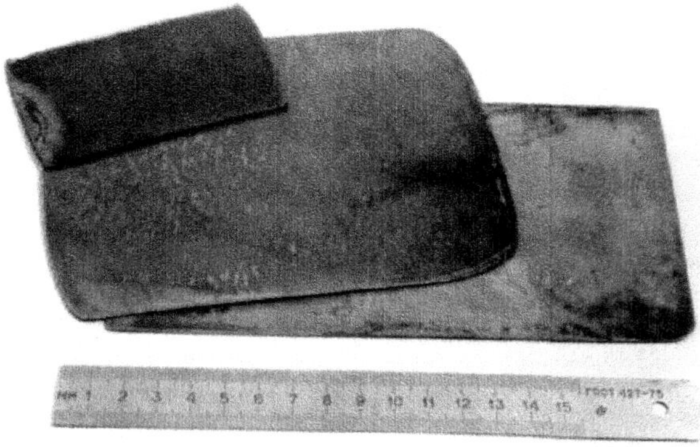

Figure 5. Sheets of the TiAl alloy produced by low-temperature pack rolling.

Analysis of grain growth in the TiAl after 30 minutes annealing revealed the less stability of the SMC condition as compared to the MC condition. It resulted in a very narrow processing window (750-850°C) for rolling of preforms with the SMC microstructure. Exceeding these limits will lead to embrittlement of the SMC preforms and their fast fracture during sheet rolling. Besides, the production of rolling preforms having the SMC structure is more expensive than that of those having the MC structure. Therefore, the MC structure appears to be more desirable as an initial condition for sheet rolling at strain rates of about 10^{-1} s^{-1}. On the other hand, sheets with SMC microstructure should possess better SP properties than MC ones [8]. It could be possible to produce such sheets via isothermal rolling at 750-850°C and strains rates of 10^{-3}-10^{-2} s^{-1}.

Along with retaining the fine-grained microstructure in the sheets, which provides superior superplastic elongations, the decrease in rolling temperature should lead to substantial cost reductions because of energy savings, use of cheaper canning materials than those currently being used, less oxidation during preheating and thus a better surface finish.

Conclusion

Analysis of the effect of grain size on the brittle-to-ductile transition temperature and microstructural thermal stability made it possible to define a temperature range for sheet rolling of stoichiometric TiAl alloy. At strain rates of about 10^{-1} s^{-1} rolling can be performed at 750-850°C in the case of preforms with submicrocrystalline structure (d=0.4 μm) and in a wider temperature interval of 750-1000°C in the case of microcrystalline (d=0.5 μm) TiAl preforms. A homogeneous microcrystalline structure was found to be an optimal microstructural condition for pack rolling at strain rates of about 10^{-1} s^{-1}.

Acknowledgements

The support of this work by the International Science and Technology Center under Project No. 2312 is gratefully acknowledged.

References

1. Clemens H., *Z. Metallkd.* 86 (1995), 814-822.
2. Clemens H., Glatz W., Schretter P., Koeppe C., Bartels A., Behr R., Wanner A., "Processing and Mechanical Properties of γ-TiAl Based Alloy Sheet Material", *Gamma Titanium Aluminides*, ed. Y-W. Kim, R. Wagner, M. Yamaguchi. Warrendale, PA: The Minerals, Metals and Materials Society, 1995, 717-726.
3. Semiatin S.L., "Wrought Processing of Ingot-Metallurgy Gamma Titanium Aluminides", *Gamma Titanium Aluminides*, ed. Y-W. Kim, R. Wagner, M. Yamaguchi. Warrendale, PA: The Minerals, Metals and Materials Society, 1995, 509-524.
4. Semiatin S.L., Ohls M., Kerr W.R., *Scripta Met.* 25 (1991), 1851-1856.
5. Seetharaman V., Semiatin S.L., "Microstructure and Mechanical Properties of Rolled Sheets of a Ti-45.5Al-2Nb-2Cr Alloy", *Gamma Titanium Aluminides*, ed. Y-W. Kim, R. Wagner, M. Yamaguchi. Warrendale, PA: The Minerals, Metals and Materials Society, 1995, 753-760.
6. Seetharaman V., Semiatin S.L., Lombard C.M., Frey N.D., "Deformation and Fracture Characteristics of a Gamma Titanium Aluminide at High Temperatures", *High-Temperature Ordered Intermetallic Alloys V*, ed. I. Baker, R. Darolia, J.D. Whittenberger, M.H. Yoo. Pittsburgh, PA: MRS Symp. Proc. Vol. 288, 1993, 513-518.
7. Lombard C.M., Ghosh A.K., and Semiatin S.L., "Superplastic Deformation of a Rolled Gamma Titanium Aluminide Alloy", *Gamma Titanium Aluminides*, ed. Y-W. Kim, R. Wagner, and M. Yamaguchi. Warrendale, PA: The Minerals, Metals and Materials Society, 1995, 579-586.
8. Imayev R.M., Salishchev G.A., Senkov O.N., Imayev V.M., Shagiev M.R., Gabdullin N.K., Kuznetsov A.V., Froes F.H., *Mater. Sci. Eng.* A300 (2001), 263-277.
9. Mukherjee A.K., Mishra R.S., *Mater. Sci. Forum* 243-245 (1997), 609-618.
10. Imayev V.M., Imayev R.M., and Salishchev G.A., *Intermetallics* 8 (2000), 1-6.
11. Salishchev G.A., Imayev R.M., Senkov O.N., Imayev V.M., Gabdullin N.K., Shagiev M.R., Kuznetsov A.V., Froes F.H., *Mat. Sci. Eng.* A286 (2000), 236-243.
12. Kaibyshev O.A., *Superplasticity of Alloys, Intermetallides and Ceramics.* Springer-Verlag, Berlin, 1992, 317.
13. V.M. Imayev, Imayev R.M., Salishchev G.A., Povarova K.B., Shagiev M.R., Kuznetsov A.V., *Scripta Mater.*, 36 (1997), 891-897.

Application of Diamond Anvil Cell Techniques for Studying Nanostructure Formation in Bulk Materials Directly during Severe Plastic Deformation

A.N.Babushkin[1*], A.A.Popov[2], O.V.Savina[1] and I.V.Sukhanov[1]
[1]Department of Physics, Ural State University, Ekaterinburg, Russia, 620083
[2]Department of Metallurgy, Ural State Technical University, Ekaterinburg, Russia, 620002

Abstract

In this paper we have applied diamond anvil cell technique to study the formation of new states in metals, alloys and nanostructured materials directly during severe plastic deformations (pressure range 15 - 50 GPa, room temperature). As sensitive properties we use thermo electromotive force and arbitrary heat transfer. We report results of the investigation of the beta titanium alloy Ti-15V-3Cr-3Al-3Sn.

Introduction

Materials processed by severe plastic deformations are obviously investigated after treatment. It is well known that changes occurring during severe plastic deformation may disappear after processing. Therefore, it is important to obtain data on occurring transformations directly during pressure treatment. Typically such data can be obtained from X-Ray studies and from changes in optical properties. We demonstrate the possibility of the application of a diamond anvil cell to study the properties of materials directly at severe plastic deformation. We present results obtained from a titanium alloy.

Experimental Procedures

High pressure has been generated in the diamond anvil cell (DAC) with anvils of the "rounded cone-plane" (Verechagin-Yakovlev) type made of synthetic carbonado-type diamonds [1, 2]. Anvils made from synthetic diamond consist of dielectric grains of diamond with layers of conducting materials. These anvils are good conductors and allow the measurement of the electrophysical properties of samples placed in the DAC. We used these anvils to study phase transitions in the large groups of dielectrics and semiconductors. The experimental details are described in [3-5].

Estimating the pressure in an opaque DAC is not easy since the pressure obtained depends on the elastic properties of the compressed layer and anvils as well as the anvil geometry. The pressures reported below have been estimated by the relation

* Corresponding Author. alexey.babushkin@usu.ru

Y.T. Zhu and V. Varyukhin (eds.), Nanostructured Materials by High-Pressure Severe Plastic Deformation, 101-106.

$$P = A\frac{F}{\pi a^2} \qquad (1)$$

where A is an empirical factor (pressure multiplication coefficient) equal to 1.51 (our experimental estimation from measurements in different thin layers). ***F*** is the force applied and a a is the radius of spot contact defined by the relation

$$a^4 - \frac{3FR(1-\nu_2^2)}{4E_2}a - \frac{0.14FR}{E_1\nu_1^{0.74}}\delta = 0. \qquad (2)$$

In (2) R is the sphere radius, E is the Young's modulus, ν is the Poisson's ratio, the subscripts 1 and 2 correspond to the layer and the anvils respectively and δ is the elastic layer thickness.

Equation (2) was primarily derived in the axially symmetric problem of elasticity theory for the case of a perfectly rigid-sphere in contact with a two-layered base - a thin layer on an infinite elastic semispace [6].

In our experiments we use a rounded cone (top anvil) with a radius of about 1 cm. The diameter of the contact spot is about 0.2 mm.

The pressure estimation procedure described above was tested on different materials (see, for example [3-5]). The error of the estimation depends on the mechanical properties of the compressed material and does not exceed 15% of the measured pressure in the range of 15 – 50 GPa.

As sensitive parameters we use the thermo electromotive force (thermoemf) and arbitrary heat transfer. In measuring thermoemf the temperature difference between cold and hot faces of a sample was about 1 K. The accuracy of the temperature measurement about 0.05 K.

The temperature difference between cold and hot faces of a sample at constant heat rate allows to carry out relative measurements of the thermal transfer through the sample. It allows to estimate the heat conductivity and the thermal capacity directly during severe plastic deformation.

We demonstrate the possibility of the application of diamond anvils on the beta titanium alloy Ti-15V-3Cr-3Al-3Sn, quenched at 850 C, foil thickness about 0.05 mm.

Results and Discussion

In Fig. 1 the dependency of thermoemf for an alloy of the pressure is shown. It was revealed that at pressures up to ~ 30 GPa the variation of thermoemf is reversible. That is after releasing the pressure the thermoemf returns to the reference value (Fig. 1a, b). At pressures higher ~ 32 GPa in a sample a new state is developed. After reducing the pressure the thermoemf differs from initial (Fig. 1c). The similar situation arises after pressure treatments above 38 and 42 GPa (Fig. 1 e – h). It means that in case of an irreversible change of the thermoemf it is possible to determine the pressure at which a new phase develops in an alloy. Thus the result turns out *in situ* at severe plastic deformations. It is possible to expect that in the beta titanium alloy Ti-15-3 at the pressure of ~32, 38 and 42 GPa new phases with different thermoemf values are formed. These phases are stable after reducing pressure to ambient pressure.

A similar result turns out of the analysis of the data on the heat transfer throughout the sample. In Fig. 2 the relative change of the heat conductivity in dependency of the pressure and the first derivative of the heat conductivity are shown. Features of thermoemf at ~ 42 GPa (Fig. 1 h) correspond to features of the heat conductivity and its first derivative. It is known that the heat conductivity at room temperature is proportional to the thermal capacity. Therefore, a first order phase transition which is accompanied by a jump of thermal capacity is possible at ~ 42 GPa.

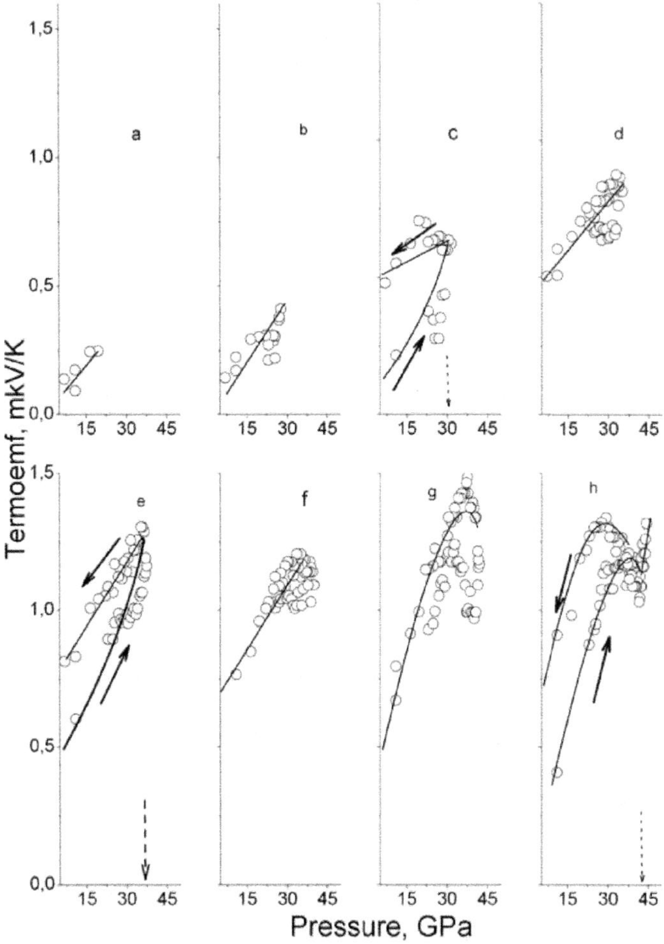

Fig.1. Thermoemf *vs* pressure dependencies. Solid arrows-indicate the direction of pressure change, dotted arrows – probable structural transformations.

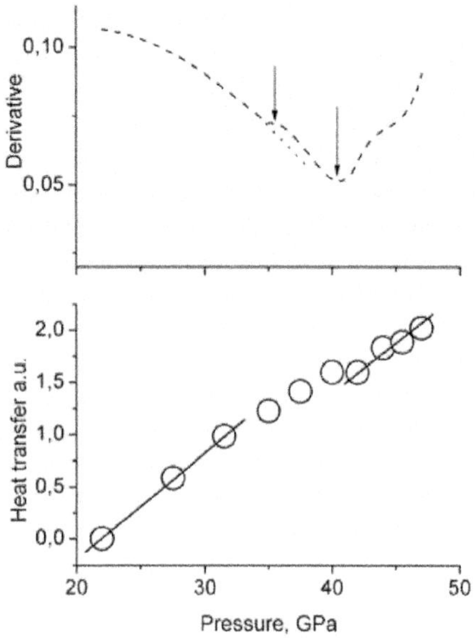

Fig.2. Relative change of heat conductivity (bottom) and the first derivative of heat conductivity *vs* pressure. Arrows specify pressure of probable structural transformations

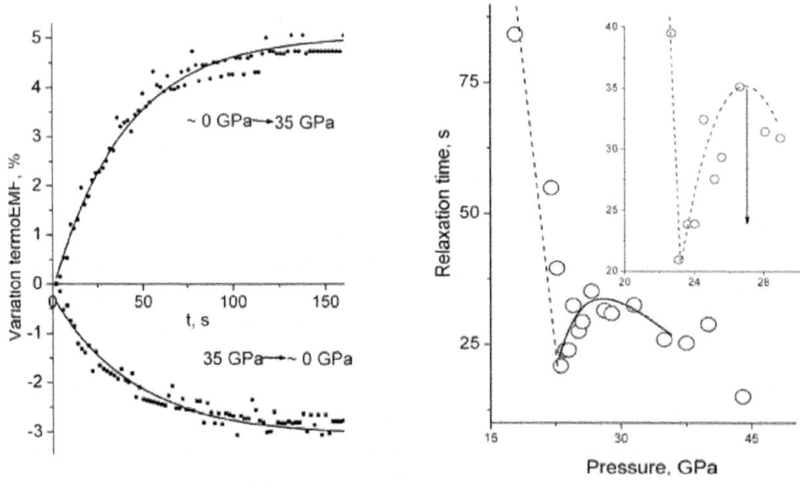

Fig.3. Time dependencies of thermoemf at increasing pressure up to 35 GPa and the subsequent reduction to ~ 0 GPa (left) and pressure dependency of relaxation time (right). The insert shows the interval 20 - 30 GPa.

Probable structural changes occur with a time lag. In Fig. 3 (left) the time dependencies of thermoemf are shown with increasing pressure up to 35 GPa and subsequent reduction to ~ 0 GPa. These curves can be well described by an exponential law. In Fig. 3 (right) the pressure dependence of the relaxation time is shown. This curve exhibits a feature precisely allocated at ~ 28 GPa. Besides there is another feature at ~ 42 GPa.

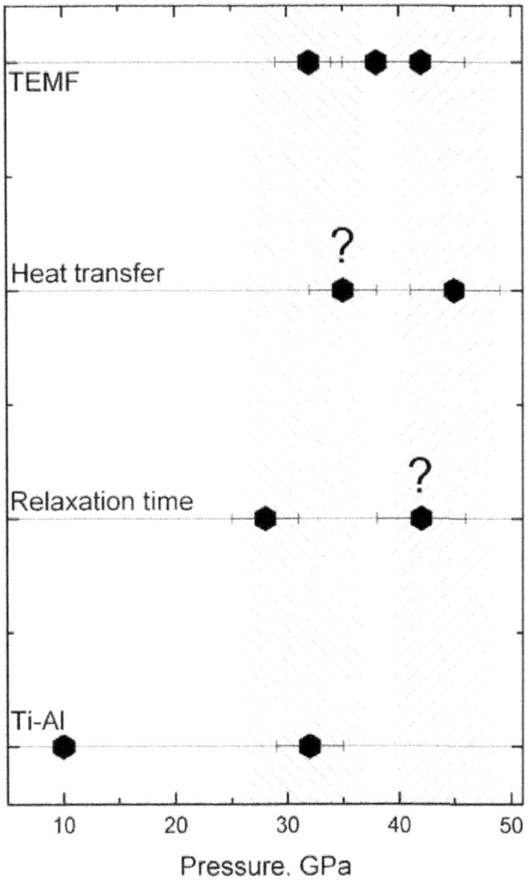

Fig. 4. Probable structural transformations in the beta titanium alloy Ti-15-3. The bottom line is given by experimental data [7]. Our experimental device does not allow to carry out reliable measurements at pressures below 10 GPa. The question indicates that the feature appears in the characteristic pressure range, but the experimental error exceeds this characteristic range

The cumulative diagram (Fig. 5) shows probable structural transformations in the beta titanium alloy Ti-15-3 obtained by different analysis methods using diamond anvils. Additionally, the transformation pressure for Ti_3Al are shown [6]. Regrettably we do not know other data on similar materials obtained directly at this high plastic deformations. The diagram convincingly shows that the different methods applied show the presence of transformations at ~ 30 and ~ 42 GPa in the alloy used.

Conclusion

The research carried out testifies the successful application of diamond anvils to study structural transformations in alloys directly during severe plastic deformations. Research can be carried out by different physical methods on one sample during one act of deformation. For the first time it is revealed that in the beta titanium alloy Ti-15-3 there are probable irreversible structural transformations at ~ 30 and ~ 42 GPa. For the first time the possibility of studying time dynamics of structural transformations in titanium alloys at room temperatures is shown.

Acknowledgements

This work is supported in part by RBRF under grant No.04-02-26702 and CRDF BRHE under grant EK-005-X1.

References

1. Verechagin L.F., Yakovlev E.N., Stepanov G.N., Vinogradov B.V. *JETF Lett.*, 16, (1972), 240-242.
2. Verechagin L.F., Yakovlev E.N., Vinogradov B.V., Stepanov G.N., Bibaev K.Kh., Alaeva T.J., Sakun V.P. *High temperatures, high pressures,* 6, (1974), 99-105.
3. Babushkin A.N., Kandrina Y.A., Kobeleva O.L., Schkerin S.N., Volkova Y.Y., in *Frontiers of High Pressure Research II: Application of High Presure to Low-Dimensional Novel Electronic Materials*. Eds. H. D. Hochheimer, B. Kuchta, P. K. Dorhout, J. L. Yarger., Kluwer Acad. Publ., Dordrecht-New York-London, 2001, p. 131.
4. Babushkin, A.N., *High Pressure Research,* 6 (1992), 349-356.
5 Babushkin A. N., Pilipenko G. I., Gavrilov F. F. *J. Phys.: Condens. Matter* 5 (1993), 8659-8664
6. Makushkin, A.P., *Sov.J. Friction and Wear,* 5 (1984), no.5, 823-832.
7. Sahu P.C., Shekar N.V.C., Yousuf M., and Rajan K.G., *Phys. Rev.Lett.*, 78 (1997), no. 6, 1054-1057.

Anomalous Nitrogen Solubility in Gradient Nanostructured Layer Formed in the Surface of Bulk Iron by Severe Plastic Deformation under Friction

A.I. Yurkova[1*], A.V. Belots'ky[1], and A.V. Byakova[1,2]

[1]Department of Metal Physics, National Technical University of Ukraine "Kiev Polytechnic Institute", 37 Prospect Peremohy, 03056 Kiev, Ukraine

[2]Phase Transformation Division, Institute for Problems of Material Science, National Academy of Sciences of Ukraine, 3 Krzhyzhanivs'ky St., 03142, Kiev, Ukraine

Abstract

By means of friction nitriding (FN) the deformation-induced structure, which includes section of nano-sized grains and that of submicro-sized grains, was formed at the top surface of pure Fe sample. Nitrogen content in bcc-Fe[N] solid solution of ultra fine grains enhanced dramatically compared to that obtained for large-grained iron subjected to conventional nitriding in furnace. When grain/cell size was limited to less than 1 μm nitrogen content reaches the value of about 0.3 wt.% that is even by 3 times higher than that appointed by binary system Fe–N for bcc–Fe[N] solid solution at the temperature 863 K at which nitrogen solubility is the highest, i. e. 0.1 wt.%.

Introduction

Nitriding the iron and steel is used widely for improvement of the tribological properties, the fatigue endurance and the corrosion resistance. The microstructural revolution due to ultra grain refining by severe plastic deformation (SPD) results in a lot of novel properties including high strength and hardness as well as other excellent customer characteristics [1], making them attractive for different engineering applications and for upgrading conventional manufacturing processes. Several SPD–based techniques capable for providing the effective grain refinement of bulk metals and alloys up to nano- and submicro-meter scales were developed nowadays and compiled in [1]. They can be grouped into two types, i.e. technique for grain refining in volume of material [1] and that in surface layer of material [2]. Nevertheless, there are only a few evidences referred to ultra grain refining the grain structure of nitriding layer [3].

This paper aims to study microstructural features of gradient surface layer on iron formed by SPD forced by directional mass transfer of nitrogen that occurs during FN process. The effect of ultra grain refining the grain structure on nitrogen solubility in the bcc–Fe[N] solid solution is the main objective too.

Experimental Procedures

An iron cylindrical–shaped sample (8 mm in diameter and 50 mm in height) with a purity 99.97 wt. % was used in the experiment. To obtain homogeneous large grains about 80 - 100 μm in size samples were annealed at the temperature about 1223 K for 180 min. Then, annealed samples were subjected to friction under strait flow of

---

* Corresponding Author. yurkova@list.ru

Y.T. Zhu and V. Varyukhin (eds.), Nanostructured Materials by High-Pressure Severe Plastic Deformation, 107-112.

ammonia-gas (NH_3). This route was developed originally for high efficiency of nitriding process and called "friction nitriding" (FN) [4]. However, it was clarified recently [3] that FN process is especially viable for ultra fine refining the grain structure up to nano-metre scale due to severe plastic deformation being forced additionally by directional mass transfer of dopant element. Therefore, new technique is named "sever plastic deformation forced by diffusion flow" (SPDD).

Fig. 1 shows a schematic presentation of the apparatus used in the present study. Main parts of this apparatus are as follow: (1) chamber that can be closed hermetically when gas atmosphere other than air was used; (2) system for gas entry and pressure control; (3) system for gas exit; (4) system for sample rotation that is equipped by spindle with chuck for holding the one side of cylindrical sample; (5) two blocks forced onto sample surface; (6) system for temperature control by chromel–alumel thermocouple. Ammonia-gas was introduced into the chamber through the hole (2) whereas other hole (3) was used for outputting NH_3-gas partly dissociated due to reaction at sample surface.

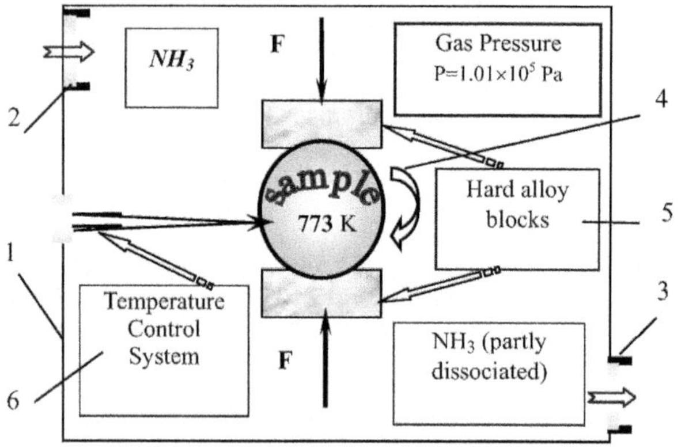

Figure 1. Schematic presentation of the apparatus for FN process: (1) chamber; (2) system for gas entry and pressure control; (3) system for gas exit; (4) system for sample rotation; (5) two blocks forced to sample surface; (6) system for temperature control.

Sample was heated by friction occurred due to its rotation between hard alloy (WC-8%Co) blocks being pressed to sample surface by certain force F, which ensured the unchanged temperature of 773 K during exposition time about to 60 min. Rotation speed was about 50 turnovers per one second. After treatment samples were cooled down to room temperature and, then, they were examines visually.

XRD line profile analysis was performed using X–ray diffractometer (20kV) with Fe radiation. Electrochemical etching the treated surface was used to remove layer-by-layer, so that the evolution of nitrogen content and microstructural features along the depth from the top surface to the strain–free matrix was studied.

Cross–section of samples was observed using powerful optical microscope Neophot–21 (resolution up to 0.4 μm). Furthermore, TEM images and selected area electron diffraction (SAED) patterns were performed to study the structure of refined surface layer. The average grain size was found by observation of TEM images.

Results and Discussion

Fig. 2 shows the structure of nitriding layer on the surface of iron treated by FN process. The results of XRD analysis and cross-sectional observation of treated Fe sample showed that thin layer of Fe_4N nitride (Fig. 2, section 1) of about 10 μm is formed at the top surface of the sample. Then, the layer of Fe_4N nitride is followed by thick section (up to 250μm) of bcc-Fe[N] that exhibits gradient-grained structure of different scale. Several structural sections (2-5) are visible within deformation region (Fig. 2). However, optical microscopy was inadequate to recognise in details the grain structure of ultra fine-grained sections (2, 3). So, the results of TEM observation were used to recognise the grain scale of sections above.

Section (4) of aquiaxed micrometer grains follows the section (3). At least at the last sections (5) adjacent to the strain-free matrix (6) exhibits the banded structure of micrometer grains with inclination to cylindrical surface (Fig. 2). Proper SPD evidences are seen by optical microscopy of this section.

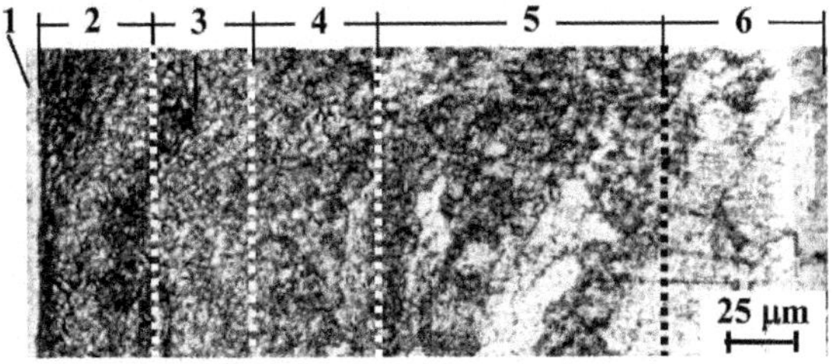

Figure 2. Optical micrograph of nitriding layer on the surface of iron treated by FN process: 1 – layer of Fe_4N nitride; 2 – nanostructured section; 3 – submicrostructured section; 4 – section of aquiaxed micro scale structure; 5 – section of micro scale banded structure; 6 – strain-free matrix.

It is noticed that processing conditions of conventional nitriding are more favourable for stimulating the nitrogen diffusion since exposition temperature about 823 K is higher and typical exposition time of 6 hours is greater than those exploited by FN route. Despite of this extension of nitriding layer formed by FN process is dramatically greater than that obtained typically by conventionally nitriding (Fig. 3).

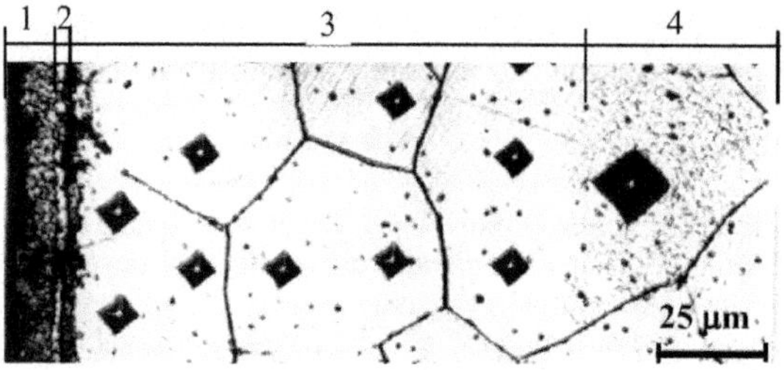

Figure 3. Optical micrograph of nitriding layer on iron sample treated by conventional nitriding in furnace: 1 – layer of $Fe_{2\text{-}3}$ N nitride; 2 – layer of Fe_4N nitride; 3 – layer of α-Fe[N]; 4 – α-Fe (matrix).

Significant acceleration of atomic nitrogen mass transfer through ultra fine grain structure is confirmed additionally by almost full disappearance of undesirable layer of $Fe_{2\text{-}3}N$-nitride being usually formed at the top surface of iron subjected to conventional nitriding, as shown in Fig. 3.

TEM images shown in Fig. 4 demonstrate the dominated features for deformation-induced structure of bcc-Fe[N] subjected to FN process.

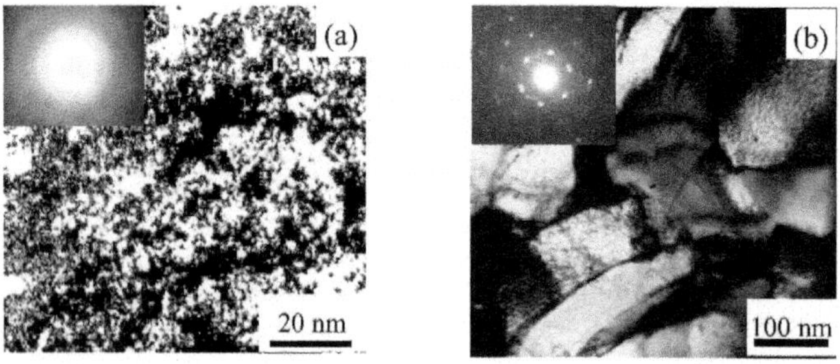

Figure 4. Bright-field TEM images obtained in Fe samples treated by FN process: (a) in nano-sized section (15 μm in deep) and (b) in submicro-sized section (50 μm in deep).

Nano-sized grain/cells with random crystallographic orientation appears in nano-sized section (2) (at the top surface) of bcc–Fe[N] (section) (up to 50 μm in deep), as indicated by SAED pattern in Fig. 4 (a). Dislocation density in this section is still high. Dislocations are preferably accumulated in tangles (DT) and dense dislocation walls (DDW), which are randomly arranged, as seen in Fig. 4 (a). In the submicro-sized section (3) grains (or subgrains) are orientated randomly and separated by DDW, as evidenced by Fig. 4 (b). It is notable that these grains are divided additionally into smaller equiaxed cells separated by DDW, DT or subboundaries. Also, inside grains one can see evident contrast of high density of lattice dislocations.

The results obtained in the present study and those published previously [3] show that grain/cell size increases gradually along the depth from 10 nm at the top surface of nano-sized section to 700 nm at the submicro-sized section. Furthermore, dislocation density decreases gradually from the 10^{12} within nano-grained section to 10^{9} cm/cm^{3} in submicro-grained section when Fe-sample was treated by friction.

Fig. 5 (a) exhibits the variation of lattice parameter "a" for bcc–Fe[N] determined along the depth from the top surface to strain-free matrix.

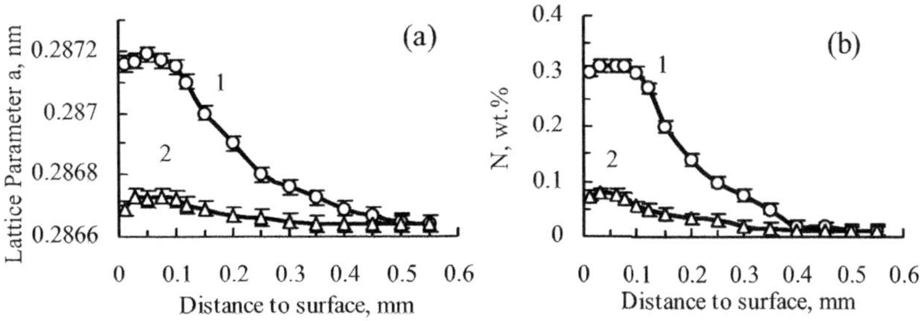

Figure 5. Data for (a) lattice parameter "a" of bcc-Fe[N] and (b) nitrogen content in bcc-Fe[N] via distance to surface treated (1) by FN process and (2) by conventional nitriding.

Following to Paranjipe et al. [5], who published the correlation *vs*. N wt. % in bcc-Fe[N] solid solution, nitrogen content in refined nitriding layer was obtained, as shown in Fig. 5 (b). Experimental results shown in Fig. 5 (b) clearly demonstrate that nano- and submicro-grained structure of high dislocation density causes nitrogen content in bcc–Fe[N] solid solution to increase dramatically. The amount of nitrogen becomes higher by 10^{2} times than that recorded in micro grained bcc–Fe[N] obtained by conventional nitriding and it is even by 3 times higher compared to that appointed by binary system Fe-N for bcc–Fe[N] solid solution at the temperature 863 K at which nitrogen solubility is the highest, i. e. 0.1 wt.%.

Conclusion

The effective refinement of nitriding layer up to nano- and submicro-meter scales has been demonstrated experimentally. The grains of bcc–Fe[N] have been refined up to 10 nm at the top of nano-sized section and they achieved 700 nm at the most within submicro-sized section. Ultra grain refining the grain structure of nitriding layer provides for anomalous high solubility of nitrogen in bcc–Fe. The amount of nitrogen becomes higher by 10^{2} times than that recorded in micro grained bcc–Fe[N] obtained by conventional nitriding and it is even 3 times higher compared to that appointed by binary system Fe-N for bcc–Fe[N] solid solution at the temperature 863 K at which nitrogen solubility is the highest, i. e. 0.1 wt.%.

Acknowledgements

This work is partly supported by the Ministry of Education and Science of Ukraine under contract #2730.

References

1. Noskova N.I., Mulyukov R.R. *Submicrostructural and Nanocrystalline Metals and Alloys.* Ekaterinburg: UrO RAS , 2003, p. 279.
2. Tao N.R., Wang Z.N., Tong W.P. et al. *Acta Mater.* 50 (2002), 4603–4616.
3. A.I. Yurkova, A.V. Belots'ky, and A.V. Byakova et al. "Nanocrystallization in Iron Alloys Induced by Friction Treatment and Nitrogen Diffusion" *Metallic Materials with High Structure Efficiency,* ed. O.N.Senkov et. al. Kluver Academic Publishers, 2004, 113–118.
4. Bilots`ky A.V., Yurkova A.I. *Technologiya i Organizatziya Proizvodstva* 2 (1988), 40–41.
5. V.G.Paranjipe, M.Cohen, M.B.Bever, and C.F.Floe. *J. Metals*., **182**, no. 2, (1950), 261–267.

Evolution of Microstructure and Mechanical Properties of Ti-5Mo-5Al-5V Alloy Processed by High Pressure Deformation

T.A. Ryumshyna*, T.E. Konstantinova

Department of Material Science, Donetsk Physical and Technical Institute NAS of Ukraine, Donetsk, 83114

Abstract

The evolution of Ti-5Mo-5Al-5V alloy microstructures after severe plastic deformation (SPD) by high pressure hydrostatic extrusion has been studied by electron and optical microscopy, and X-ray analysis. It has been established that the deformation mechanism depends on the phase composition of the alloy. The mono-phase 100% β-state is deformed by formation of deformation package martensite which contains disoriented fragments having sizes of about 20 nm. At early stages of deformation, a translation mechanism is observed in the two-phase alloy. At severe deformations this mechanism changes into a bending one. Reorganization of structure during the aging heat treatment of the as-deformed alloy causes an increase of the alloy strength by 35-40% while maintaining plasticity.

Introduction

Due to their high structural efficiency, titanium alloys are widely used in different industries, especially in aerospace engineering. Despite a considerable number of publications on titanium alloys, the possibilities for property improvement by structural minimization have not been exhausted yet. To improve the properties, severe plastic deformation together with heat treatments have been used in recent years. One can comprehensively control the microstructure and consequently mechanical properties only if the deformation mechanisms are known.

Experimental Procedures

Deformation mechanism effects on the microstructural state of an industrial titanium alloy (Ti-5Al-5Mo-5V-1Ni-1Fe) have been studies in this work. Two initial microstructural states have been studied: the mono-phase β-state (the BCC-structure) obtained after annealing at 900 °C (for 1.5 h), and quenching in water and a two-phase β + 55%α – state (the BCC + HCP -structure) obtained by a two-stage annealing consisting of a heat treatment at 850 °C for 1.5 h, air cooling, a furnace heat treatment at 750 °C for 3 h, and then air-cooling. The samples prepared were subjected to deformation with the use of a high hydrostatic pressure (HP) up to ε = 50%. The influence of deformation on the character of precipitates after aging at 750 °C for 4 h (for single-phase β -state) and aging at 515 °C for 4 h (for two-phase β + α -state) on the microstructure and mechanical properties was also considered. Observations of the microstructural states were performed with optical microscopy (OM), transmission electron microscopy (TEM), and X-ray diffraction analysis.

* Corresponding Author. ryumshyna@donapex.net

Y.T. Zhu and V. Varyukhin (eds.), Nanostructured Materials by High-Pressure Severe Plastic Deformation, 113-119.

Mechanical properties ($\sigma_{0.2}$, σ_B, ψ, δ) were determined with standard samples with diameter 5 and 10 mm under tension to failure at a rate of 10 mm/min .

Results and Discussion

Microstructure of the mono-phase alloy in 100% β - state after HP-deforming.

After the 900°C annealing and quenching, the OM observation shows the formation of equiaxial large grains with dimensions of 0.15 mm in diameter in the alloy. The TEM observations reveal a streaked contrast in the β – phase (Fig.1a). The latter can be conditioned by a regular distribution of doping elements by the type of modulated structure [1].

Deformation results in the emergence of sliding bands (Fig.1b) in favorably oriented grains. Deformation increases result in an increase in the total number of bands in grains, deformation band appearance in new grains, and band distortion at the grain boundaries. Previously, it was determined by X-ray analysis [2] that at strains of 5-15% an orthorhombic α′′- martensite appears, with the martensite orientationally contacting with the initial β – matrix. Deformation leads to a considerable decrease in the region of coherent scattering (CSR), from 80 μm in the initial state down to 2 μm at a strain of 15%. TEM images show the martensite in the form of plates and stacks, which, as a rule, have a complex internal "stack" structure (Fig.1c,1d).

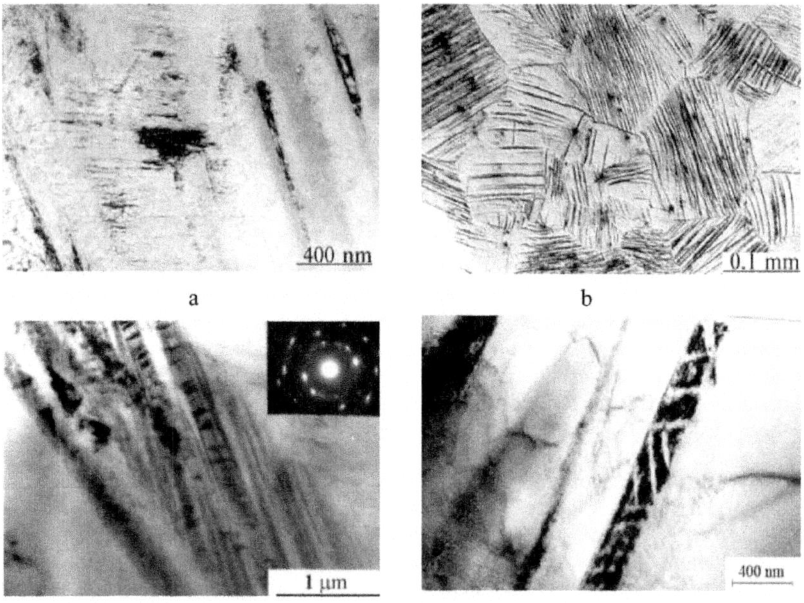

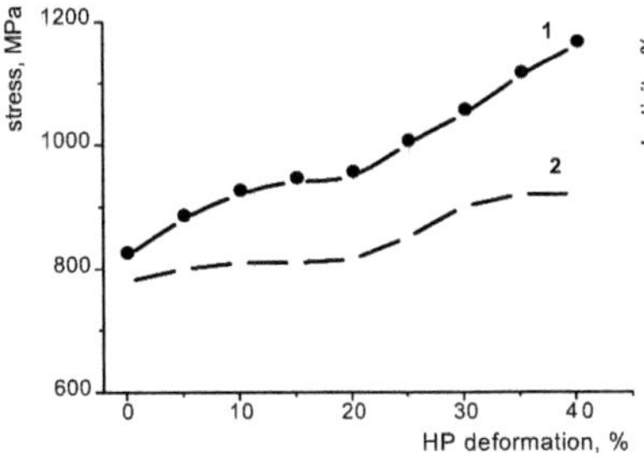

e

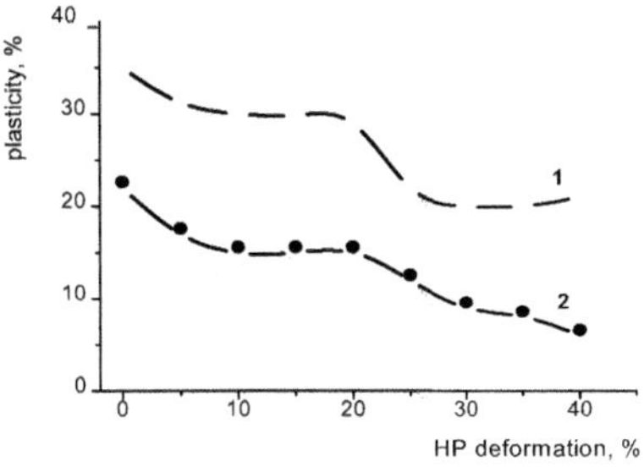

f

Figure. 1. Microstructure of the mono β-phase state before (a) and after a hydrostatic deforming (b,c,d) and the mechanical properties (e,f)
a) $\varepsilon = 0\%$ (TEM), modulation of precipitation in β – grain (TEM);
b) $\varepsilon = 15\%$ (OM), traces of α′′-martensite;
c), d) $\varepsilon = 30\%$ (TEM), a internal structure of martensite.
e) tensile stress σ_b (1) and yield stress $\sigma_{0.2}$ (2)
f) Necking ψ (1) and ultimate elongation δ (2)

When strains are higher than 30%, a roentgen graphic martensite is not observed, even though the CSR size still keeps decreasing. At developed deformations ($\varepsilon > 30\%$) an α′′- martensite formation likely forms, but then a quick reverse α′′→ β transformation takes place, and as a result of this transformation β – grains significantly fragment [1].

An increase in HP-deforming up to strains of 50% leads the tensile strength (σ_b) to rise from 820 up to 1100 MPa, the yield strength ($\sigma_{0.2}$) from 790 to 900 MPa. But decreases in plasticity are observed. Necking (ψ) changes from 35% to 22%, and ultimate elongation (δ) changes from 22% to 10%.

So, when deforming a single-phase alloy stored energy dissipation take place as a martensitic phase transformations produce boundaries with discretely different orientations within initial grains. According to [3] such a deformation mechanism should be thought of as *rotational*.

As the reverse α′′→ β transformation occurs during deformation, the initial orientation in the β –phase is not preserved, and as a result β –grains are greatly fragmented. This provides new possibilities to control the alloy microstructure. following annealing in a two phase region (750°C aging heat treatment) an initial α - phase precipitations appear on the boundaries of fragments, thus creating a peculiar frame structure (Fig. 4a,b).

Microstructure of the two-phase 45% β + 55% α alloy after HP-deforming

An initial structure of a two-phase 45% β + 55% α alloy is more stable in comparison with single-phase state. The primary α – phase is present in colonies of large non-equiaxed particles (Fig.2a). In the β – phase a streaked contrast is observed, which is indicative of structure modulation (Fig.2b). X-ray analysis in the (200) reflex reveal the presence of two maxima which are 0.6° apart, suggesting that there are two BCC-solid solutions in the matrix. Their solutions differ by a lattice period which points to alloy division by chemical composition [1].

When acted on by the deformation processing, the β –grains elongate. Also, the orientation and α – plate sizes change. The grains are fragmented along the deformation axis (Fig.2c). The TEM image reveals random orientation growth within grains. But in this case, the occurrence of extinction contours (Fig. 2d) is indicative of a continuous changing of random orientations within grains [4].

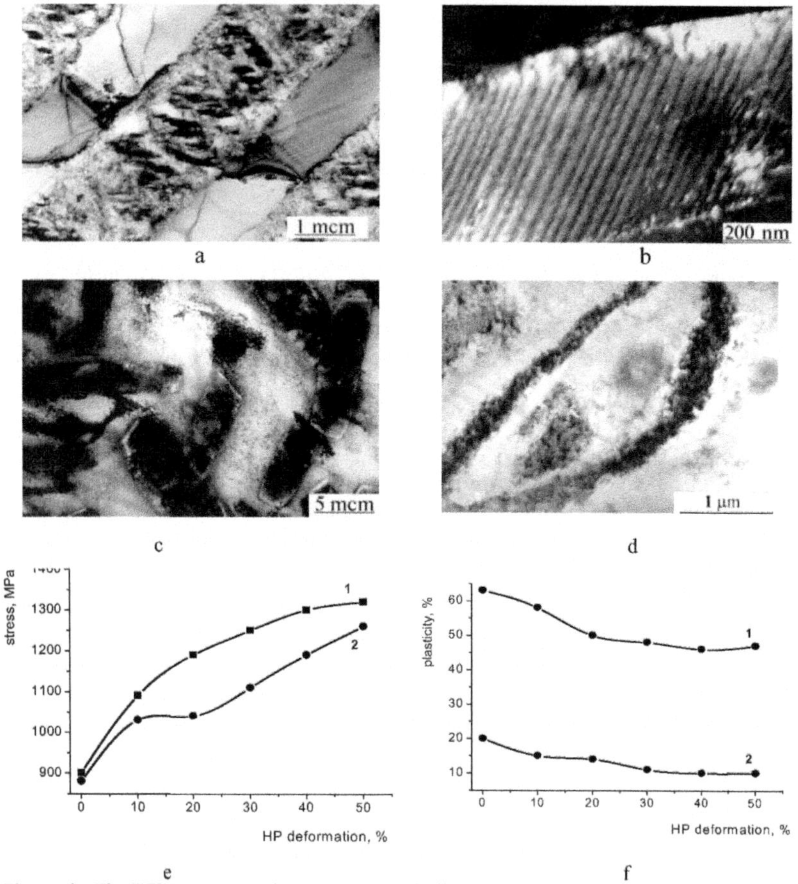

Figure. 2. The TEM structure of a two – phase Ti-alloy (45% β + 55% α):
a) an initial structure (850 °C, 1.5h →750 °C, 3 h);
b) a streaked contrast in β – phase (a modulated structure);
c) a HP- deforming (ε = 30%);
d) banding extinction contours in β – phase.

e) tensile stress σ_b (1) and yield stress $\sigma_{0.2}$ (2)
f) necking ψ (1) and ultimate elongation δ (2)

The formation of regions with a high crystal lattice bending are likely due to the hindering of dislocation translational motion by the high density of dislocations which come into existence during the deformation. The crystal lattice bending may occur as a result of the formation of dislocation piles-up in the form of two dislocation rows of opposite sign, with the rows lying within one plane or being located at an angle to each other [5].

Mechanical characteristics of the two-phase alloy change to a greater degree than the mono-phase alloy. When HP-deforming increases to $\varepsilon = 50\%$ the σ_b increase from 900 MPa to 1.30 GPa, the $\sigma_{0.2}$ increases from 880 MPa to 1.25 GPa (fig.2e), and plasticity decreases: ψ from 62% to 45%, elongation δ - from 20% to 10% (fig.2f). If in the mono-phase state, sample elongation is due to homogeneous deformation of the whole sample, but in the two-phase state, the elongation is greatly enhanced by the localization of plastic flow in the sample neck region.

The modulus of elasticity E of the two-phase state is twice that of the mono-phase state (130 GPa and 65 GPa, respectively) (Fig.3). Such a change in the modulus of elasticity has been noticed in [6], where a low-modulus state is named a "rubber" state. The increase of the modulus of elasticity in the two phase state might be connected with the rearrangement of doping elements in the process of α -phase precipitation, that might manifest itself in a different degree of the β – state modulation

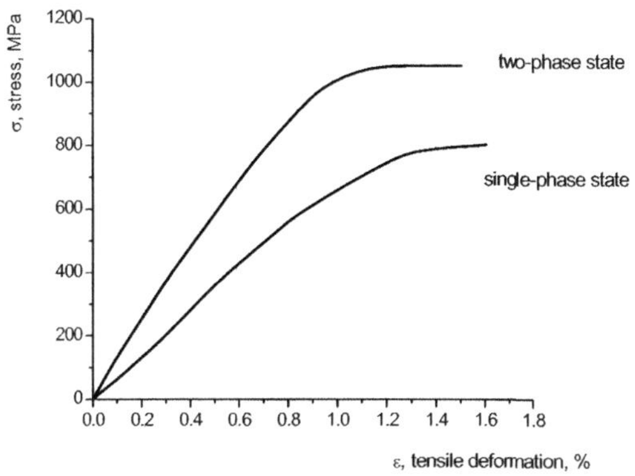

Figure. 3. The starting stage of a deformation curve when the starting single state and two-phase state samples (diameter 10 mm) are being extended

Changes in the composition and fine structure of the mono- and two-phase alloy states are responsible for the differences in the modulus of elasticity, and as a result, for variations in the deformation mechanism. The state with a high rigidity (a high-modulus state) allows the deformation bending mechanism to occur. At low values of the modulus (a "rubber" state) the rigidity decreases. The alloy shift instability increases, and the formation of deformation martensite proves to be possible by the rotation mechanism.

The ideas concerning the mechanisms of plastic deformation obtained have allowed the selection of processing conditions which get a good combination of strength and plasticity. It was found that such treatment conditions are a stepwise annealing together with an intermediate high pressure deformation, followed by an aging heat treatment. The combination of annealing and the following deformation providing conditions for deformation bending mechanism, which make it possible to form a structure with continuously changing random orientations and a high dislocation density. After aging, a precipitation α–phase is found to be highly dispersed (Fig.4c.d). This processing (850 °C → 750 °C + HP-deforming 50% + aging 515 °C) increases the alloy's tensile strength σ_b to 1.59 GPa and preserves the plasticity (δ = 9%, Ψ = 24%) relative to the standard processing (without intermediate deformation 850 C → 750 C + aging 515 C) when TS σ_b = 1.27 GPa, δ = 11%, Ψ = 36 %.

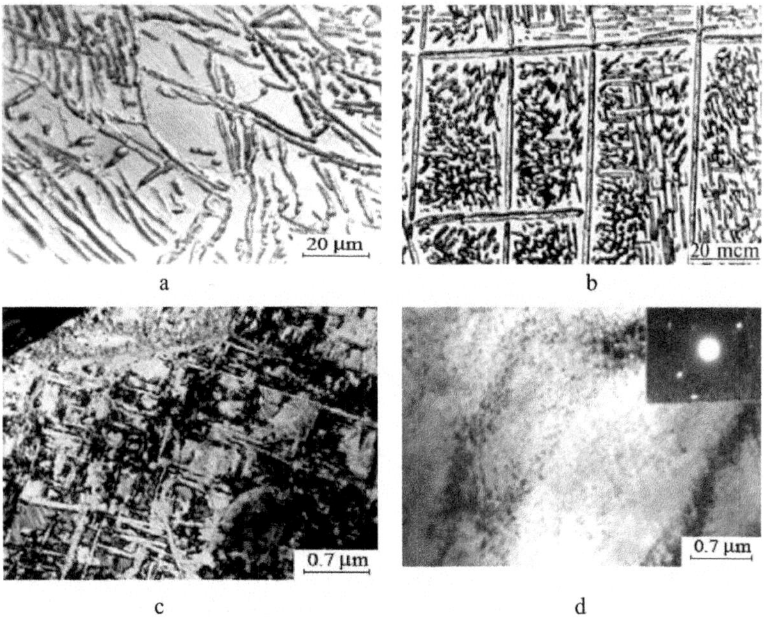

Figure. 4. The structure of a mono-phase alloy (a, b) and two-phase alloy (c, d), precipitation of α phase in β-grains:
a) 900°C, 1h + 750 °C, (OM);
b) 900°C, 1h + HP deforming 20% +750 °C, (OM);
c) 850 °C, (1.5h) →750 °C,) + aging 515 °C;
d) 850 °C, (1.5h) →750 °C,) + HP deforming 50% + aging 515 °C.

The high pressure deforming is important to increase the strength of the alloy by decreasing of size of microstructural elements. Special conditions during HP-deformation allow for the creation of large gradients of strains and stresses which result in the accumulation the SPD in the material.

Conclusion

The effects of a phase composition on the structure and mechanical properties of the Ti-5Mo-5Al-5V alloy have been studied in the present work. It was established that in the mono-phase β –state, the alloy is being deformed by the rotary mechanism. In the two phase state the alloy is deformed by a bending mechanism. Using the HP-deforming allowed for the creation large strain gradients in the material, which lead to the accumulation of SPD and to the encouragement of a new possibility for controlling the alloy microstructure and properties.

Acknowledgements

Authors express thanks to G.K.Volkova for useful discussion.

References

1. T.E.Konstantinova. *Mesostructure of deformed metals.* Donetsk, DonFTI NANU, 1997, p. 168 (in Russian) .
2. T.E.Konstantinova, G.K.Volkova, A.A.Adamets. *Proceedings of the Conf. on Martensite Transformation in solids.* Kiev: IMF AN Ukraine, 1992, p.294 (in Russian).
3. V.I.Vladimirov, A.E.Romanov. *Disclynations in crystals.* Leningrad, Nauka,1986, (in Russian).
4. T.E.Konstantinova, N.V.Tokij, V.B.Primisler, A.A.Dobrikov. *Electron Microscopy and Material Strength.* (Kiev: IPM NANU: 1994), p.60 (in Russian).
5. N.V.Tokij, T.E.Konstantinova, V.N.Varyuchin. *Met. Phys. Adv. Tech.,* 20 (1998), no 11, p. 71-76 (in Russian).
6. Takashi Saito, Tadahiko Furuta, Junh-Hwan Hwang and other. *Science,* 300 (18 April 2003), p. 464-467.

Martensitic Transformations in Nanocrystalline Fe-Cr-Ni and Fe-Mn Alloys

B.M. Efros[1*], V.P. Pilyugin[2], A.M. Patselov[2], S.V. Gladkovskii[3], N.B. Efros[1] and L.V. Loladze[1]

[1]Physics & Technology Institute of NASc of Ukraine, Donetsk, 83114, Ukraine
[2]Institute of Metal Physics of the RAN, Ekaterinburg, 620219, Russia
[3]Ural State Technical University,, Ekaterinburg, 620002, Russia

Abstract

Structures of Fe-Cr-Ni and Fe-Mn alloys subjected to severe plastic deformation under pressure have been studied using different methods. It is shown that in FCC γ-Fe-18%Cr-10%Ni steel the nanocrystalline structure formed increases the amount of high-pressure HCP ε-phase. It is also shown that with the formation of nanocrystalline structure there is an increase of forward $P^{\gamma\to\varepsilon}$ and a decrease of reverse $P^{\varepsilon\to\gamma}$ martensitic transformation in FCC γ-Fe-Mn alloys. It has been found that pressure increase to 20 GPa reduces the $P^{\varepsilon\to\gamma}$ martensitic transformation in nanocrystalline Fe-Mn alloys in FCC γ-field by DAC in such a way that a 100% stabilization of high-pressure HCP ε-phase under normal conditions and a long endurance after pressure removal are possible.

Introduction

The problem of severe plastic deformations (SPD) applied to numerous intensive technologies of plastic forming the metals and alloys and resulting in a series of structural and phase transformations with the purpose of imparting the improved mechanical and service properties to the materials is of special interest [1,2]. However, despite a wide practical use of the SPD, many mechanisms of the nanocrystalline (NC) structure formation and their relation to the processes of phase formation in metals and alloys remain unclear.

In this respect, the purpose of this work is to study the influence of deformation under pressure on structural and phase states in Fe-Cr-Ni and Fe-Mn alloys which are the basis for the production of high-strength nonmagnetic wear-and corrosion-resistant steels and alloys with stable and metastable FCC γ-phase.

Experimental Procedures

The objects studied were FCC γ-Fe-Cr-NI alloys, Fe-18%Cr-10%Ni steel and FCC γ-Fe-Mn (C_{Mn}=20-55%) alloys after thermoplastic treatment and water quenching from 1100°C. Samples were in the form of discs of 5 mm diameter and 0.3 mm thick. For the SPD of the given materials a device for deformation by high-pressure torsion (HPT) and a diamond anvil cell (DAC) were used [3,4]. Pressure value P and the deformation degree e were approximately ranging from ≈ 0 - 20 GPa and ≈ 0 - 10 unit of true deformation, respectively.

* Corresponding Author. efros@hpress.dipt.donetsk.ua

Y.T. Zhu and V. Varyukhin (eds.), Nanostructured Materials by High-Pressure Severe Plastic Deformation, 121-126.

The degree of true deformation was estimated by the formula:

$$e = \ln\{1 + [\phi(R_i/h_i)]\} + \ln(h_0/h_i),$$

where ϕ is torsion angle (in rad), R_i is distance from the axis of rotation, h_i is thickness of the sample before and after the SPD with a corresponding radius R.

To study structural and phase states of samples the methods of x-ray diffraction analysis, optical microscopy, transmission electron microscopy and Mössbauer spectroscopy were used.

The grain size of studied samples was estimated as mean linear intercepts to an accuracy of ± 5%. The strength properties of the deformed samples were estimated by Vickers microhardness measurements HV with the help of a hardness tester "Leitz miniload 2".

Results and Discussion

Extreme influences as a method of producing new materials and of improving the important properties of materials utilized in technology specify substantial progress in this field. At a combined action of high pressure (value P of the order of 1 - 5% of compression modulus) and SPD, in materials, there occur different phase the transitions resulting in the formation of new metastable phases.

Stainless steel $Fe_{72}Cr_{18}Ni_{10}$

The initial state of studied steel is FCC γ - state with the structure of hot deformation, HV microhardness ≈ 1.8 GPa (Fig.1,a). Boundaries of coarse grains and of twin-structure elements are seen. One can also observe sites of dislocation accumulation with a comparatively low dislocation density.

In the case when the steel is only treated with pressure (P = 8 GPa) it is seen that structure elements of the sample become finer, microtwins are seen and dislocation cells are formed (Fig.1,b). Deformation is non-uniform, there are bands of localized deformation, microdiffraction image from band zone is characterized by smeared out reflexes in the form of arcs, not by point reflexes. After the high-pressure treatment the HV value was ≈ 2.2 GPa (Fig. 2).

In the case of HPT with strain e = 1.8 (P = 8 GPa) there are, apart from twins and stacking faults, regions of fine-disperse structure, HCP ε-phase is in the form of twins in FCC γ-matrix (Fig.1,c), value HV abruptly increase up to 3.5 GPa (see Fig.2). It should be noted that this structure is similar to that after shrinkage deformation, but in the latter case there are no regions of fine-disperse structure.
With the increase in deformation degree to e = 3.2, there are the same two types of structure, but share of the fine-disperse component increases (phase composition: $\varepsilon+\gamma+\alpha$) (Fig.3). The average size of substructure elements makes about 110 nm. Diffraction of two types is present: point one-from areas with twins, and the arc-like one-from areas with cellular structure. Moreover, the twins are subdivided into fragments. The change in microhardness is not so abrupt; it was approximately equal to 3.9 GPa (see Fig.2).

For HPT with e = 3.9 there are microfragments of broken twins. There, the fragments are coarse,r of about 200 nm. The structure is not uniform, in areas of fine-disperse structure, the size of substructure elements makes 30 - 40 nm with the average size being ≈ 90 nm (Fig.1,d). Insignificant increase in HV value (~ 4.1 GPa) is accompanied by a significant increase of scatter-from 5.0 to 3.7 GPa (see Fig.2).

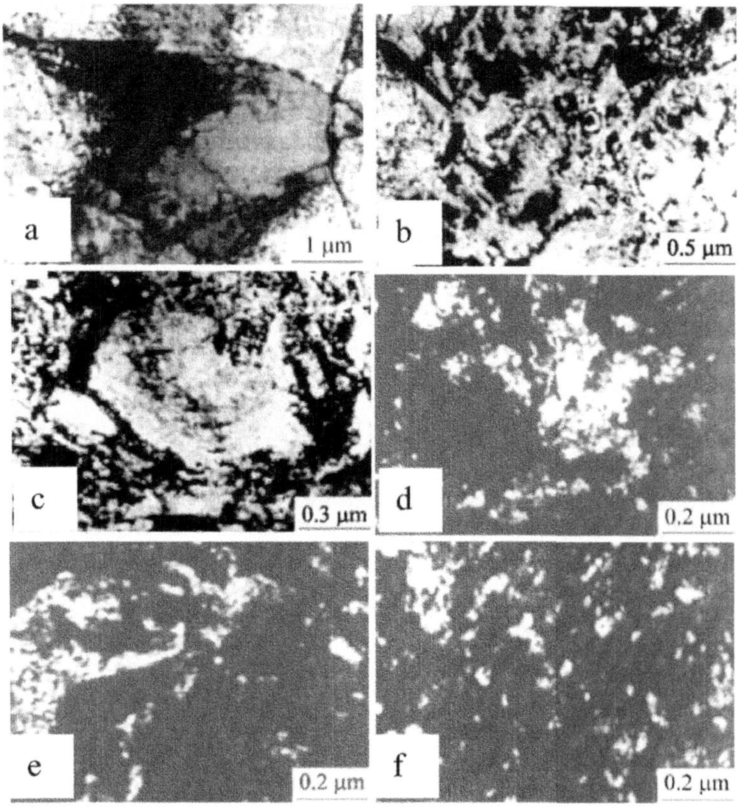

Figure.1. Structure of steel Fe-18%Cr-10%Ni samples in the initial state (a) and after SPD under pressure P = 8 GPa to a degree e: b – 0; c – 1.8; d – 3.9; e – 5.5; f – 6.2

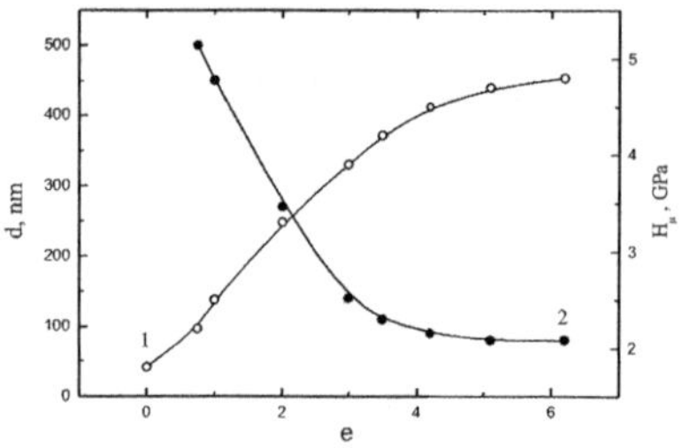

Figure. 2. Change of Vickers microhardness HV (1) and average grain size d (2) of stainless steel versus degree of SPD by HPT (P = 8 GPa)

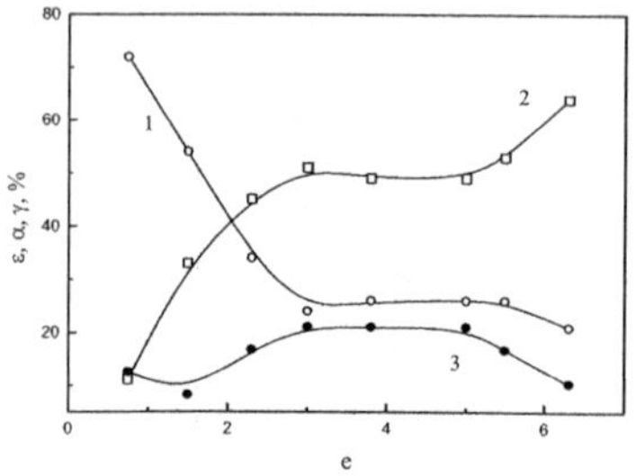

Figure. 3. Change of phase components: γ (1), ε (2) and α (3) in stainless steel versus degree of SPD by HPT (P = 8 GPa)

With the increase in deformation degree to e = 5.1 the fragmented structure is more well-defined (the average size of fragments makes 80 nm), but there is some non-uniformity for the same phase composition (see Fig.3). Diffraction is circular and consisting of many reflexes.

For HPT with e=5.5 structure fragments become finer, averaging ~ 70 nm. Areas are come across consisting of fragments of close orientation and of typical size of the order of 500 nm (Fig.1,e). Circular diffraction is also seen, but it is more smeared, as structure was more finer as compared to shearing strain e = 5.1. For the sample after HPT with e = 5.5, the scatter in values of HV is low (the average HV value being equal to 4.8 GPa) (see Fig.2), in contrast to sample after shearing strain with e = 3.9.

For the highest, attainable in this work, degree of deformation e = 6.2. In this case, the average crystallite size makes ~ 60 nm (Fig.1,f). After the analysis of dark-field image, in reflex of the $(200)_\gamma$ type, the crystallite size points to a similar decrease in fragments of γ -and ε - phase (see Fig.3). The most coarse substructure elements are 140 nm in size (see Fig.2).

After the analysis of the evolution of X-ray diffraction patterns of the investigated steel subjected to HPT (P = 8 GPa) it is seen that with the increase in shearing strain e (pressure level being constant) in the stainless steel, the quantity of high-pressure HCP ε-phase increases (see Fig.3). It has been also revealed that past the treatment, in the samples, the quantity of BCC α-phase was not large (~ 5 %). Diffraction peak $(200)_\alpha$ typical of the BCC α-phase makes it possible to assume that the BCC α-martensite of unloaded is formed upon pressure decrease.

FCC γ-Fe-Mn alloys

The influence of C_{Mn} on C_ε high-pressure HCP ε-phase content in Fe-Mn alloys depending on parameters of deformation under pressure is illustrated in Fig.4. It is known that after quenching the high-pressure HCP ε phase is present in the Fe-Mn alloys with $C_{Mn} \approx 10$ - 30 % at room temperature [5].

In the initial state the structures of studied alloys with the high-pressure phase, for example (γ+ε)-Fe-20%Mn alloy, consist of plates of HCP ε-phase ($C_\varepsilon \approx 45$ %) with the habit plane $\{111\}_\gamma$ and residual FCC γ-phase ($C_\gamma \approx 55$ %). The characteristic feature of the studied alloy, which determines its the capability to plastic deformation, is the development of strain/pressure-induced (γ + ε) → ε, α martensitic transformations [5].

After the SPD by HPT in the BCC α-Fe-Mn alloys with low C_{Mn} (to 7 wt. % Mn) the high-pressure HCP ε-phase does not occur even after deformation of e = 6.5 at pressure P = 10 GPa (see Fig.4) [6].

Plastic deformation of (γ+ε) - and γ-Fe-Mn alloys with e ≈ 1 at pressure P ≈ 1.5 GPa (method of hydroextrusion), and the shock waves (P = 8 - 16 GPa) also a little bit expand ΔC_{Mn}, in which the high-pressure HCP ε-phase is observed (see Fig.4) [5,6].

At the initial stages of the SPD by HPT the change in structure of Fe-Mn alloys with stable and metastable FCC γ-phase, the appearance of stacking faults, deformation twins and plates of HCP ε-phase is pointed out [5,6]. The increase of deformation degree to e ≈ 4 results in the appearance of banded type structures formed the basically from the twins and stacking faults. Further increase of e from 4 up to ≈ 10 causes the NC state formation with the average grain size d ≈ 80 nm.

After SPD by HPT of FCC γ-Fe-Mn alloys with C_{Mn} ≈ 40 - 55 wt.%Mn the formation of high-pressure phase - the HCP ε-phase is fixed. After pressure removal the C_ε value decreases with the increase of C_{Mn}.

The increase of pressure to 20 GPa at the SPD by HPT with deformation degrees e = 4 - 5, results in practically 100 % formation of high pressure HCP ε-phase in FCC γ-Fe-Mn with C_{Mn}≈ 40 - 55 wt. %Mn (see Fig.4). The high-pressure HCP ε-phase remains stable after pressure removal and a prolonged holding under normal conditions (atmospheric pressure and room temperature).

The research of temperature stability of the so-obtained HCP ε-phase has shown that the heating to temperature exceeding ≈ 250°C, initiates the reverse HCP ε → FCC γ-phase transformation. For obtaining a single-phase state (FCC γ-phase) at heating in an interval of temperatures 340 - 390°C, a 5 minute endurance is enough (Fig.5).

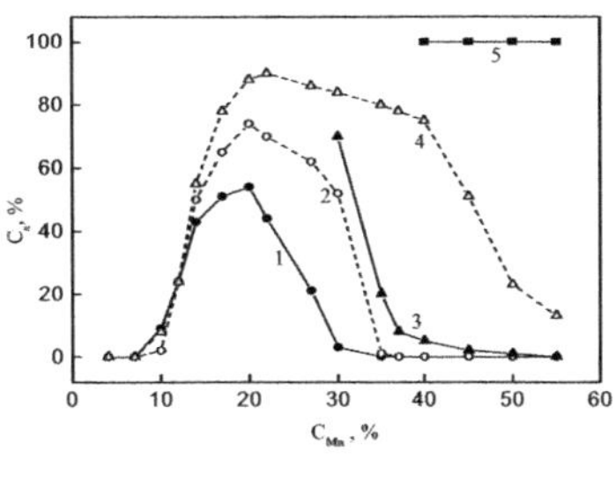

Figure. 4. Influence of different treatments on concentration dependence of HCP ε-phase in Fe-Mn alloys versus concentration C_{Mn}: 1 – initial (quenching) state; 2 – hydroextrusion (e = 1, P = 1.5 GPa); 3 – shock waves (P = 8-16 GPa); 4 – HPT (e = 6.4, P = 10 GPa); 5 – HPT (e = 5, P = 20 GPa)

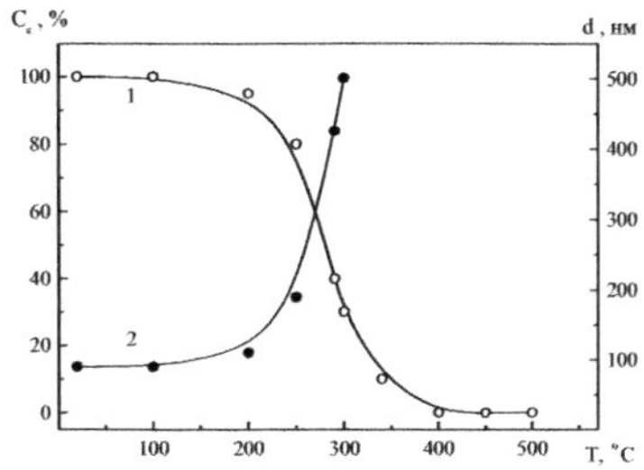

Figure. 5. Influence of tempering temperature on concentration dependence of high-pressure HCP ε-phase C_ε (1) and average grain size d (2) of FCC γ-Fe-45%Mn alloy after HPT (e = 5, P = 20 GPa)

The adequacy of investigation results depends substantially on the correct choice of the parameters unambiguously describing the stressed state, for example such as η as well as μ_σ [1,7]. The parameter of the stressed state η ($\eta = \tilde{\sigma}/T$, where σ = - P is the hydrostatic component of the stress tensor and T is the tangential stress intensity) characterizes the «rigidity» of the stressed state scheme, i.e. tensile-to-compression stresse ratio: if $\eta > 0$, the tensile stresses, and if $\eta < 0$, the compressing stresses are prevailing during the deformation. The Lode-Nadai parameter μ_σ determines the type of the stressed state scheme. The values of $\mu_\sigma = -1$ and 0 relate to the stressed state of compression and torsion accordingly. The calculated values of η at the SPD by HPT and the DAC show that with a growth of deformation degree e and a decrease of η, characterizing the «softening» of the stressed state (increase of P role).

At SPD by HPT and DAC in (γ+ε)-and γ-Fe-Mn alloys with high-pressure HCP ε-phase it is possible to designate two characteristic substructures differing by amount and thickness of plates of HCP ε-phase and morphology of BCC α-martensite crystals depending on deformation degree e. Thin plates of HCP ε-phase in FCC γ-matrix along two or three-intersection plane $\{111\}_\gamma$ characterized the first (I) type (P > 5-10 GPa). The second (II) one is characterized by availability of large plates of HCP ε–phase usually along one plane from the system $\{111\}_\gamma$ in which BCC α–crystals of lath shape are arranged, having alike orientation or forming structural complex of the "frame"- type (P ~ 0.5 – 1.5 GPa). In this case the amount of BCC α–phase is increased with the increase of deformation degree e.

Increasing of the degree e at SPD by HPT and DAC as well as increasing of the pressure in the deformable region of (γ+ε)-and γ-Fe-Mn alloys results in the growth of share of the type I-structure (P > 10 GPa). Besides, in the deformed substructure at $e \geq 0.5$ together with the reduction of BCC α–deformation martensite phase as a result of suppression of martensitic $(\gamma + \varepsilon) \rightarrow \alpha$ transformation, new plates of HCP ε–phase appear as a result of shift of the dynamic balance $\gamma \leftrightarrow \varepsilon$ in direction of activation of $(\gamma + \varepsilon) \rightarrow \varepsilon$ - deformation martensitic transformation.

A change in the amount of phase components in structure of the investigated γ-Fe-Cr-Ni steel and (γ+ε), γ-Fe-Mn alloys unambiguously testifies that with a decrease of η (increase of P) the intensity of deformation $(\gamma, (\varepsilon + \gamma) \rightarrow \alpha$ - martensitic transformations falls, and the intensity of $(\gamma, (\varepsilon + \gamma) \rightarrow \varepsilon$ - martensitic transformations increases (see Fig.3 and 4).

Conclusion

The obtained results of complex research of FCC γ-Fe-Cr-Ni steel and FCC γ-Fe-Mn alloys have revealed that their structural and phase state depends on parameters of the SPD by HPT and DAC and initial phase composition and chemical concentration, that, in the end, determines the level of mechanical and service properties of materials.

References

1. Beygelzimer Y., Varyukhin V., and Efros B., *Physical Mechanics of Hydrostatic Processing of Materials*. Donetsk: DonFTI Publ., 2000, p. 198.
2. Efros B.M., Pilyugin V.P., Gladkovskii S.V., *Defects and Diffusion Forum* 208 (2002), 263-266.
3. Kuznetsov R.I., Bykov V.I., Chernyshov V.P., Pilyugin V.P., *Pribory i Technika Eksperimenta*, 1988, no 1: 246-254.
4. Beresnev B.I. and Efros B.M., *Physica*, 139-140B (1986), 910-916.
5. Efros B.M., *Fizika i Technika Vysokich Davlenij*, 8 (1998), 82-94.
6. Adadyrov G.A., Dremin A.N., Shekhiman V.S., *Metals*, 1968, no 6: 135-142.
7. Beygelzimer Y.Y., Efros B.M., Shishkova N.V., *Metals*, 1995, no 1: 121-126.

III. PROCESSING, MICROSTRUCTURES AND MECHANICAL PROPERTIES

Precipitation Behavior in Age-Hardenable Alloys after Severe Plastic Deformation

Z. Horita[1*], T. Fujita[1], K. Kaneko[1] and T.G. Langdon[2]
[1]Department of Materials Science and Engineering, Faculty of Engineering Kyushu University, Fukuoka 812-8581, Japan
[2]Departments of Aerospace & Mechanical Engineering and Materials Science University of Southern California, Los Angeles, CA 90089-1453, USA

Abstract

In this study, equal-channel angular pressing (ECAP) is conducted on age-hardenable Al-based alloys and the precipitation behavior is examined in association with the subsequent aging process. The microstructures are examined using transmission electron microscopy and the tensile properties including hardeness are measured. It is shown that severe plastic deformation through ECAP introduces unusual phenomena in the precipitation process and there should be a potential for enhancement of strength over the conventional age-hardening process.

Introduction

Equal-channel angular pressing (ECAP) is a processing technique to introduce severe plastic strain into metallic materials [1]. The application of ECAP to the materials results in a significant grain refinement [2,3] and a subsequent increase in the strength at room temperature through the Hall-Petch relationship [4-7] or an introduction of the superplastic ductility when tested in tension at high temperatures [3,8-10]. In addition to the capability of grain refinement, the ECAP process is also effective to control the morphology and distribution of second-phase particles in two-phase metallic materials [11-14]. Since dispersion of fine particles is an important process to enhance the strength, it is anticipated that the strengthening of materials may be achieved not only with the grain refinement but also with the particle dispersion when the ECAP process is applied to the materials containing second-phase particles.

Aging is the most common process to produce a fine dispersion of precipitate particles. It was shown that ECAP accelerated the aging process to form stable phases and/or gave rise to the formation of unexpected particles when aging was conducted following ECAP. An example was the formation of the Si phase in the Al-0.9wt%Mg_2Si alloy despite the composition of no excess Si [12]. There are also reports that aging of ECAP-processed Al alloys led to an increase in the hardness [13,14]. These results suggest that there should be a great potential for the enhancement of the strength over the value achieved by conventional aging.

In this study, ECAP is applied to three different Al-based alloys, Al-Ag, Al-Mg-Si and Al-Si-Ge, all of which are well known to exhibit age-hardening. ECAP is conducted on these alloys and subsequent aging behavior is examined in terms of microstructures and mechanical properties. The microstructures are observed by transmission electron microscopy.

---

* Corresponding Author. horita@zaiko.kyushu-u.ac.jp

Y.T. Zhu and V. Varyukhin (eds.), Nanostructured Materials by High-Pressure Severe Plastic Deformation, 129-136.

Experimental Materials and Procedures

The following alloys were prepared through melting and casting procedures: Al-10.8wt%Ag, Al-1.0wt%Si-2.2wt%Ge, Al-1.0wt%Mg_2Si-0.4wt%Mg (excess Mg), and Al-0.7wt%Mg_2Si-0.7wt%Si (excess Si). All alloys were homogenized and solution-treated followed by quenching into iced water. The alloys were cut into rods with dimensions of 10 mm in diameter and 60 mm in length and they were subjected to ECAP at room temperature. ECAP was conducted through route B_C for 8 passes using a die having a channel angle of 90° and an angle of 20° for the outer arc of curvature at the channel corner [15]. Aging was conducted for the ECAP samples including solution-treated samples at temperatures of 373, 423 or 473 K for periods of up to 1080 ks (300 hours). For comparison, aging at 373 K was also undertaken with the samples subjected to rolling at room temperature with an area reduction of 70% (corresponding to equivalent strain of 1.4). Microstructures were examined using transmission electron microscopy (TEM) and scanning transmission electron microscopy (STEM) equipped with an energy dispersive X-ray spectrometer. Thin specimens for microstructural observations were prepared using a twin-jet electro-polishing technique in a solution of 20%HNO_3 and 80%CH_3OH for the Al-Ag alloy and 20%$HClO_4$, 10%$C_3H_8O_3$ and 70%C_2H_5OH for the other alloys. Vickers microhardness was measured by applying a load of 25 g for 10 s and measurements were repeated 10 times at different positions. Tensile specimens with a gage length of 5 mm and cross sectional dimensions of 2x3 mm^2 were cut with the tensile axes parallel to the longitudinal directions of the rods. Tensile tests were conducted at room temperature with a strain rate of 1.0×10^{-3} s^{-1}.

Results and Discussion

Al-10.8wt%Ag alloy

Figure 1 shows variations of the Vickers microhardness with respect to the aging time for the samples subjected to ECAP and for the solution-treated sample without ECAP. The hardness increases with aging for the as-solution-treated sample and exhibits maxima for both aging temperatures of 373 K and 473 K but with a rate which is faster at 473 K than at 373 K. The maximum hardness was obtained after aging for 10 hours at 473 K but after 100 hours at 373 K. However, the aging behavior is very different for the ECAP sample. The hardness continuously decreases from the initial stage with aging at 473 K, but it gradually increases with aging at 373 K and reaches a maximum after aging for 100 hours. The maximum hardness obtained by aging the ECAP sample at 373 K is well above the maximum hardness for the as-solution-treated samples.

Microstructures are shown in Fig.2(a) for the ECAP sample aged at 373 K for 100 hours leading to the maximum hardness and in Fig.2 (b) for the ECAP sample aged at 473 K for the same aging period. It is confirmed that there is a significant difference in the microstructures. Three features are noted in association with the microstructure shown in Fig.2 (a): (1) small grains having sizes of less than 1 μm but containing dislocations with the grains as marked A, which is a typical feature observed in ECAP samples [16], (2) very small particles which are finely dispersed within the grains as marked B, and (3) some large particles which are mostly present on the grain boundaries as marked C. These large particles are considered to be the stable γ–phase as they are not coherent with the matrix. For the small particles within

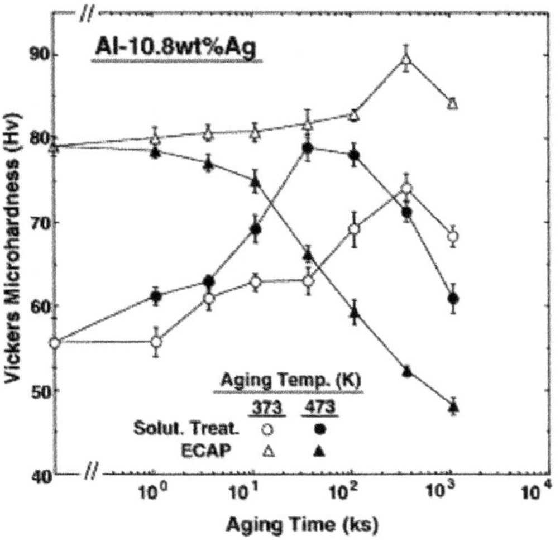

Fig.1 Vickers micro-hardness plotted against aging time for samples subjected to 8 passes of ECAP and for solution-treated samples without ECAP in Al-10.8wt%Ag alloy.

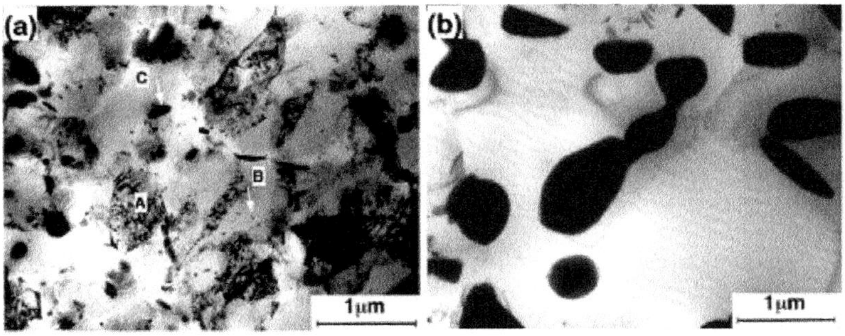

Fig.2 TEM micrographs for ECAP samples aged for 100 hours at (a) 373 K and (b) 473 K.

Fig.3 Higher magnification image within grain, showing two types of small particles.

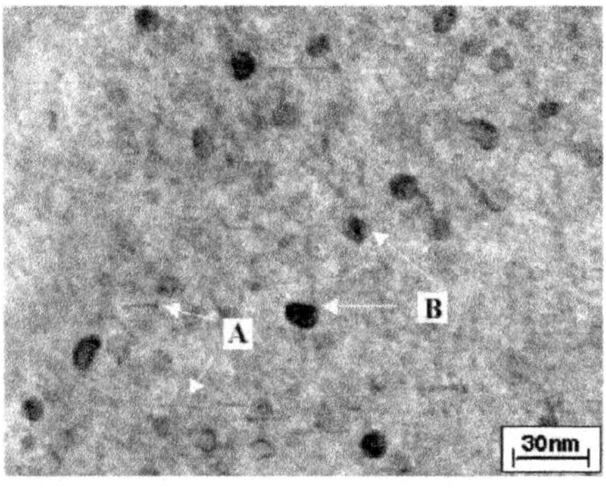

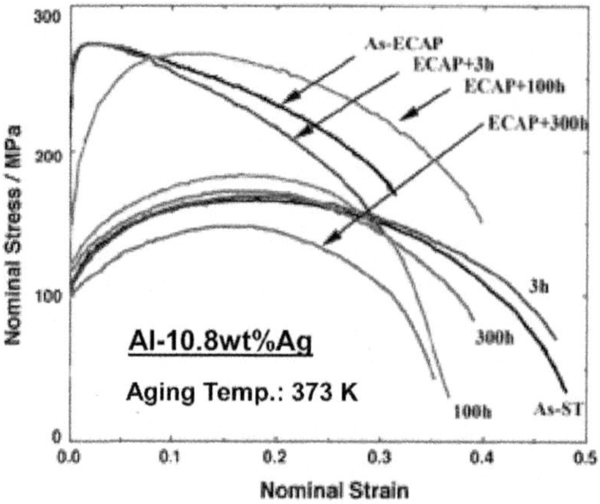

Fig.4 Stress-strain curves are shown in Fig.3 and the effect of aging at 373 K is compared for the specimens with and without ECAP

grains, a higher magnification image is shown in Fig.3. It is revealed that there are two types of particles in the image: one with elongated shapes as marked A and the other with spherical shapes as marked B. The former appears to be the plate-like γ' phase as often observed in samples subjected to aging directly following solution treatment. The latter should be due to the formation of η-zones according to a further analysis using electron diffraction. It is apparent from Fig.2 (b) that the microstructure after aging at 473 K for 100 hours contains neither small grains nor very small particles within the grains but large γ-phase particles are present. It is considered that the hardening observed in the sample aged at 373 K is attributed to the formation of the very small particles within the grains and the hardness remains the highest because of the simultaneous effect of the grain refinement due to the ECAP process and the hardening due to the low temperature aging which does not lead to grain growth.

Stress-strain curves are shown in Fig.4 and the effect of aging at 373 K is compared for the specimens with and without ECAP. Whereas the effect is minor on the specimens without ECAP, aging significantly affects the level of flow stress including the shape of the stress-strain curve for the specimens subjected to ECAP. An important feature is that the elongation is improved without decreasing the tensile strength when the ECAP specimen is aged for 100 hours. It is concluded that the 100-hour aging is the optimum condition for the enhancement of the tensile stress as well as the uniform elongation: the improvement in the tensile stress is a factor of 1.5 when compared with the solution-treated specimen without ECAP while keeping a similar level in the uniform elongation.

Al-Mg-Si alloys

Figure 5 shows (a) a TEM bright field image and (b) a dark field image including (c) an SAED pattern of the sample containing excess Si after annealing for 1 hour at 473 K. The dark field image was taken using a diffraction spot indicated by the arrow. Many rod-like precipitates are visible which are elongated along the <100> direction of the matrix. A lattice image of a precipitate is shown in Fig.6(a) and a diffractogram taken from the precipitate is in Fig.6(b). Close inspection of the image

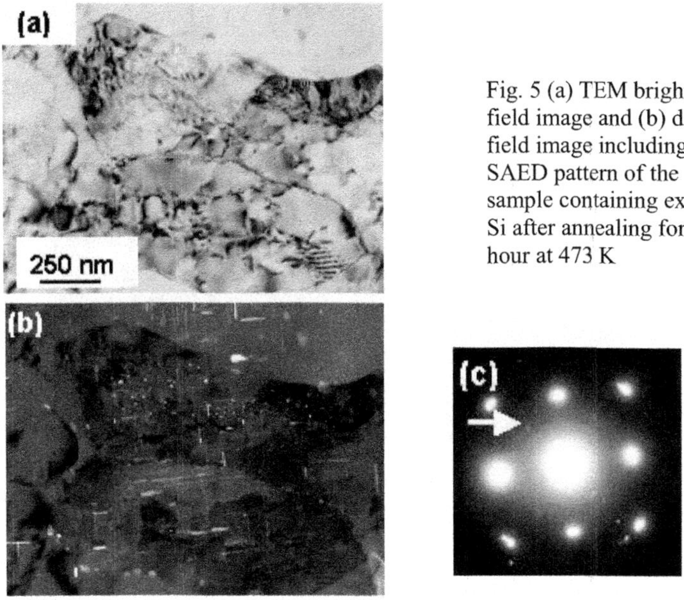

Fig. 5 (a) TEM bright field image and (b) dark field image including (c) SAED pattern of the sample containing excess Si after annealing for 1 hour at 473 K

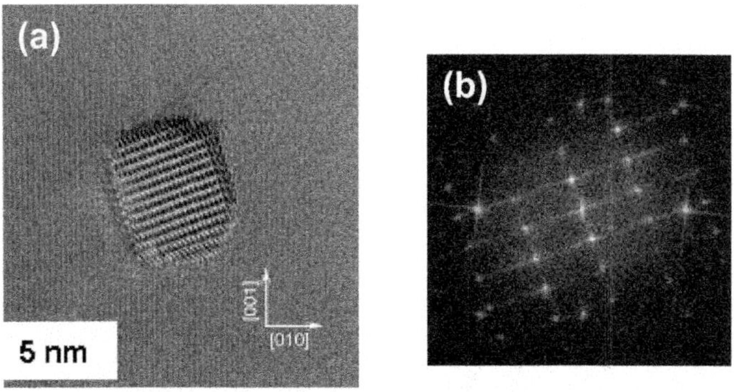

Fig. 6 (a) Lattice image of type A precipitate and (b) diffractogram taken from the precipitate

and diffractogram reveals that this particle is a type A precipitate. Matsuda and Ikeno reported [17] that the formation of type A precipitates occurred in a solution-treated sample after prolonged aging such as for 1000 hours at 473 K. Thus, the aging for 1 hour for the present ECAP sample is very short and it is considered that the aging was accelerated by severe plastic strain introduced by the ECAP process. This result is consistent with an earlier observation [12] that precipitation process proceeds at a faster rate than in the normal strain-free condition.

Figure 7 shows (a) a STEM image and (b)-(d) X-ray mappings of the

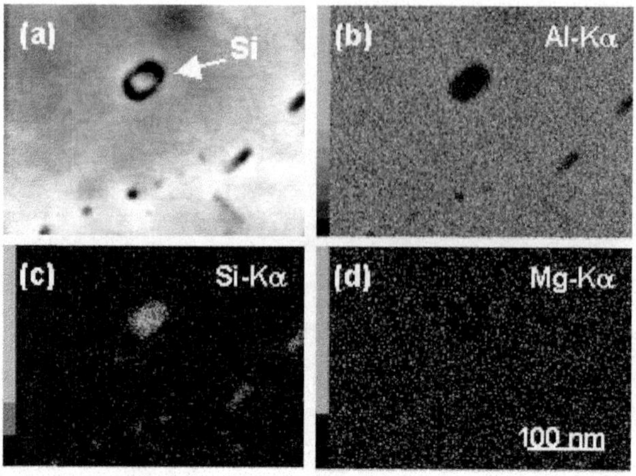

Fig. 7 (a) STEM image and (b)-(d) X-ray mappings using Al K_α, Si K_α and Mg K_α lines for excess-Mg. alloy

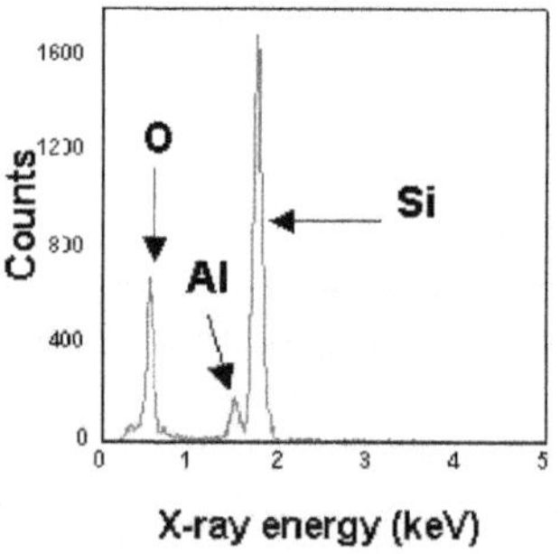

Fig. 8 X-ray spectrum from particle indicated by arrow in Fig.7 (a)

containing excess Mg. The sample was aged for 30 hours at 473 K. The X-ray analysis reveals that the particle indicated by the arrow in Fig.7 (a) is rich in Si. This is also confirmed with an X-ray spectrum shown in Fig.8. There is no peak of Mg and it is concluded that a Si phase formed in the sample despite the composition which is excess of Mg. Although an oxygen peak is present, this must be due to an oxide layer formed during electro-polishing for electron microscopy. The presence of the Si particle is attributed to the introduction of severe plastic strain through the ECAP process. The presence of Si particles was also observed in an earlier study of a balanced Al-0.9%Mg2Si alloy after ECAP process [12].

Al-1.0wt%Si-2.2wt%Ge alloy

Figure 9 shows variations of the Vickers microhardness with respect to the aging time for samples subjected to ECAP and for a solution-treated sample without ECAP. As in Fig.1, the hardness increases with aging for the as-solution-treated sample with a rate which is faster at 423 K than at 373 K. The maximum hardness is reached after aging for ~30 hours at 423 K but it keeps increasing even after 300 corresponding

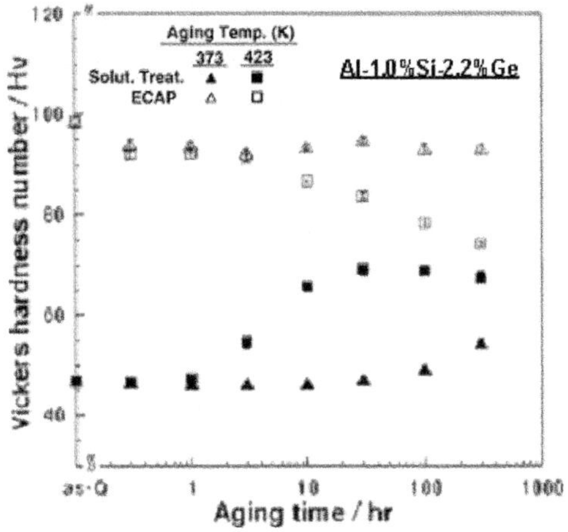

Fig.9 Vickers micro-hardness plotted against aging time for samples subjected to 8 passes of ECAP and for solution-treated samples without ECAP in Al-1.0wtSi-2.2wt%Ge alloy.

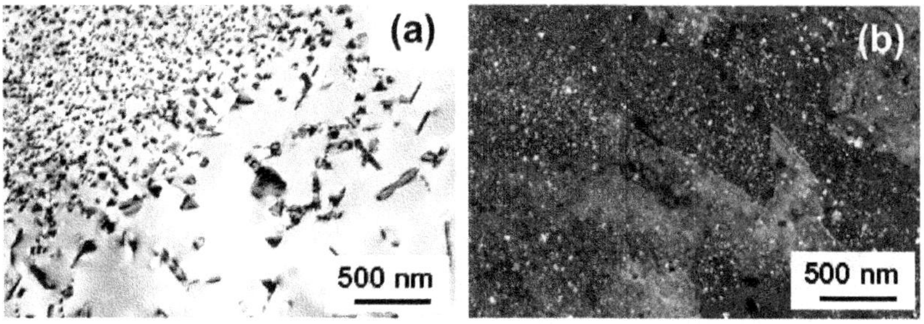

Fig.10 Microstructures after aging at 423 K for 100 hours (a) for the as-solution-treated sample and (b) for the ECAP sample.

area taken with the Al K_α, Mg K_α and Si K_α lines for the sample hours at 373 K. However, the hardness continuously decreases from the initial stage with aging at 423 K and it remains almost unchanged for aging at 373 K. Microstructures after aging at 423 K for 100 hours are shown in Fig.10 (a) and (b) for the as-solution-treated sample and for the ECAP sample, respectively. Very small particles are finely precipitated throughout the ECAP sample but a coarse distribution of particles is partially visible in the as-solution-treated sample. It is considered that the homogeneous distribution of fine particles in the ECAP sample may be attributed to a homogeneous distribution of nucleation sites introduced by severe plastic strain through the ECAP process. This difference in particle distribution should be responsible for the higher hardness in the ECAP sample than in the as-solution-treated sample after 100-hour aging.

Summary and Conclusions

1. *Al-10.8wt%Ag alloy*: (1) The hardness of the ECAP sample gradually increased

with aging at 373 K and reached a maximum after aging for 100 hours. (2) Small grain sizes less than 1 μm were retained and very small particles of η zones and plate-like γ' phase were precipitated within the grains after the maximum aging condition. (3) The precipitation of such very small particles within the fine-grained matrix was responsible for the age hardening. (4) While the tensile stress remained the same level as that for the as-ECAP sample, the uniform elongation was significantly improved and reached a level similar to the solution-treated specimen without ECAP.

2. *Al-Mg-Si alloys*: (1) Precipitates of A type were observed in the excess-Si alloy after very short time aging as 1 hour when aging was performed on the ECAP sample. (2) Precipitates of Si phase were formed even in the excess-Mg alloy after aging of the ECAP sample.

3. *Al-1.0wt%Si-2.2wt%Ge alloy*: (1) The hardness continuously decreased from the initial stage with aging at 423 K but remained almost constant for aging at 373 K. (2) Very small particles were finely precipitated throughout the ECAP sample but a coarse distribution of particles was partially present in the as-solution treated sample. (3) The homogeneous distribution of fine particles in the ECAP sample should be attributed to a homogeneous distribution of nucleation sites introduced by severe plastic strain through the ECAP process.

Acknowledgements

This work was supported in part by a Grant-in-Aid for Scientific Research from the Ministry of Education, Science, Sports and Culture of Japan, in part by the Light Metals Educational Foundation of Japan and in part by the National Science Foundation of the United States under Grant No. DMR-0243331.

References

1. Segal V.M., Reznikov V.I., Drobyshevskiy A.E. and Kopylov V.I., *Russian Metall.* 1 (1981), 99.
2. Valiev R.Z., Krasilnikov N.A., Tsenev N.K., *Mater. Sci. Eng.* A137 (1991), 35.
3. Valiev RZ, Islamgaliev RK and Alexandrov RK. *Prog. Mater. Sci.* 45 (2000), 103.
4. Hall E.O., *Proc. Phys. Soc.*, B64 (1951), 747.
5. Petch N.J., *J. Iron Steel Inst.*, 174 (1953), 25.
6. Furukawa M., Horita Z., Nemoto M., Valiev R.Z., Langdon T.G.,, *Acta Mater.*, 44 (1996), 4619.
7. Horita Z., Fjinami T., Nemoto M., Langdon T.G., *Metall Mater Trans A*, 31A (2000), 691.
8. Komura S., Horita Z., Furukawa M., Nemoto M., Langdon T.G., *Metall. Mater. Trans A*, A32 (2001), 707.
9. Neishi K., Uchida T., Yamauchi A., Nakamura K., Horita Z.., Langdon T.G., *Mater. Sci. Eng.*, A307 (2001), 23.
10. Matsubara K., Miyahara Y., Horita Z., Langdon T.G., Acta Mater., 51 (2003), 3073.
11. Murayama M., Horita Z., Hono K., *Acta Mater.*, 49 (2001), 21.
12. Oh-ishi K., Hashi Y., Sadakata A., Kaneko K., Horita Z., Langdon T.G., *Mater. Sci. Forum*, 396-402 (2002), 333.
13. Kim J.K., Jeong H.G., Hong S.I., Kim Y.S., Kim W.J., *Scripta Mater.*, 45 (2001), 901.
14. Ohashi K., Fujita T., Kaneko K., Horita Z., Langdon T.G , *Mater Sci. Forum*, 426-432 (2003), 2637.
15. Furukawa M., Iwahashi Y., Horita Z., Nemoto M., Langdon T.G *Mater. Sci. Eng.* A257 (1998), 328.
16. Iwahashi Y., Horita Z., Nemoto M., Langdon T.G., *Metall. Mater. Trans.*, 29A (1998), 2503.
17. Matsuda K., Ikeno S., *J. Japan Inst. Light Metals*, 50 (2000), 23.

Understanding the Textural and Microstructural Development in Two-Pass ECAP by Route A of a Commercial Aluminium Alloy

J.C. Werenskiold[1,*] and H.J. Roven[1]
[1]Department of Materials Technology, Norwegian University of Science and Technology, N-7491 Trondheim, Norway

Abstract

The textural and microstructural development throughout the deformation zone in ECAP has been investigated by EBSD and XRD and coupled with measured strains. The microstructural development during the first ECAP pass is dominated by deformation banding and the results presented are discussed in terms of stable texture components and their effect on the microstructural development.

Introduction

Numerous earlier works on ECAP have mainly been focused on the very fine grain structure obtainable by SPD. In the present work, focus is set on deformation mechanisms and the earlier stages of grain subdivision. A detailed study of the microstructural and textural development throughout the deformation zone in the first and second ECAP pass by route A, has been made. The early stages of grain subdivision seem to be dominated by deformation banding (DB). The DB mechanism is discussed in terms of the texture development and stable texture components.

Experimental Procedures

The material used in the present study is a commercial AA6082.50 aluminium alloy supplied by Hydro Aluminium AS. The chemical composition is Al-0.1Si-0.6Mg-0.5-Mn-0.2Fe [wt%].

Billets of dimensions 19.5×19.5×100mm^3 were machined from the received material and was ECAP'ed in the as cast and homogenized (03-temper) state at room temperature with a constant ram speed of approximately 3 mm/s, following route A [1]. The billets were coated with the petroleum and graphite based lubricant, OMEGA99® before pressing. The ECAP die has an upper channel cross section of 20×20mm^2, die angle φ= 90°, outer curvature angle ψ=20.6° and an exit cross section of 19.5×19.5mm^2.

Texture and microstructural data through the deformation zone were obtained by electron backscatter diffraction (EBSD) in a Hitachi S-4300SE FE-SEM, equipped with a Nordif EBSD detector and the TSL OIM EBSD software. Samples were mechanically grinded and polished to 1μm, followed by polishing in a diluted OPS solution and finally electro chemically polished in a 30% HNO_3 – 60% methanol solution at -30°C.

Standard texture measurements were also made by XRD in a Siemens D5000 difractometer after 1-4, 6 and 8 passes by route A. ODF sections were calculated by the Van Houtte software.

The texture data is presented as ODF sections (φ_1,Φ,φ_2)=(0-180°, 0-180°, 45°) in

* Corresponding Author: Jens.Christofer.Werenskiold@material.ntnu.no

Y.T. Zhu and V. Varyukhin (eds.), Nanostructured Materials by High-Pressure Severe Plastic Deformation, 137-144.

accordance to the Bunge notation, covering all stable texture components. Crystal orientations are presented in a *(hkl)*/θ notation where *(hkl)* is the crystal plane normal in the Z-direction and θ is the CCW rotation (X towards Y) about the <*hkl*> crystal axis, i.e. equivalent to rotation about Z. The applied coordinate system is given in Figure 1.

The strain introduced by ECAP was measured by applying a grid marking technique from Lectroetch® and analyzed using the commercially strain analysis software ASAME® and some in-house software. The grid size was 1×1mm^2 with a line thickness of ~0.1mm.

Results and Discussion

Identification of the ideal texture components

The typical ECAP texture (at the end of each pass, route A) has been elaborated by the present author, and is found to be a combination of a few stable orientations which are, in the present work, referred to as the ideal ECAP texture components. The identified components are listed in Table 1. The first seven components in Table 1 are equal to the components presented by Tóth et al. in [2].

Table 1.The ideal ECAP texture components in the present case for FCC aluminum. The components are given as crystal directions in the global X, Y and Z directions, by the corresponding Euler angles and in accordance with the *(hkl)*/θ notation.

Texture component	Crystal directions in the global X, Y and Z directions			Euler angles			*(hkl)*/θ
	X	Y	Z	φ1	Φ	φ2	
A_{2E}	[-3 3 25]	[25 -25 6]	[1 1 0]	99.74	90	45	(110)/105
A_{1E}	[-25 25 6]	[3 -3 25]	[1 1 0]	170.37	90	45	(110)/165
C_E	[71 -71 100]	[71 -71 -100]	[1 1 0]	**45**	90	**45**	(110)/45
A_E	[2 -20 9]	[20 -2 -9]	[1 1 2]	**45**	35.26	**45**	(112)/45
$-A_E$	[-2 20 -9]	[-20 2 9]	[1 1 2]	**45**	144.7	**45**	(112)/225
B_E	[27 -100 73]	[100 -27 -73]	[1 1 1]	**45**	54.74	**45**	(111)/45
$-B_E$	[-73 -27 100]	[73 -100 27]	[1 1 1]	105	54.74	45	(111)/105
D_E	[13 -12 7]	[8 4 -7]	[2 5 5]	**45**	105	**45**	(255)/120
Cube	[100]	[010]	[001]	**45**	0/180	**45**	(001)/45
Z-rot. cube	[-5-11 0]	[11-5 0]	[001]	**70**	0/180	**45**	(001)/70

Texture development: first pass

The starting texture from the homogenized billets was very close to a random distribution (Figure 3a). The texture strength increases continuously through the deformation zone during the first pass, as shown in Figure 3a-h), and becomes more well-defined with increasing strain. All texture components are observed to rotate ~15° about the Z-direction (in the global XYZ-frame) from the X-direction towards the Y-direction (positive rotation). This means that the physical principal shear direction starts at ~30° to the X-direction (at 20% strain), and rotates ~15° about Z-direction and ends up at ~45° to the X direction (at 100% strain). The angle of 45° corresponds to the intersection of the two ECAP channels. The gradual 15° rotation about the Z-axis suggests that the "shear plane" or more correct, the physical principal shear direction is not a single defined direction or plane, rather the deformation occurs over a finite "fan shaped" zone, as shown in Figure 1. This observation is also supported by the strain measurements, which shows that the width of the actual deformation zone in the specimen centerline is about ~5mm for this specific die design and high friction conditions.

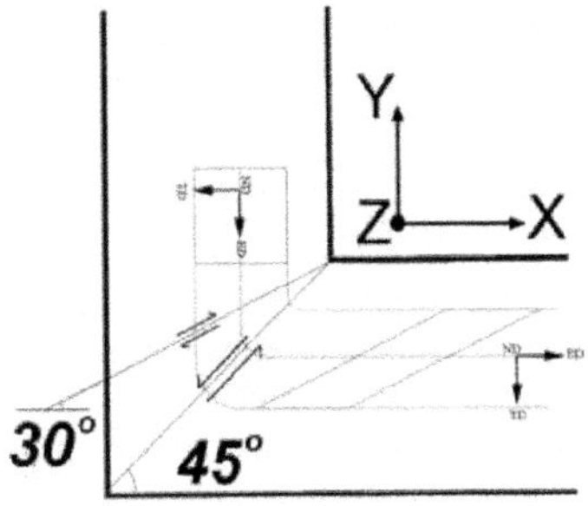

Figure 1. Schematic presentation of the physical principal shear direction, starting at 30° to the X-axis.

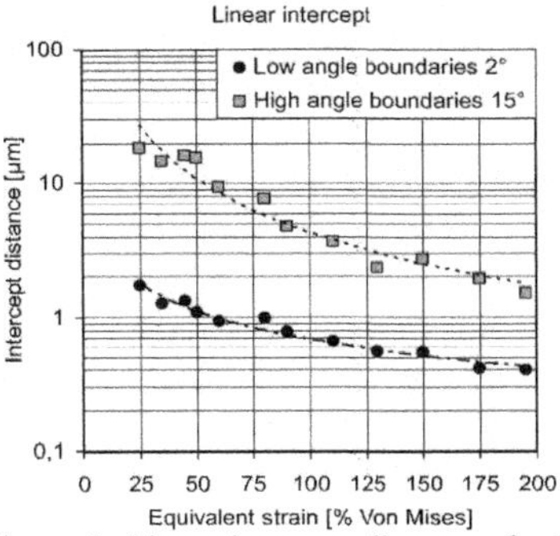

Figure 2. Linear intercept distances for low- and high-angle boundaries (>2°) and (>15°) through the deformation zone. The equivalent strain is accumulated up to two passes; first pass from 0-100%, second pass from 100-200%.

The actual width of the deformation zone is dependent upon the outer corner angle of the die, the friction in the die channels (hence also the back pressure) and the work hardening of the material [3].

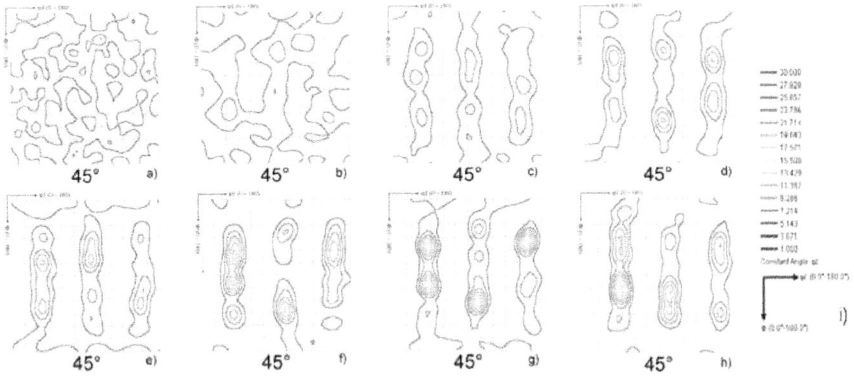

Figure 3. Measured EBSD ODF sections for φ_2=45° for the first pass, for a) 10%, b) 20%, c) 35%, d) 55%, e) 65%, f) 75%, g) 85%, h) 95% strain and i) applied scale (times random).

The measured strain tensor components show that the shear component increases linearly with the equivalent strain, while a material element will experience compression in the ED direction and tension in the TD direction up till ~50% strain. Here the tension / compression is reversed, i.e. the material element will experience tension in the ED direction and compression in the TD direction until the element has passed through the deformation zone where the final distortion of the element can be described by simple shear only.

It is believed that this mixed deformation mode, generated as a result of the fan-shaped deformation zone and ECAP geometry, is the reason behind the gradual texture component rotation. This is also supported by the texture modeling by Gholinia et al. [4], where they compared the texture generated from the mixed deformation mode generated by ECAP, and texture generated from simple shear. By using the actual observed strain tensor, they succesfully reproduced the observed ECAP texture.

Texture development: second pass

As expected, the starting texture in the second pass (Figure 4a) is almost identical to the final texture from the first pass, but rotated -90° about the Z-axis.

When the sample is rotated -90° about the Z axis, according to route A, the texture components get new orientations in relation to the XYZ-frame, and are no longer in their stable orientations. The new orientations will try to realign to the closest stable orientation. During this process, some components will diminish while others continuously increase in strength. We may assume that crystal rotations will occur mainly as rotations about the Z-axis (in the XY-plane) due to the observed plane strain deformation mode.

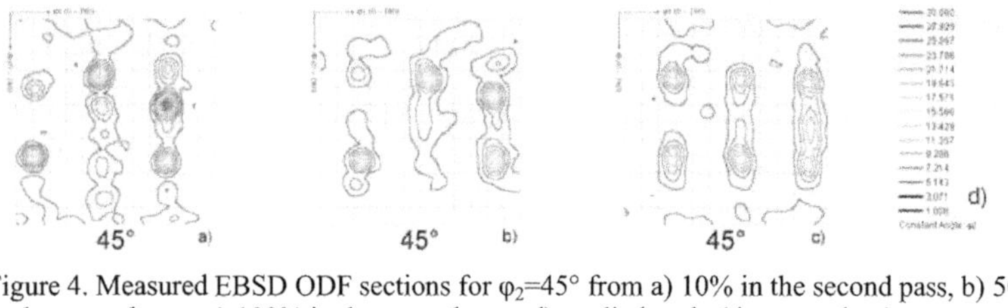

Figure 4. Measured EBSD ODF sections for φ_2=45° from a) 10% in the second pass, b) 50% in the second pass, c) 100% in the second pass, d) applied scale (times random).

By comparing the intensities of the different texture components at the end of the first and second pass (Table 2), one find support for the above assumption.

The $\mathbf{B_E}$/-$\mathbf{B_E}$ components are observed to continuously increase in strength by increasing number of passes. These components will also increase due to the –$\mathbf{A_E}$/$\mathbf{A_E}$ and $\mathbf{D_E}$ components, which are assumed to rotate into the -$\mathbf{B_E}$/$\mathbf{B_E}$ components.

The $\mathbf{A_{1E}}$ component is observed to increase in strength in the second pass, to be stable in third pass and thereafter decrease slightly by increasing number of passes.

The $\mathbf{C_E}$ and $\mathbf{D_E}$ components may rotate into $\mathbf{A_{1E}}$ in the second pass and may therefore contribute to the $\mathbf{A_{1E}}$ texture strength.

The $\mathbf{C_E}$ component decreases slightly in the second pass. The $\mathbf{A_{1E}}$ and $\mathbf{A_{2E}}$ may both rotate into the $\mathbf{C_E}$ component in the second pass and therefore contribute to the $\mathbf{C_E}$ texture strength.

The $\mathbf{A_E}$ and -$\mathbf{A_E}$ components decrease in the second and third pass and is thereafter stable at intensity ~1 times random. This is probably because none of the other components rotate into the $\mathbf{A_E}$/ -$\mathbf{A_E}$ components. This is also the case for the $\mathbf{D_E}$ component, which only seems to be stable in the first pass.

The $\mathbf{A_{2E}}$ component is weak in the first pass and decreases slightly in the second pass. Here, the $\mathbf{A_{2E}}$ component rotates into the $\mathbf{C_E}$ component and only the $\mathbf{A_{1E}}$ component is likely to rotate into the $\mathbf{A_{2E}}$ component.

Please note that the texture intensities observed by EBSD is much stronger than intensities observed by XRD. This is most likely a result of the different measuring techniques. Therefore, the absolute texture intensities should not be compared, but rather the relative variations in intensities from each measurement method. In spite of a severe plastic deformation, texture intensities in general tend not to increase with strain.

Table 2. Texture component intensities from XRD measurements, route A, pass 1-8.

Intensities from XRD measurements (times random)						
Texture component	1st pass	2nd pass	3rd pass	4th pass	6th pass	8th pass
A_{2E}	2	1.4	4	2	2.5	1.4
A_{1E}	6.4	8	8	5	5.6	4
C_E	5	4	2.8	2.5	4	2.8
A_E	3.2	2.8	1	1	1.4	1.2
$-A_E$	4	2.8	1	1	1.4	1.2
B_E	3.2	5.6	5.6	6.4	8	8
$-B_E$	3.2	4	5.6	6.4	8	8
D_E	4	2.8	2	1.6	2	2
Cube	2.5	1.4	1	1.3	1.4	1.4
Z-rotated Cube	1.6	2	2.8	2.5	2	2

Microstructural development

The microstructural development during the first and second ECAP pass by route A can in general be described by grain subdivision where the formation of deformation bands (DB's) seem to be the dominating mechanism. Deformation banding can be described in terms of LEDS. For a detailed review on deformation bands, see e.g. [5]. Grains not oriented in any stable orientation with respect to the subjected strain field will subdivide by deformation bands rotating into stable orientations. For the formation of a deformation band to be energetically favorable, the reduction of elastic strain energy due to a reduction of simultaneously acting slip systems must be larger than the energy created in the DB boundaries and the excess elastic strain energy on account of accommodation stresses.

The deformation bands are crystalographic in nature and form an alternating lamellar structure, separated by high angle boundaries, typically higher than 25°, where finally both DB's and the original grain matrix rotate towards the stable texture components. The DB size is dependent on many factors such as the reduction of the effective Taylor factor, the DB boundary energy and the average cell misorientation in the matrix where the DB's form. However, the most important factor controlling the equilibrium DB size is the shear stress at the time of DB formation. The shear stress is again dependent on the grain size of the material, following a Hall-Petch type relation, i.e. higher shear stress leads to smaller deformation bands which again defines the new grain size which leads to an increase in shear stress etc.

However, the deformation bands do not form in a grain already oriented within the stable texture components, i.e. a new DB will not form inside an old DB, already in a stable orientation, independent of the shear stress. However, by rotating the ECAP sample according to route A (or any other route changing the macroscopic shear plane), orientations which were stable during the previous pass will now become unstable, increasing the potential for new deformation bands to form. In addition, the average cell misorientation increases, leading to increased potential for formation of smaller DB's. Due to the grain refinement during the previous pass, the shear stress has increased and new, smaller deformation bands will form, crossing the structure formed during the previous pass, resulting in an even more refined structure.

At some finite accumulated strain (typically after 4 to 6 passes), the potential for DB formation will become too small because the grain size has become equal to the minimum DB size which is governed from equilibrium considerations. In the present case, this size is ~0.4μm in width, and ~2μm in length.

Grains may however become smaller than this equilibrium size, but not by formation of new deformation bands, which is a crystallographic process. By introducing the work piece to even higher strains by ECAP, the geometric nature of shear deformation will flatten and elongate the already established structure, and deformation bands will become divided by low to medium angle boundaries into more equiaxed grains (similar to what is observed in cold rolling).

There is however other deformation mechanisms leading to the formation of high angle, sub micron sized grains.

A secondary mechanism is the formation of very small, elongated grains, inside larger grains, separated from the grain matrix by high angle boundaries. These grains form by collapse of orientation gradients due to reordering of dislocations into high angle boundaries, typically ~20° misorientation, thereby reducing the internal stresses by formation of low energy dislocation structures (LEDS). These grains have non-crystalographic orientation, i.e. not oriented within the stable texture components. The observed size of such grains formed during the first ECAP pass is typically ~0.1 to 0.2μm in width and ~0.5 to 1.5μm in length.

The deformation of grains oriented within the stable texture components occur by crystalographic slip, leading to a cell structure, aligned in the direction of shear. The cell size decreases with increasing accumulated strain while the average cell misorientation increases steadily but seem to saturate below ~10°.

Linear intercept values for low and high angle boundaries are shown in Figure 2 as a function of accumulated strain to a total strain of 2, corresponding to two passes by route A. Low angle boundaries correspond to the cell size evolution and high angle boundaries correspond to the reduction of grain size due to the formation of deformation bands.

A few examples of grain subdivision by deformation banding are shown in the following. At 20% strain in the first pass, a grain with orientation (335)/60 is subdivided by crossing DB's with orientations (335)/30 and (335)/90, both orientations having a <110> crystal direction aligned in the principal shear direction (which at 20% strain is located at ~30°to the X-axis). The grain is shown in Figure 5 with the corresponding (111) and (110) pole figures.

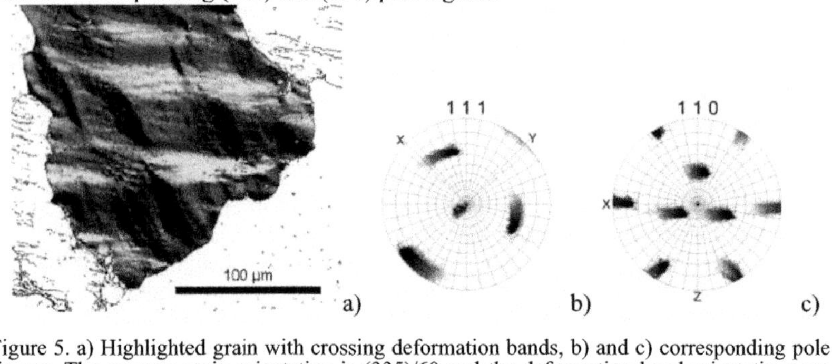

Figure 5. a) Highlighted grain with crossing deformation bands, b) and c) corresponding pole figures. The average grain orientation is (335)/60 and the deformation band orientations are (335)/30 (yellow) and (335)/90 (black). Position 25% strain in the first pass.

At 60% strain in the first pass, a grain oriented as (001)/90 (Figure 6) is observed to subdivide by DB's with orientations 60°apart, i.e. (110)/105 and (110)/165, corresponding to the $\mathbf{A_{2E}}$ and $\mathbf{A_{1E}}$ texture components. Some observed subdivisions from the first pass are given in Table 3 as initial grain orientation, DB orientation and the closest texture component along with the deviation from the given texture component (MOA).

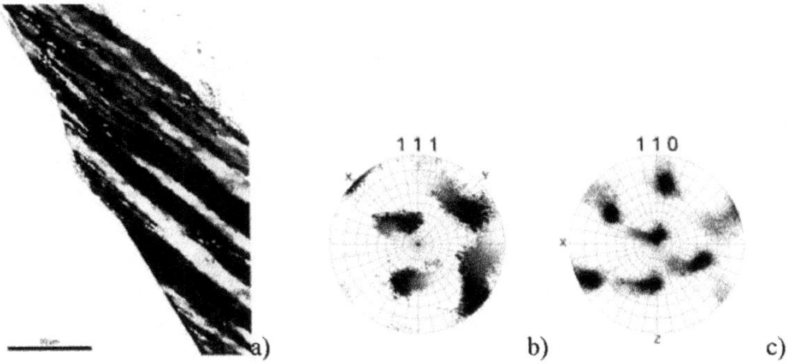

Figure 6. Example of deformation banding in a "near cube" oriented grain. a) Highlighted grain, b) and c) corresponding pole figures. The original grain orientation was close to (001)/90. The grain is subdivided into two new orientations, both stable texture components; $\mathbf{A_{2E}}$ (110)/105 (yellow) and $\mathbf{A_{1E}}$ (110)/165 (black). Position 60% strain in the first pass.

Table 3. Observed subdivisions by deformation banding during 1st pass.

Grain orientation: $(\varphi_1,\Phi,\varphi_2)$	DB orientation: $(\varphi_1,\Phi,\varphi_2)$	Texture comp/MOA
(85,40,45)	(45,40,45) + (105,40,45)	$\mathbf{A_E}$ /5° + $\mathbf{-B_E}$ /15°
(135,103,45)	(105,103,45) + (155,103,45)	$\mathbf{A_{2E}}$ /14° + $\mathbf{A_{1E}}$ /20°
(45,90,45): $\mathbf{C_E}$	(45,90,45) + (85,90,45)	$\mathbf{C_E}$ /0° + Σ9 to $\mathbf{C_E}$
(45,0,0)	(100,90,45) + (170,90,45)	$\mathbf{A_{2E}}$ /0° + $\mathbf{A_{1E}}$ /0°
(70,90,45)	(45,90,45) + (100,90,45)	$\mathbf{C_E}$ /0° + $\mathbf{A_{2E}}$ /0°
(45,0,0)	(45,0,45)	**Cube**/0°
(160,35,45)	Stable	N.A.
(105,145,45)	(45,145,45)	$\mathbf{-A_E}$ /0°
(105,35,45)	(45,35,45)	$\mathbf{A_E}$ /0°
(20,11,124)	(96,88,73)	**Cube**/18°

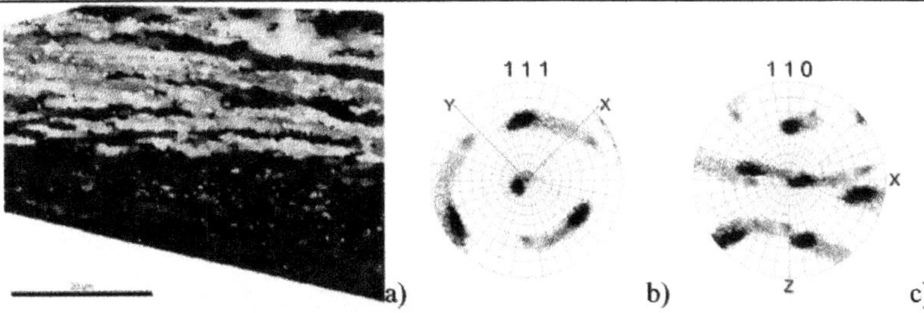

Figure 7. Example of a grain which most probably was subdivided by deformation bands during the first pass ((112)/165 oriented grain with (111)/105 oriented deformation bands) and is now realigning towards the new stable orientations (111)/45 at 30% strain in the second pass.

At 30% strain into the second pass, a grain subdivided by DB's in the first pass (DB1), is now observed to realign one of its former components towards a new stable component (Figure 7). The grain orientation at the end of the first pass was ~(112)/165 with deformation bands oriented ~(111)/105 (DB1), i.e. the $\mathbf{-B_E}$ component. After the -90° rotation about the Z-axis according to route A, the DB's from the first pass got a (111)/15 orientation which is realigning towards the (111)/45 orientation (DB2), i.e. the $\mathbf{B_E}$ component.

A few observed subdivisions / realignments occurred during the second pass are given in Table 4 as the assumed grain orientation from the first pass, the DB orientation from the first pass (DB1), the new grain orientation in the second pass (rotated according to route A), the new DB (DB1) orientation in the second pass (rotated according to route A) and the realigned / new DB orientation (DB2).

Table 4. Observed realignment of grains carrying three different grain and deformation band orientations during the 2nd pass.

Grain orientation, 1st pass. (φ_1,Φ,φ_2): component	DB1 orientation, 1st pass. (φ_1,Φ,φ_2): component	Grain orientation, 2nd pass. (φ_1,Φ,φ_2): component	DB1 orientation, 2nd pass. (φ_1,Φ,φ_2): component	DB2 orientation, 2nd pass. (φ_1,Φ,φ_2): component
(165,35,45)	(105,55,45): -$\mathbf{B_E}$	(75,35,45)	(15,55,45)	(45,55,45): $\mathbf{B_E}$
(115,35,45)	(45,35,45): $\mathbf{A_E}$	(25,35,45)	(315,35,45)	(45,35,45): $\mathbf{A_E}$
(45,90,45): $\mathbf{C_E}$	N.A.	(125,90,45)	N.A.	(105,90,45): $\mathbf{A_{2E}}$

Conclusion

The typical ECAP texture starts to develop already at ~25% strain and increases in strength during the first pass. The strongest texture components observed in the first pass are $\mathbf{A_{1E}}$, $\mathbf{C_E}$, $\pm\mathbf{A_E}$, $\mathbf{D_E}$, $\pm\mathbf{B_E}$ and $\mathbf{A_{2E}}$ in decreasing order. In the second pass, $\mathbf{A_{1E}}$ and $\pm\mathbf{B_E}$ increase in strength while all other components decrease. This is believed to be a result of a realignment of the different texture components to the closest stable component in the second pass. At higher strains, up to 8 passes, the $\pm\mathbf{B_E}$ component shows a continuous increase in strength while all other components, including $\mathbf{A_{1E}}$ decrease in strength. The $\pm\mathbf{B_E}$ component is stable in route A because it has three slip directions aligned in the XY-plane and may easily rotate about the Z-direction.

The observed deformation banding can be explained in terms of LEDS theory, and has been shown to be an important mechanism in the early stages of grain subdivision, and is further believed to be the main source of high angle grain boundary formation by gradual grain subdivision down to a HAGB size of approximately ~0.4μm, when other deformation mechanisms may be energetically more favorable.

Acknowledgements

The authors wish to thank the Norwegian Research Council under the project COMPFORM and Hydro Aluminum AS for financial support.

References

[1] Segal V.M, *Materials Science and Engineering A,* 197 (1995), 157-164.
[2] Toth L.S., Massion R.A., Germain L., Baik S.C., Suwas S., *Acta Mater.* 52 (2004), 1889.
[3] Roven H.J., Dumoulin S., Werenskiold J.C., *Ultrafine Grained Materials III* ,ed.Y.T. Zhu, T.G. Langdon, R.Z. Valiev, S.L. Semiatin, D.H. Shin, T.C. Lowe (Eds), TMS, The Minerals, Metals & Materials Society, 2004, pp. 117-124.
[4] Gholinia A., Bate P., Prangnell P.B., *Acta Mater.* 50 (2002), 2121-2136.
[5] Kuhlmann-Wilsdorf D., *Acta Mater.* 47 (1999), 1697-1712.

Microstructural Refinement of Bulk Nb and Ta by Severe Plastic Deformation for Composite Superconductor Applications

S.N. Mathaudhu, R.E. Barber and K.T. Hartwig*

Department of Mechanical Engineering, Texas A&M University, College Station, TX 77843-3123

Abstract

Fine grained and homogeneous recrystallized microstructures in precursor Nb and Ta are necessary to avoid filament sausaging and layer fracturing during the production of Nb_3Sn superconductor wires. In this study, the effectiveness of multi-pass equal channel angular extrusion (ECAE) to break up large as-cast microstructures (>5mm diameter columnar grains) in vacuum arc remelted Ta and electron beam remelted Nb is discussed. Extrusions were performed at room temperature in 90° tooling via ECAE with routes C, B_c, and E. Results show significant refinement to average grain sizes of less than 30μm after four extrusions. Four passes via routes E and B_c showed the best potential for refinement due to deformation of the as-cast microstructure on orthogonal planes.

Introduction

Composite multifilamentary Nb_3Sn superconductors are often fabricated with pure Nb and pure Ta precursor components. The Nb begins as a small diameter rod (3-6 mm) surrounded by pure Cu, and ends up after wire drawing and a reaction heat treatment as a small diameter (4 μm diameter) filament of Nb_3Sn. The Ta starts out as a thin sheet (0.38 mm thick) and is reduced to a 3-5μm diffusion barrier layer during wire drawing. Advanced superconducting magnets for high energy physics accelerators and plasma containment for fusion will require low temperature superconductors with finer filaments (submicron diameter) and thinner Ta diffusion barriers (1-3μm thickness) [1,2]. It is extremely difficult to fabricate such conductors without encountering Nb filament sausaging (1) and Ta layer fracture. Filament sausaging leads to a lower critical current density in the composite superconductor, and Ta layer fracture leads to wire breakage and shorter piece lengths during wire manufacture.

The problems of Nb filament sausaging and Ta diffusion barrier fracture are most likely due to 1) a large starting grain size, 2) a non-uniform starting grain size, and/or 3) significant texture variations in the starting microstructure. These causes arise because of the extremely large grained and highly textured microstructure of the original cast metal, and the subsequent ineffectiveness of conventional metal working methods (forging and rolling with intermediate heat treatments) to refine and homogenize the microstructure [3]. Strong texture gradients and banded recrystallized microstructures are common in Nb and Ta wire and sheet products.
Preliminary work on the microstructural breakdown of large grained cast Nb and Ta using severe plastic deformation (SPD) methods has been quite successful [4,5]. In

* Corresponding Author. thartwig@mengr.tamu.edu

Y.T. Zhu and V. Varyukhin (eds.), Nanostructured Materials by High-Pressure Severe Plastic Deformation, 145-150.

the case of Nb, bulk material strained to 4.6 by ECAE at room temperature gave a recrystallized grain size of 25 microns but the microstructure contained bands of larger grain size [5]. For vacuum arc remelted (VAR) Ta, an average recrystallized grain size of 15 microns was achieved after an ECAE strain of 4.6 [4]. The as-worked Ta (strained to 4.6) had a microstructure composed of 100-350 nm wide by 200-800 nm long dislocation cells and subgrains containing a high density of dislocations. Comparison of the ECAE processed Nb microstructure with similar fully annealed commercial material shows that the average grain size of the ECAE processed and recrystallized Nb appears to be as fine as, and perhaps finer than, the best commercial material.

The motivation behind the work reported here is the belief that SPD processing of bulk Nb and Ta via ECAE will give a finer and more uniform microstructure than can be produced easily by conventional metal working processes. This improvement could enable the fabrication of better performing and more cost effective low temperature composite superconductors. Supporting evidence for the benefits of processing bulk metal via ECAE to achieve better microstructural refinement already exists in the sputtering target field [6] and encourages further work on refinement of cast microstructures.

Experimental Procedures

The materials used for this research were commercially pure (99.7%) electron beam remelted (EB) Nb and 99.98% pure vacuum arc remelted (VAR) Ta ingots. Both materials consisted of ingots with initial columnar grain structures and large as-cast grains (>5mm diameter for the Ta and >20mm for the Nb). Billets with nominal dimensions of 25mm x 25mm x 150 mm were cut from cast ingots with their long axes parallel to the ingot long axis.

Extrusions were performed at room temperature in sliding wall 90° tooling with a punch speed of 5.1 mm/sec. Multipass processing following routes C, B_c and E was used. Routes C and B_c are well researched, while route E is a hybrid route that involves extrusions via route 2C followed by a 90° rotation, then extrusion via route 2C again so as to deform the microstructure on three orthogonal planes. Post-extrusion anneals (60 min for Nb, 90 min for Ta) were performed under a vacuum of at least 10^{-6} torr. Vickers microhardness measurements were taken with a 300g load for 13 s on a Buehler Micromet. Microstructural characterization was done using a Leica MEF4M stereo wide-field metallograph. Grain sizes were determined using the Image Processing Toolkit (IPTK) plug-in for Adobe Photoshop.

Results and Discussion

Fig. 1 shows the as-worked Vickers microhardness for both Ta and Nb to four passes. It is clear in both cases that significant hardening occurs after one extrusion with continued increases being less for each subsequent extrusion. In the processed Ta, it appears that hardness saturation occurs at four extrusions, which corresponds to an equivalent strain of about five. Different regions of the starting billets will experience different amounts of plastic strain due to their different crystallographic orientations to the ECAE shear plane. Whether or not grains have favorable slip systems will influence how much a localized region will deform [3]. This results in some regions of the billet being severely deformed, accumulating a higher density of dislocations, and work-hardening more than others. The decreasing standard deviation of hardness with increasing number of extrusions indicates that, with an increasing number of extrusions, additional slip systems are activated, thereby homogenizing the strain within the grains composing the billet.

The variations between the initial texture/orientation of the few large grains composing the original billets probably accounts for the differences seen in the Vickers microhardness of the four-pass materials (Ta: 4C and 4E, Nb: 4C, 4E and $4B_c$).

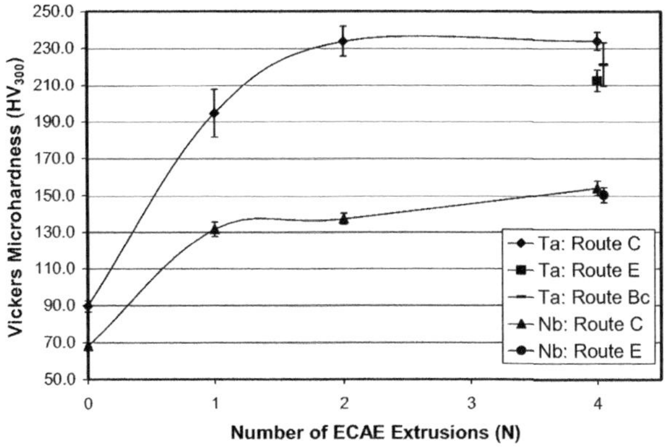

Figure 1. Vickers microhardness measurements as a function of number of extrusions for EB Nb and Ta subjected to multi-pass ECAE at room temperature. Error bars are measurement standard deviation.

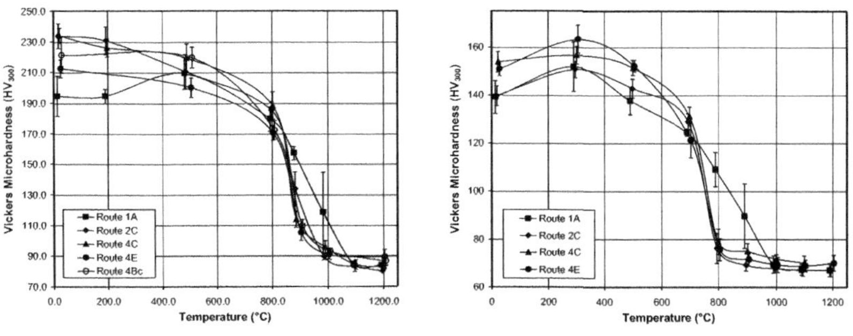

(a) (b)

Figure 2. Recrystallization curves (HV_{300} vs. Temperature) for a) VAR Ta and b) EB Nb ECAE processed at room temperature. Error bars are measurement standard deviation.

The recrystallization temperatures (See Fig. 2) are near 1100°C for the Ta processed by routes 1A and 2C, and the same for Nb processed by route 1C. The rest of the routes recrystallized completely near 1000°C. Observe that the materials processed by route 1A have large variations in strain even near the recrystallization temperatures, as demonstrated by the large standard deviation in the localized microhardness values. These materials have experienced non-uniform strain and therefore do not recrystallize uniformly, which results in significant grain growth in some regions and higher temperatures for full recrystallization. The maximums occurring on the recrystallization curves near 300ºC for Nb are characteristic of either polygonization or Cottrell atmospheres [7].

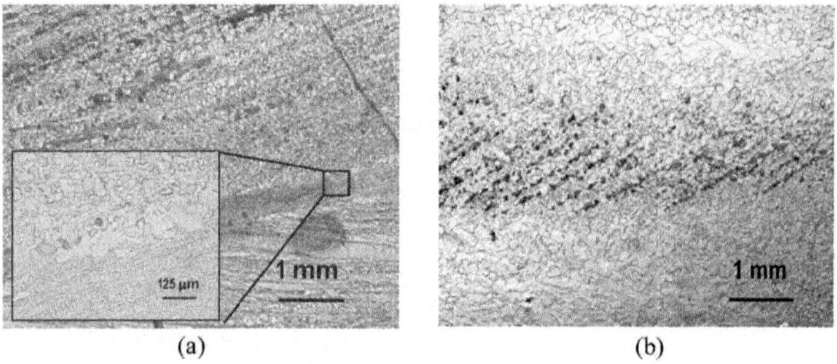

Figure 3. Flow plane optical micrographs of heat treated Ta processed via route 1A at a) 990ºC (55% recrystallized) and b) 1100ºC showing banded and non-uniform microstructure after only one extrusion.

Fig. 3 gives verification of the observations concerning the material processed via route 1A. Fig. 3a shows partial recrystallization occurs due to the non-uniform strain. Full recrystallization has occurred at 1100ºC (Fig. 3b), but banding and grain growth can result in a non-homogeneous distribution of grains in the sample. With an increasing number of extrusions, the average grain size decreases and microstructural homogeneity increases (Fig. 4). It also appears that routes B_c and E result in a finer grain size than route C for an equivalent amount of strain.

As expected, the average grain size decreases with increasing strain (Fig. 5), with microstructural uniformity increasing as well. For both the Ta and Nb, after two extrusions via route C, there are regions that contain the minimum average recrystallized grain size attainable with additional extrusions. These must occur in regions where the initial grain structure is oriented with the shear plane for easy break-up and deformation. The effectiveness of using routes E and B_c over route C involve rotating prior grain regions by 90° that might not have been oriented well to the shear plane, thereby offering other orientations that favor microstructural deformation. It appears that using routes which incorporate billet deformation on three orthogonal planes results in better breakup of large cast microstructures into more fine-grained, equiaxed and homogeneous recrystallized microstructures.

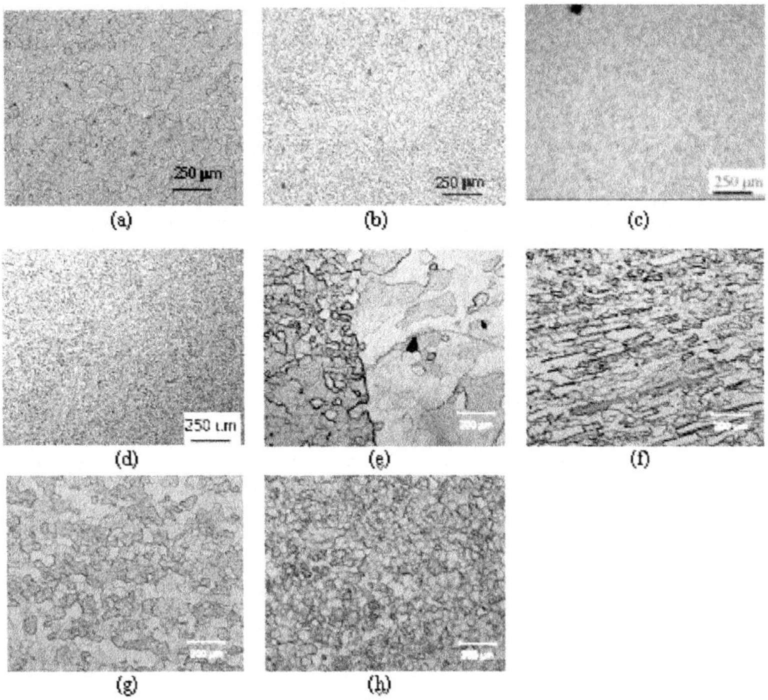

Figure 4. Flow plane optical micrographs of fully recrystallized ECAE processed material: a) Ta: 2C and annealed at 1100ºC, b) Ta: 4C and annealed at 1000ºC, c) Ta: $4B_c$ and annealed at 1000ºC, d) Ta: 4E and annealed at 1000ºC, e) Nb: 1A and annealed at 1100ºC, f) Nb: 2C and annealed at 1000ºC, g) Nb: 4C and annealed at 1000ºC and h) Nb: 4E and annealed at 1000ºC.

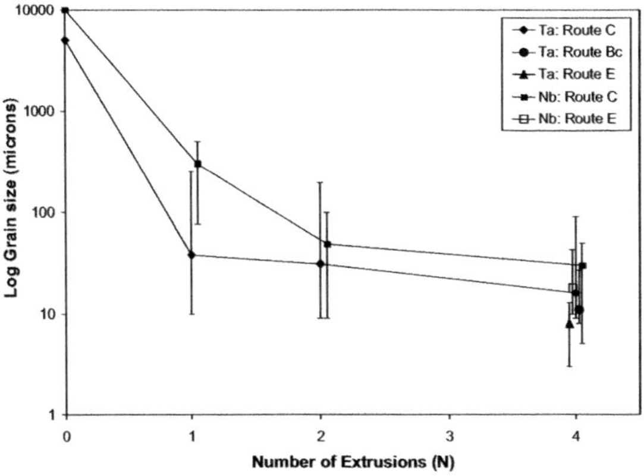

Figure 5. Variation in the fully recrystallized grain size as a function of extrusion route and number of extrusion passes for Ta and Nb processed by ECAE at room temperature.

Conclusions

The effectiveness of ECAE processing for the purpose of breaking up, grain refining and homogenizing VAR Ta and EB Nb has been demonstrated. The results of the study indicate that:

1. ECAE is effective in breaking up as-cast metals allowing the production of micron scale recrystallized microstructures.
2. The small grains observed in some regions of two pass ECAE processed and recrystallized material are an indication of the smallest grain size attainable with further increases in number of extrusions.
3. The intersecting shear planes utilized by routes E and B_c cause multiple slip planes to be activated, thereby increasing the uniformity of deformation and the homogeneity of the recrystallized microstructures.

Acknowledgements

The State of Texas Higher Education Coordinating Board and U.S. Department of Energy (contract #DE-FG02-03ER83773) are acknowledged for funding the work presented. H.C. Starck, Inc. and Reference Metals Inc are thanked for supplying the Ta and Nb used in this study.

References

1. High, Y.E., Lee, P.J., McKinell, J.C. and Larbalestier, D.C., *Adv. Cryo. Eng.* 38 (1992), 647-652.
2. Scanlan, R.M., Royet, J. and Hannaford, R., *IEEE Trans. Mag.*, MAG 23 (1987), 1719-1723.
3. Sandim, H.R.Z., Lins, J.F.C., Pinto, A.L. and Padilha, A.F., *Mat. Sci. Eng. A.*, 354 (2003), 217-228
4. Hartwig, K.T., Mathaudhu, S.N., Maier, H.J. and Karaman, I., *Ultrafine Grained Materials II*, ed. Y.T. Zhu et al., Warrendale, PA: The Minerals, Metals and Materials Society, 2003, 151-160.
5. Hartwig, K.T., Bryant, D.O., Mathaudhu, S.N., Barber, R.E. and Im, J.T., *Adv. Cryo. Eng.* 50 (2003), 441-449.
6. Ferrasse, S, Segal, V.M. and Alford, F., *Mat. Sci. Eng. A.*, 372 (2004), 44-55.
7. Hoelzer, D.T., West, M.K., Zinkle, S.J. and Rowcliffe, A.F., *J. Nuc. Mat.* 283-287 (2000), 616-621.

Modeling Hardening Effects and Grain Size Evolution in Metals Induced by Severe Plastic Deformation

Z.Mróz[1*] and A.Baltov[2]

[1] Institute of Fundamental Technological Research, Warsaw, Poland
[2] Institute of Mechanics, Bulgarian Academy of Sciences, Sofia, Bulgaria

Abstract

The present paper is concerned with the analysis of some cyclic deformation processes aimed at improved metal forming and generation of fine grain structure. Next, the model of microstructure evolution is discussed and the evolution rules are proposed. The shear banding process accompanying the continuing recrystallization is analyzed and the respective constitutive equations are presented.

Introduction

Bulk nanocrystalline materials (metals and alloys) with grain size less than 100 nm exhibit numerous interesting material propagates, different from those of usual crystalline materials [1]. The microstructure of such materials is characterized by nanodimensional crystalline grains with high-angle grain boundaries. The occurrence of triple point junctions between adjoining crystals is frequently observed [4] and small pores are generated near triple points. The grain boundary layers of small thickness are developed with concentrated dislocations while the inner grain portion remains practically without dislocations. The plastic deformation occurs through slip and rotation at grain boundaries with elastic response within the inner grain portion. The diffusion process also occurs across the boundary layer. The grain size distribution in the representative material element is of lognormal character [5]. The well known Hall-Petch relation between the yield limit and the mean grain size is valid for grain sizes exceeding the limit value below which this relation does not apply. The mechanical and thermal properties of nanomaterials have been tested by numerous researchers. Hardness, yield limit, fatigue life, temperature and strain rate sensitivity were exhibited to grow with the decreasing grain size.

There are different technological procedures for generation of nanocrystalline materials, such as

i) powder metallurgy methods
ii) crystallization from amorphous phase [2]
iii) severe plastic deformation methods, (SPD) [3]

In this paper our attention will be paid to SPD techniques which possess many advantages. Let us mention such techniques as equal channel angular extrusion, twist extrusion, accumulative roll bonding process, successive 3D forging, constrained groove processing, et al. All these methods have a common feature of multi-step plastic deformation with changing loading orientation at each step. Locally a material element undergoes plastic deformation with changing orientation of principal stress axes, thus inducing at each step a new flow mode. The shear band pattern is generated with localized deformation distributed in a system of bands whose intersections are the potential nucleation centers for recrystallization and development of ultra fine grain structure. The SPD processes usually occur at increased hydrostatic pressure to prevent fracture and allow for large plastic straining. The metal forming method with fundamental process assisted by cyclic deformation (KOBO-technique) [4,5] belongs to this class of processes. Extrusion,

* Corresponding author. zmroz@ippt.gov.pl

Y.T. Zhu and V. Varyukhin (eds.), Nanostructured Materials by High-Pressure Severe Plastic Deformation, 151-159.

forging and rolling processes are accompanied by imposed cyclic straining, thus leading to distributed shear band patterns with the resulting microstructure evolution inducing cyclic softening and eventually ultrafine grain structure.

Progressive plastic deformation assisted by cyclic loading

The response of metals under controlled monotonic and cyclic deformation was analyzed by numerous researchers. Under monotonic straining the material hardening increases monotonically until damage effects reduce the yield stress and generate localized flow. However, under cyclic loading imposed on a heavily cold worked state the effect of cyclic softening is observed with a limit value of stress amplitude similar to that of an annealed material subjected to the same cyclic deformation. The existence of steady cyclic states independent of the initial hardening state was observed by numerous investigators, starting from the early pioneering tests by Coffin [6] who observed this property on aluminium. The constitutive models of cyclic response of metals with account for anisotropic hardening were discussed by numerous researchers [7]. The importance of knowledge of limit cyclic states is related to prediction of low cycle fatigue life of elements when plastic deformation at notch roots occurs at the steady state. For non-symmetric stress cycles the effect of accumulating strain may lead to incremental failure of a structural element.

Besides structural aspects of cyclic response of metals, there is a possibility to implement technological processes such as extrusion or drawing, rolling and forging by superposing plastic strain in order to reduce load level required for process, increase formability of the material and also to generate fine grain structure. Korbel and Bochniak [4,5] developed the so called KOBO method or the approach SBDMFO (Structure Based Design of Metal Forming Operation) aimed at cyclic strain assisted metal forming with microstructure control.

Figure 1 illustrates schematically the extrusion process through a die executing cyclic rotation and thus inducing cyclic shear to the extruded work piece. Table 1 provides the comparative data of parameters for the extruded aluminium by conventional and KOBO methods

Figure 2 presents the extruded wire of CuSn8 alloy by applying KOBO method and the conventional method. After high reduction the ultrafine structure was generated with the grain size of order of 200nm. The strength properties are slightly affected, namely the yield stress $Y_s=185.0$ *MPa* for conventional and $Y_s = 149.7$ and KOBO methods, *UTS=588.9 MPa* and *533.0 MPa* for two methods, however the uniform elongation was considerably increased from *A=12.2%* for conventional and *A=21.6%* for KOBO methods.

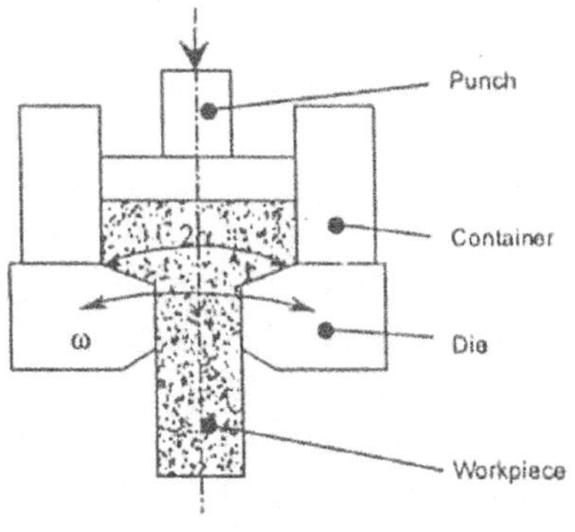

Fig. 1. Extrusion process through the die inducing cyclic torsion

To analyze the deformation process coupled with cyclic torsion, consider a cylinder subjected to monotonic axial compression or tension and cyclic torsion with the specified frequency and amplitude of the angle of twist An analytical treatment can be provided for the case of a thin walled tube. The solution for a cylinder can be considered by superposing solutions for a set of tubes with the neglect of radial stress resulting from tube interaction.

Table 1: Mechanical properties of extruded aluminium rods after conventional and cyclic extrusion: A_r – total elongation, A – elongation at maximum stress

Processing conditions	Yield stress Y_s [MPa]	Ultimate tensile strength UTS [MPa]	A %	A_r %
Conventional extrusion	183.2	190.1	1.8	11.2
Extrusion by KOBO method $\alpha = \pm 3.5^o$	123.1	162.0	7.1	27.1
Extrusion by KOBO method $\alpha = \pm 11^o$	102.8	122.4	6.2	28.9

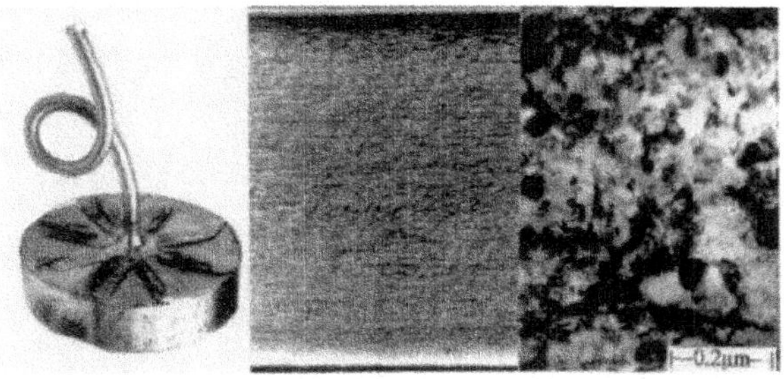

CuSn8 wire obtained as a result of applying KOBO-method (variant 700) and its nanostructure (optical and transmission electron microscopy techniques).

Procedure	Y_s [Mpa	UTS [MPa]	A [%]
Variant 500	**185.0**	**558.9**	**12.2**
Variant 700	**149.7**	**533.0**	**21.6**

Fig. 2. Mechanical properties of CuSn8 wires extruded using the KOBO method at room temperature.

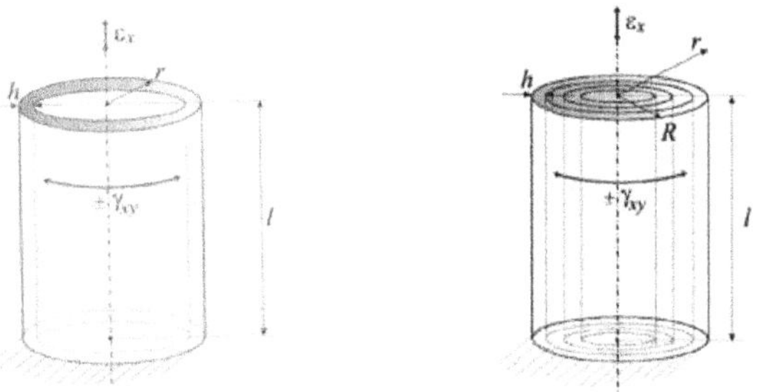

Fig. 3. A thin walled tube and cylinder under uniaxial tension/compression and cyclic torsion

Assume the axial strain rate to be specified and constant. The alternating torsion is imposed in order to reduce the axial stress and applied axial force. Denote by ε_x and γ_{xy} the axial and shear strain components, with their rates denoted by $\dot{\varepsilon}_x, \dot{\gamma}_{xy}$. The Cauchy stress components referred to the actual configuration are σ_x, τ_{xy}, Fig. 3.

For uniform length variation the axial strain and its rate are $|\varepsilon_x| = \alpha t$, $|\dot{\varepsilon}_x| = \alpha$, where $\dot{\alpha} = \dot{l}l / l_0$. For the logarithmic strain measure there is $\varepsilon_x = \ln(1 \pm l_o \dot{\alpha} t)$ and $|\dot{\varepsilon}_x| = \dot{l}/l = \dot{\alpha}\, l_0 / l$. The shear strain is assumed to oscillate with the amplitude $2\gamma_m$ and the period T. For the piecewise linear oscillation, Fig. 3b we have $\dot{\beta} = |\dot{\gamma}_{xy}| = 4\gamma_m / T$. Denote the ratio of rates of shear and axial strains by η, thus

$$\eta = \frac{\dot{\beta}}{\dot{\alpha}}, \quad \dot{\beta} = \frac{4\gamma_m}{T}, \quad \dot{\beta} > 0, \quad \dot{\alpha} > 0 \tag{1}$$

For the harmonic variation of γ_{xy}, Fig. 1c, we can write

$$\gamma_{xy} = \gamma_m \sin(\frac{2\pi}{T}t), \quad \dot{\gamma}_{xy} = \dot{\beta}\frac{\pi}{2}\cos(\frac{2\pi}{T}t), \quad \dot{\beta} = \frac{4\gamma_m}{T} \tag{2}$$

and the ratio of strain rates is

$$\frac{\dot{\gamma}_{xy}(t)}{|\dot{\varepsilon}|} = \eta\frac{\pi}{2}\cos(\frac{2\pi}{T}t) = \eta\frac{\pi}{2}\cos(\frac{2\pi}{T}\frac{|\varepsilon_x|}{\dot{\alpha}}) = \eta\frac{\pi}{2}\cos(2\pi\frac{|\varepsilon_x|}{E_x}) \tag{3}$$

where $E_x = 4\gamma_m / \eta$ is the strain period corresponding to time period T. Let us note that the time measure can be replaced by the axial strain, $|\varepsilon_x| = \dot{\alpha} t$.

To understand the constitutive model assumptions in the analysis of cyclic response of tube, consider the combined hardening model for which the back stress tensor α_{ij} specifies the deformation anisotropy and the yield stress $\sigma_p = \sigma_p(\lambda)$ specifies the isotropic hardening.

The yield condition is expressed as follows

$$F = \sqrt{(\sigma_x + \alpha_x)^2 + 3(\tau_{xy} - \alpha_{xy})^2} - \sigma_p(\lambda) = 0 \tag{4}$$

where $\sigma_p(\lambda) = \sigma_0 + R(\lambda)$ and σ_0 is the initial yield stress.
The elasto-plastic incremental (or rate) equations take the form

$$\dot{\varepsilon}_x = \frac{\dot{\sigma}_x}{E} + \dot{\lambda}\frac{\sigma_x - \alpha_x}{\sigma_p}, \quad \dot{\gamma}_{xy} = \frac{\dot{\tau}_{xy}}{G} + \dot{\lambda}\frac{3(\tau_{xy} - \alpha_{xy})}{\sigma_p} = \eta\dot{\varepsilon}_x \tag{5}$$

where E and G denote the Young and Kirchhoff moduli. The plastic multiplier $\dot{\lambda}$ is obtained from the consistency condition

$$\dot{F} = (\sigma_x - \alpha_x)(\dot{\sigma}_x - \dot{\alpha}_x) + 3(\tau_{xy} - \alpha_{xy})(\dot{\tau}_{xy} - \dot{\alpha}_{xy}) = 0 \tag{6}$$

The back stress evolution is specified by the relations

$$\begin{aligned} \dot{\alpha}_x &= C_1\dot{\varepsilon}_x^p - \dot{\lambda}C_2\alpha_x = \dot{\lambda}(C_1\frac{\sigma_x - \alpha_x}{\sigma_p} - C_2\alpha_x \\ \dot{\alpha}_x &= C_1\dot{\gamma}_{xy}^p - \dot{\lambda}C_2\alpha_{xy} = \dot{\lambda}(C_1\frac{3(\tau_{xy} - \alpha_{xy}}{\sigma_p} - C_2\alpha_{xy} \end{aligned} \tag{7}$$

where the first term represents the linear hardening (Prager rule) and the second term represents the plastic recovery (Frederick-Armstrong rule) proportional to the value of back stress. The hardening and recovery parameters C_1 and C_2 are assumed as constants but they are usually functions of the deformation history.

A more general back stress evolution rule is generated by assuming the integral representation, proposed by Mróz and Rodzik [7], namely

$$\alpha_{ij} = \int_0^z \rho(z - \xi)\frac{d\varepsilon_{ij}^p}{d\xi}d\xi \tag{8}$$

where $\rho(z)$ is a Kernel function to be specified and ξ varies between 0 and the actual value of $z(\varepsilon^p)$, where

$$\dot{z} = \frac{\dot{\lambda}}{g(\lambda)}, \qquad \dot{\lambda} = (\dot{\boldsymbol{\varepsilon}}^p \cdot \boldsymbol{\varepsilon}^p)^{\frac{1}{2}} \tag{9}$$

and $g(\lambda)>0$ specifies the non-uniform dependence of z on the length of plastic strain trajectory. The incremental or rate form of (8) can be written as follows

$$\frac{d\alpha_{ij}}{dz} = \rho(0)\frac{d\varepsilon_{ij}^p}{dz} + \int_0^z \frac{\partial\rho(z-\xi)}{\partial z}\frac{d\varepsilon_{ij}^p(\xi)}{d\xi}d\xi = \rho(0)\frac{d\varepsilon_{ij}^p}{dz} - h_{ij}(z) \tag{10}$$

or in the alternative form

$$\dot{\alpha}_{ij} = \rho(0)\dot{\varepsilon}_{ij}^p - h_{ij}(z)\dot{z}, \quad h_{ij} = -\int_0^z \frac{\partial\rho(z-\xi)}{\partial z}\frac{d\varepsilon_{ij}^p}{d\xi}d\xi \tag{11}$$

The form (11) of evolution rule is more general than (7) as the recovery function is now a functional of the past plastic strain history. The kernel function $\rho(z)$ occurring in (11) is an exponentially decaying function of z, so that

$$\rho(z) = \sum_{i=1}^{n-1} A_i e^{-\gamma_i z} + A_n \tag{12}$$

For the parameters A_i constant the convergence property occurs when using (13) and the steady cyclic states are not affected by prior deformation history. However in general these parameters depend on maximal plastic prestrain ξ_m, plastic strain amplitude ξ_a, and curvature of the deformation path θ, $A_i = A_i(\xi_m, \xi_a, \theta)$.

Figure 4 presents the experimental and predicted response curves for a cylinder under combined compression and cyclic torsion. The Frederick-Armstrong recovery model specified by (7) and the integral recovery description specified by (11) are used in the simulation analysis. The annealed copper cylinders were tested by Grossman and Pawlicki [8]. The axial compression strain rate was $\dot{\varepsilon}_x = 0.0211/s$ and the angle of torsion was $\alpha = \pm 2^0, \pm 5^0, \pm 8^0$ with three frequencies $f=0.3$ Hz, $f=1$ Hz, $f=1.8$ Hz. The following non-dimensional parameters are specified:

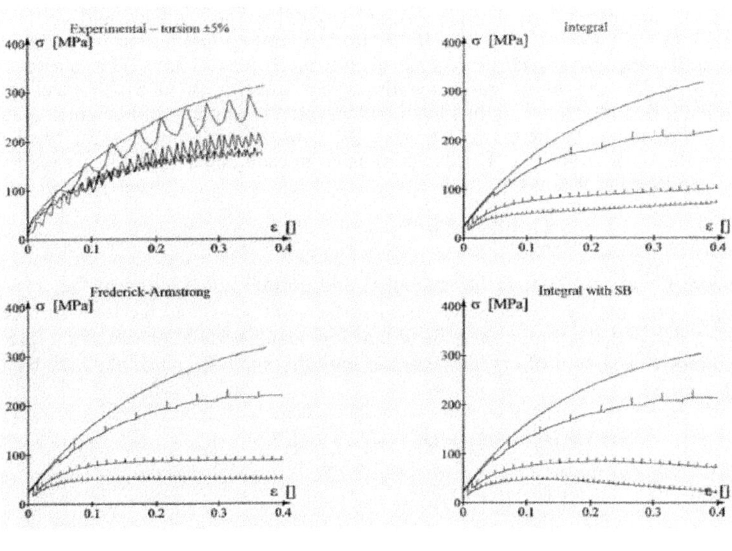

Fig. 4 . Experimental and predicted response curves for cylinder and tube.

Table 2. Cyclic loading parameters at the external cylinder surface: $\eta = 4\dot{\gamma}_{mf} / E_x$

α	γ_m	f=0.3 Hz	f=1 Hz	f=1.8 Hz
$\pm 2^0$	0,03431	η=1.96	η=6,54	η=11,76
$\pm 5^0$	0,08367	η=4,78	η=15,94	η=28,69
$\pm 8^0$	0,1307	η=7,46	η=24,89	η=44,81

Assume $E=126000$ MPa, $G=48500$ MPa, $\sigma_p = \sigma_0 = 20MPa$, and the hardening-recovery parameters $C_1=1500$ MPa, $C_2=4.8$. For the integral back stress evolution we assume $n=2$, $A_1=1600$ MPa, $A_2=60$ MPa, $\gamma_1=6$, $\gamma_2=0$. These parameters are calibrated in order to fit the uniaxial compression curve. The axial stress evolution predicted by the models is much larger than that observed experimentally. However, this prediction is related to the external thin walled layer. Considering the cylinder we have to reduce η at each layer proportionally to the ratio of its radius to the external radius, and average the axial stress

$$\eta_i = \frac{r_i}{R}\eta_e, \qquad \sigma_x = \frac{\sum \sigma_{xi} A_i}{A} \tag{13}$$

where r_i, A_i, are the radii and cross-sectional areas of particular layers, and R, A denote the external radius and the total cross-sectional area. This averaging procedure provides prediction of axial force evolution close to experimental curves.

Analysis of shear banding

The available results of metallographic observations reveal that in heavily deformed metals or even at small strains if they are preceded by a change of deformation path, a multiscale pattern of shear localization modes progressively replaces the crystallographic multiple slip or twinning. The term micro shear band is understood as a long and very thin (of order 0.1 μm) sheet- like region of concentrated plastic shear crossing grain boundaries. The micro shear bands are usually not parallel to densely packed crystallographic planes and form clusters (packets) of thickness of order (10-100 μm). It appears that the successive generation of active shear bands is responsible for advanced plastic flow. The intersections of shear bands and twins generates the strain localization and the mechanism of formation of large angle grains leading consentively to grain refinement process.

Following Pecherski and Korbel [9] consider the double shear band system with the shear strain $\dot{\gamma}_{SB}^{(i)}$, I=1.2 accumulated in each band. The velocity gradient can be expressed as follows

$$L_{SB} = \sum_{i=1}^{2} \dot{\gamma}_{SB}^{(i)} \mathbf{s}^{(i)} \otimes \mathbf{n}^{(i)}, \quad i = 1,2 \tag{14}$$

where $\mathbf{s}^{(i)}$ and $\boldsymbol{n}^{(i)}$ denote the tangential and normal unit vectors to the shear band. Fig. 5 presents the Huber-Mises yield surface $F(\tau',k)=0$ and the limit surface $F_l(\tau', R)=0$ associated with the shear band development. The total plastic strain rate $\mathbf{D}^p$ is expressed as a sum of crystallographic multiple slips $\mathbf{D}_c^p$ and the shear band slips $\mathbf{D}_{SB}$ so that

$$\mathbf{D}^p = \mathbf{D}_c^p + \mathbf{D}_{SB}^p = \mathbf{D}_c^p + \frac{\sqrt{2}}{2}\sum_{i=1}^{2} \dot{\gamma}_{SB}^{(i)} \mathbf{N}_i \tag{15}$$

where

$$\mathbf{N}_i = \frac{\sqrt{2}}{2}(\mathbf{s}^{(i)} \otimes \mathbf{n}^{(i)} + \mathbf{n}^{(i)} \otimes \mathbf{s}^{(i)}) \tag{16}$$

and the crystallographic slip is described by the usual flow rule

$$D_c^p = \frac{\sqrt{2}}{2}\boldsymbol{\mu}_F \dot{\gamma}_s, \quad \dot{\gamma}_s = \frac{\sqrt{2}}{2}\frac{\dot{\boldsymbol{\tau}}' \cdot \boldsymbol{\mu}_F}{h}, \quad \boldsymbol{\mu}_F = \frac{1}{\sqrt{2k}}\boldsymbol{\tau}' \tag{17}$$

The yield surface and the limit surface are

$$F = \sqrt{\frac{1}{2}\boldsymbol{\tau}' \cdot \boldsymbol{\tau}'} - k = 0, \quad F_l = \sqrt{\frac{1}{2}\boldsymbol{\tau}' \cdot \boldsymbol{\tau}'} - R = 0 \tag{18}$$

where k is t he yield shear stress $\boldsymbol{\tau}'$ is the stress deviator and $R>k$ is the radius of the limit surface specifying the initiation of shear band deformation. The tensor $\boldsymbol{\mu}_F$ represents the unit normal vector to the yield surface. The total strain rate can be decomposed into the normal and tangential components to the limit surface

$$\mathbf{D}^p = \frac{\sqrt{2}}{2}\dot{\gamma} * \boldsymbol{\mu}_F + \frac{\sqrt{2}}{2}\dot{\varepsilon}_{SB}\mathbf{t} \tag{19}$$

where

$$\dot{\gamma}^* = \dot{\gamma}_s + \dot{\gamma}_{SB}, \quad \dot{\gamma}_{SB} = \cos 2\beta(\dot{\gamma}_{SB}^{(1)} + \dot{\gamma})_{SB}^{(2)}), \quad \dot{\varepsilon}_{SB} = \sin 2\beta(\dot{\gamma}_{SB}^{(1)} - \dot{\gamma}_{SB}^{(2)}) \tag{20}$$

and $\mathbf{t}$ is the unit tangent vector to the limit surface and the angle β is shown in Fig. 5. Introduce the scalar factors representing contributions of the shear bonding system to the total plastic shear strain rate γ *, so that

$$\dot{\gamma}_{SB}^{(1)} \cos 2\beta = f_{SB}^{(1)}\dot{\gamma}^*, \qquad \dot{\gamma}_{SB}^{(2)} \sec 2\beta = f_{SB}^{(2)}\dot{\gamma} * \tag{21}$$

Denoting $f_{SB} = f_{SB}^{(1)} + f_{SB}^{(2)}$, $\Delta f_{SB} = f_{SB}^{(1)} - f_{SB}^{(2)}$, the flow rule can be expressed as follows

$$\mathbf{D}^p = \frac{\dot{\tau}\mu_f}{2h(1 - f_{SB})}\boldsymbol{\mu}_F + \frac{\dot{\tau} \cdot \boldsymbol{\mu}_f}{2h(1 - f_{SB})}\frac{\Delta f_{SB} \tan 2\beta}{\sin \delta'}\mathbf{t} \tag{22}$$

where

$$\mathbf{t} = \frac{\dot{\boldsymbol{\tau}}' - (\boldsymbol{\tau}' \cdot \boldsymbol{\mu}_F)\boldsymbol{\mu}_F}{\|\tau'\|}$$

is the unit tangent vector to the limit surface and δ' denotes the angle between the normal $\boldsymbol{\mu}_F$ to the limit surface and stress rate vectors.

Grain evolution rule

Now, let us discuss the grain evolution rule during the alternating plastic deformation. It can be assumed that the process of grain refinement can proceed as continuing recrystallization or grain fragmentation modes. The intersections of shear bands or of twins and shear bands are the potential nucleation sites. The strain concentration generates high value of local free (or strain) energy which is relaxed in a form of new grain boundary and new grain structure. To provide an elementary treatment of grain refinement process, consider a representative volume regarded as a sphere of radius R_0 and the initial free energy Φ_0 per unit volume. Assume this volume to be transformed into a new structure with

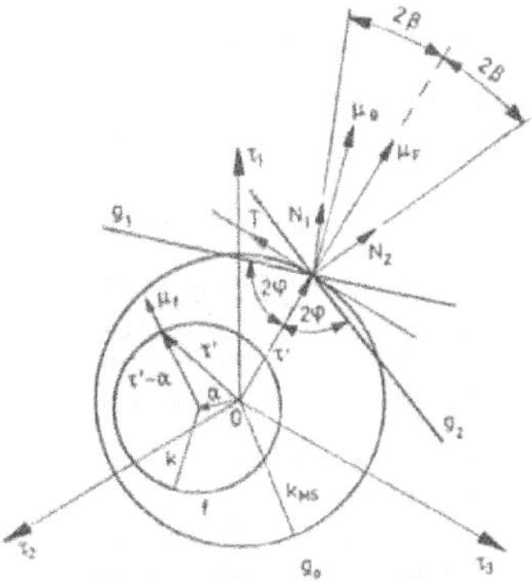

Fig. 5 .Yield and the limit surface for shear banded material.

relaxed strain concentration and with the specific free energy Φ_v and surface energy Φ_s. Assume the transformation to satisfy the energy balance, so that

$$\Phi_0 V_n = \Phi_v V_n + \Phi_s S_n \tag{23}$$

where $V_n = \frac{4}{3}\pi R_0^3$, $S_n = 4\pi R_0^2$. Equation (24) provides the radius of the transformed domain

$$R_0 = \frac{3\Phi_s}{\Phi_0 - \Phi_v} \tag{24}$$

On the other hand, when the increment of the energy is considered, we have

$$\Delta\Phi = -(\Phi_0 - \Phi_v)V_n + \Phi_s S_n \tag{25}$$

and the extremum condition $\partial(\Delta\Phi)/\partial R = 0$ provides

$$R_{0m} = \frac{2\Phi_s}{\Phi_0 - \Phi_v} \tag{26}$$

In fact, the value R_{0m} corresponds to the initiation of transformation process but R_0 specifies the nucleation size at advanced stage. Assume now that new structure grows from R_0 to R_f which is a final value when the transformation process is terminated. The value of R_f can be assessed by assuming the number of nucleation sites N to be dependent on the value of non-proportional plastic deformation. For the total volume of the element V_0 and the volume associated with R_f, we have $NV_f = V_0 = const$ and the value of R_f can be assessed in the form

$$R_f = AN^{-\frac{1}{3}}, \quad \dot{R}_f = -\frac{1}{3}AN^{-\frac{4}{3}}\dot{N} \tag{27}$$

Assuming that rate $\dot{N}$ depends on non-proportional plastic strain rate

$$\dot{N} = c\dot{\varepsilon}_p^{\frac{1}{m}} \tag{28}$$

the evolution rule of grain size can be expressed as follows.

$$\dot{r} = \dot{\varepsilon}_p^n\left[\frac{r_f - r}{r_f - 1}\right], \quad r_f = \frac{R_f}{R_0}, \quad r = \frac{R}{R_0}, \quad \dot{r}_f = -\frac{1}{3}\bar{A}N^{-\frac{4}{3}}\dot{N} \tag{29}$$

These relations can now be generalized to the multiaxial stress and strain state by writing

$$\dot{r} = \left[\frac{r_f - r}{r_f - 1}\right](\sigma_{ij}\cdot\dot{\varepsilon}_{ij}^p)^n(1+\alpha\sin^2\theta), \quad \dot{N} = c(\dot{\varepsilon}_e)^{\frac{1}{m}}(1+\alpha\sin^2\theta) \tag{30}$$

where $\dot{\varepsilon}_e$ the effective plastic strain θ is the non-proportionality angle between total plastic strain and strain rate,

$$\cos\theta = \frac{\dot{\boldsymbol{\varepsilon}}^p \cdot \boldsymbol{\varepsilon}^p}{\left\|\dot{\varepsilon}^p\right\| \cdot \left\|\varepsilon^p\right\|} \tag{31}$$

The parameters $A, \bar{A}, n, m, \alpha$ occurring in the formulae are the material parameters. The present formulation provides the foundation for more detailed description of grain size evolution and its relation to plastic strain.

References

1. Valiev,R. Z., .Islamgaliev I. V., .Aleksandrow I. V.,, *Progr, Mat. Sci.* 45, (2000),103.
2. Bochniak W.,Korbel A., (2000*), Mat. Sci. Technol.*, 16, (2000), 664-674
3. Bochniak W., Korbel A. , *Journ. Mat. Process Techn.*, 134, (2003), 120-134
4. Gleiter H., *Acta Mat.,* 48, (2000) , 1-29

5. Weertman J.R. et al. (1999), *MRS Bulletin*, pp. 44-60
6. Coffin L. F., *Appl. Mat. Res.*,1, (1962), 129-141.
7. Mroz Z., Rodzik P., *Europ. Journ Mechanics, A/Solids*, 15, (1996), 1-28.
8. Grosman F., Pawlicki J., *Silesian Techn. Univ. Rep.* ,52, (2004).
9. Pęcherski R., Korbel K. , *Arch. Mech.*, 54, (2002), 603-620

Nanocrystalline Structure Formation under Severe Plastic Deformation and its Influence on Mechanical Properties

Yu.M. Podrezov*
Institute for Problems of Materials Sciences, NAS of Ukraine, 3, Krzhizhanovsky Str, Kiev, 03142

Abstract

Consideration is given to the distinction between nanocrystal strengthening and Hall-Petch grain size strengthening which varies linearly with $d^{-1/2}$. The important role of grain boundary structure in the strengthening formation of nanomaterials is emphasized. Static and dynamic recoveries are the main reasons limiting the minimal size of structural elements under deformation in high deformed materials. Structural relaxation proceeding by the recovery mechanism both during plastic deformation and during unloading causes loss of strengthening in high deformed materials. At the initial stage of repeated deformation the recovered cell structure interacts with moving dislocations in a special way. In the micro-deformation level hardening stress is practically independent on previous deformation degrees in wide interval of deformation. The usual increase of strengthening with rise of deformation is observed only from the yield point.

Introduction

The creation of new high deformation methods offers ample scope both for strain hardening theory and for deformed materials structure engineering. Equal canal angel pressure (ECA) method was created by Segal [1]. Simple shift uniform deformation of high intensity can be achieved on a macro sample 15x15x150 mm without exchanging of its sizes (Fig.1). Other method based on the deformation by twist extrusion (TE) scheme was proposed by Beygelsimer [2] (Fig.2). The creation of shift uniform deformation without sizes exchanging allows to carry out repeated deformed treatment in the different (or the same) direction of deformation and as the result to control structure formation process under high deformation degrees.

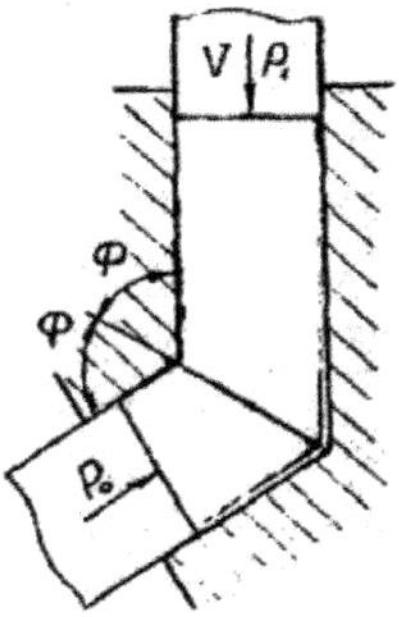

Figure 1. ECA pressure scheme [1]

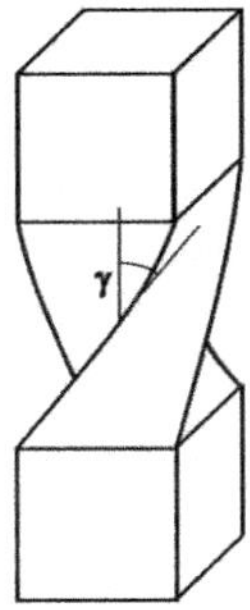

Figure2. Deformation by twist extrusion (TE) [2]

* Corresponding Author. podrezov@materials.kiev.ua

Y.T. Zhu and V. Varyukhin (eds.), Nanostructured Materials by High-Pressure Severe Plastic Deformation, 161-168.

The achievement of high strength in deformed materials with nanostructure is essentially more complicate task then it follows from microscopic theory of strength. According to the models of this theory strength of materials is connected with structural size dimension by Hall - Petch equation $\sigma_T = \sigma_o + \kappa_y d^{-1/2}$ or its Holt variation for cells materials $\sigma_T = \sigma_o + \kappa_y d^{-1}$. The extrapolation to nano scale of grain or cell size dimension theoretically predicts a very high strength of nano-crystalline materials. But experimental results demonstrate essentially worse situation. A. W. Tompson [3] reviewed experimental data from variety of investigation obtained on high deformed Fe - Armco wires. Relation between flow stress increment $\sigma - \sigma_o$ and substructure size is shown in Fig.3. In this graph we add our data obtained on Fe - Armco deformed by (ECA) pressure.

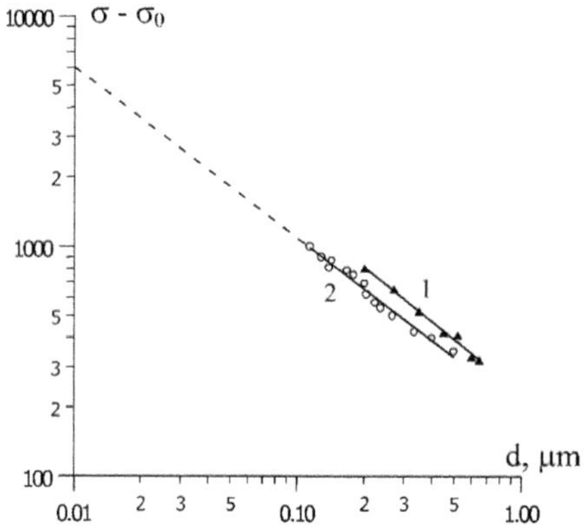

Figure 3. Effect of cell size on yield point of high deformed Armco-Fe after ECA pressure (1) and wires drawing (2) [4].

Important conclusions about interconnection between structural evolution and strengthening of high deformed materials is followed from these results. Firstly, it is the change of strengthening mechanism from grain size sensitivity (Hall - Petch equation) to cell size sensitivity (Holt equation) at critical grain size $d < 0{,}4$ μm. Secondly, it is the theoretical possibility to obtain super high strengthening in nanocrystalline Fe- Armco. It follows from experimental data (see dashed line) that extrapolation of experimental data to nanostructure sizes (10 nm) gives the value of yield point for such material approximately 6000 MPa. But the thread conclusion restricts such possibility. Difficulty is amplified due to the fact that cells or subgrain size in Armco-Fe obtained by different methods of severe plastic deformation (rolling, drawing, EC-pressure), can not be less than 100 nm. As the result, the yield point for such material is only 1000 Mpa [3-6].

Experimental Procedure

In order to explain the restriction of structural dispersion in deformed materials careful investigations of deformed substructure by TEM were carried out. More important experimental work was published by Langford and Cohen [6] who investigated high deformed Fe-Armco wires. Practically the same results were obtained in our work where severe plastic deformation in Fe- Armco was produced by rolling and ECA - pressure.

Results and Discussion

Following expression for f, the fraction of the initial number of cells in a cross section at any given deformation power can be write:

$$f = \frac{N}{N_i}\exp[-(e - e_i)] \qquad (1)$$

where N is the initial number of cells per unit cross-sectional area as formed at some early strain e_i and N is the number of cells per unit cross -sectional area at some subsequent strain e. Measurements of N have been made during the deformation and corresponding values of f vs e are plotted in Fig.4. As it can be deduced from Fig. 4, f is near unity both for wire drawing and for rolling deformation under low and middle deformation degrees. This result is in a good agreement with Tailor - Pallany law: both bulk material and structural elements (grains, cell) have the same shape under deformation. Since ECAE is a process capable of producing plastic deformation without causing substantial change in geometric shape of billet, cells size of ECAE deformed iron at the middle deformation degrees is substantially higher.

Under high deformation power principal changing in f vs. e dependence takes place. There is the large decrease in f throughout the subsequent elongation, signifying that substantial dynamic recovery is operative during the cellular refinement. This means that many cells are being lost from the structure simultaneously with cross -sectional reduction of the remaining cells. It is clear that cell wall migration constitutes very important aspect of the structural changes and critical deformation degree e_c under which changing in f vs. e dependence takes place is very important value. Cell size evolution under deformation is accompanied by increasing of cellular misorientation.

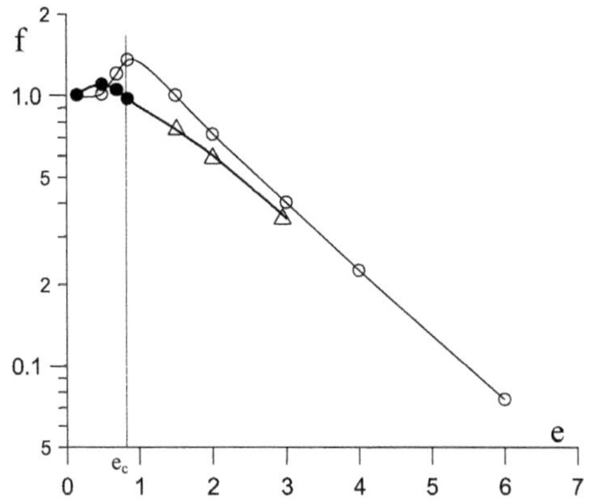

Figure 4. The dependence of *f*-parameter vs. deformation degree: • - rolling, Δ - ECAP, o - wire-drawing ([4].

According to Tompson`s model [3]of structural evolution, static and dynamic recovery are the main reasons limiting the minimal size of structural elements under deformation in high deformed materials. From one hand many cells are being lost from the structure during this process, from another hand recovery process promotes boundary perfection and increases misorientation of cells. But this model cannot explain the existence of critical deformation degree e_c, which characterizes start or strong activation of this process.

Rybin [7] proposed disclination model of severe plastic deformation. This model is based on analysis of micromechanical stress arising around a dislocation during plastic deformation. According to this model plastic deformation in crystals is developed by moving of dislocation at a low and middle deformation. But after transformation of substructure from dislocation forest to cells at $e_i \sim 0,2$ a change of deformation mechanism takes place and severe deformation is evoked by generation of disclination modes without diminution in subgrain size.

Calculation in the framework of disclination model gives critical cell size ~ 0,2 μm for BCC metals. As far as critical cell size depends from some physical constants (Burgers vector, elastic modulus, fault energy), for some FCC metals its value is essentially lower. It is in a good agreement with last experimental data, obtained on ECAE deformed Ni and Al where average size of structural element was 30-50 nm [8]. Disclination theory predicts monotonic rise of misorientation with increasing of deformation power and does not imply any spatial features of mechanical behaviour of high deformed materials after critical deformation e_c.

The change of mechanical behavior of deformed materials near critical deformation e_c occurs [9]. The parabolic strain hardening which was observed at the low and average deformation degrees was followed by the linear stage of deformation at high deformation degrees. The growth of the fracture toughness for specimens with cracks introduced into the plane perpendicular to the plane of deformation was shown.

Mechanical behavior near critical deformation e_c we observed in high deformed Fe-Armco after ECA pressure. The growth of the fracture toughness for specimens with cracks introduced into the plane perpendicular to the plane of deformation was shown, and the decreasing of the fracture characteristic for specimens with cracks introduced into the plane parallel to the plane of deformation was observed (Fig. 5). Increasing of the deformation degree ($e>e_c$) promotes the change of the failure mechanism as: quasi-cleavage → quasi-cleavage with delamination. On the contrary, transition from the parabolic stage of strain hardening to the linear one is not related with the change of the deformation mechanism. Therefore the processes occuring near critical deformation e_c are very important and must be investigated more carefully.

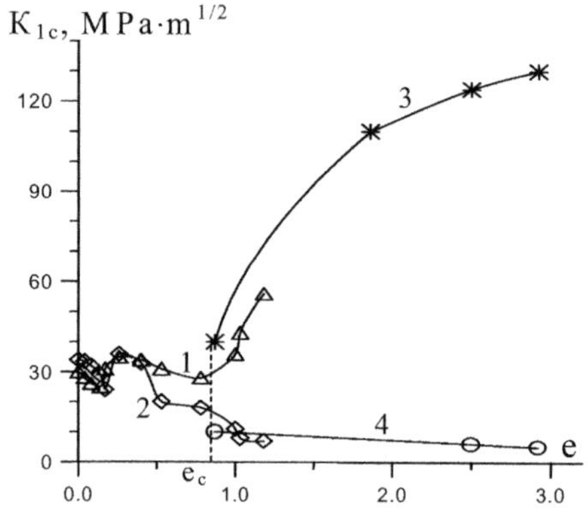

Figure 5. Fracture toughness of strained iron. 1, 2 - rolling, 3,4 - ECAP (1,3 - crack introduced into the plane perpendicular to the plane of deformation; 2,4 - crack introduced into the plane parallel to the plane of deformation).

The noted features of structure formation in deformed materials predetermine regularities of designing their mechanical properties. Analysis of strengthening curves for a deformed material is made usually based on structural changes occuring in the material at its continuous loading. At that an assumption is made that each next structural condition is obtained by an evolution of previous one, and change of deforming stress is a consequence of the evolution. Structure sensitive models of strain hardening are described in detail by many authors [10-12].

Practically the same approaches are used to analyze structural sensitivity of mechanical properties of pre-strained materials [4]. Widely used at that a known postulate of mechanics that deforming stress being achieved in a material at repeated deformation corresponds exactly to deforming stress achieved in it at unloading moment at initial deformation This postulate does hold well in range of low and middle deformation degrees for materials which have not susceptibility to recovery process [13]. For example it verified enough reality in Al-Ti-Cr intermetalic specimens in our experiments under uniaxial compression tests, Fig.6a. But this postulate does not hold for materials which have inclination to recovery. Such results were obtained in aluminum (Fig.6b) which have inclination to recovery. Strain hardening curve obtained by single loading compression test (Fig. 6b, curve 1) was compared with curves obtained after repeating deformation (second deformation - curve2, third deformation - curve3 and forth deformation - curve 4) . Experimental data show that pre-strained materials have both lower yield point and lower strain hardening coefficient than it follows from mechanic postulate.

Unfortunately, due to the fact that conditions of uniaxial deformation are limited by friction and size effects under compression and by neck formation process under tension, it is impossible to verify the postulate at high deformation powers (more than e=0,4). Therefore, in order to obtain strain hardening curve of high-deformed aluminium in unified loading condition we have made repeated grinding of lateral faces of a specimen deformed by single compression. As the result the specimen was put into shape of rectangle and tested by repeated compression. This strain hardening curve (curve 5) lays essentially lower than single one (curve 1) but its start position coincides with the end of last curves obtained by repeated compression (curve 4). Since summary hardening curve of pre deformed materials could be obtained by mating results of consecutive loading cycles, but because of recovery process its strain hardening coefficient may be essentially lower then one obtained under continuous loading.

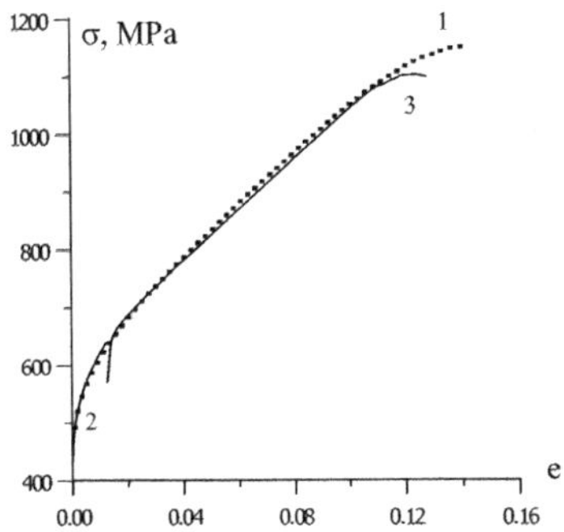

Figure 6a. Stress-strain curves of Al-Ti-Cr alloys (1,2 – continuous deformation, 3 – repeating deformation)

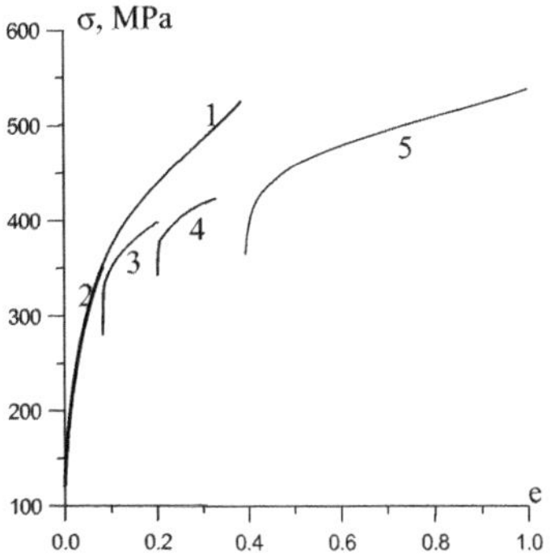

Figure 6b. Stress-strain curves of clean Al (1,2 – continuous deformation, 3-5 - repeating deformation)

Similar effects we obtained in titanium specimens [14]. Strain hardening curves obtained by uniaxial tension (subject to neck formation moment) and data of hardening of the material after high deformed rolling were compared (Fig. 7). In this case the hardening is stronger essentially for the specimens tested in continuous loading than that for the specimens tested in reloading of pre-deformed material. In particular, failure stress 1600 MPa is achieved under uniaxial tension of pure titanium. In high deformed rolled titanium the stress never exceeds 600-700 MPa. Undoubtedly a difference of both loading conditions and of specimen texture should be taken into account in this case. However only these factors in any case do not explain essential difference in titanium properties in conditions of continuous loading and pre-deformation states.

Larikov [15] explains recovery process by increasing of mobility of screw dislocation components and as the result activation of cross slip in plastic deformation mechanism. The processes of structural relaxation proceeding by recovery mechanism both during plastic deformation and during unloading seems to be the most important factors to promote loss of strengthening in high deformed materials.

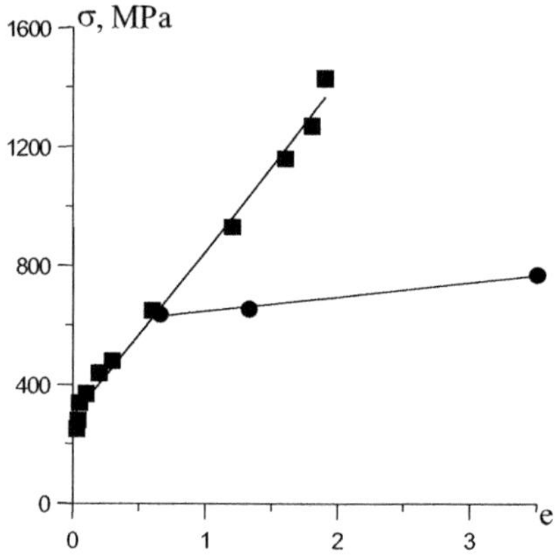

Figure 7. Stress –strain curves for Ti:(■) – tension, (●) – rolling

We analyzed micro-plasticity curves of pre-deformed rolled molybdenum specimens (range of deformation degrees 9-75%). It was found (Fig. 8a) that at the initial stage of repeating deformation micro-plasticity stress-strain diagrams are non-sensitive to pre-deformation degrees (and therefore to structural state of deformed materials). In the micro-deformation level at 10^{-5}, 10^{-4} and 5 10^{-4} deformation degrees (Fig. 8b) deformation stress is practically the same for all investigated materials. Differences are starting to be noticeable at 2 10^{-3} (yield point). And only at comparatively high repeating deformation degree $5 \cdot 10^{-3}$ stable increase of stress with the rise of deformation is observed. Obviously that hardening curves obtained on micro-deformation levels can not be explained in the framework of traditional structural model. In the initial stage of repeated deformation the recovered cells structure interacts with moving dislocation by spatial way.

These data are important for creation of an adequate model of mechanical behaviour of high-deformed materials, and may be useful in explaining of anomalous dependence of fracture toughness on deformation power of materials, tested in ductile-brittle transition temperature region (Fig. 5). In this case the laws of interaction of dislocations being emitted from a crack tip with primary structure in conditions when number of the dislocations is not too large should be accounted for.

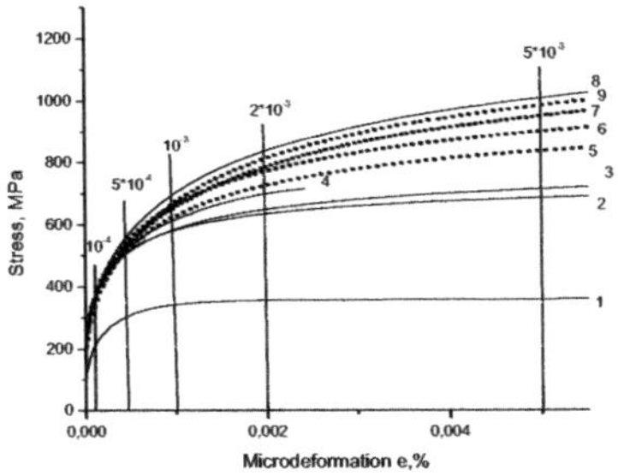

Figure 8a. Microplastisity curves of molybdenum deformed by rolling up to different deformation degrees: 1 initial, 2 – 9%, 3 –13%, 4-23%,5- 39%, 6-53%, 7 – 63%, 8-68%, 9 –73%

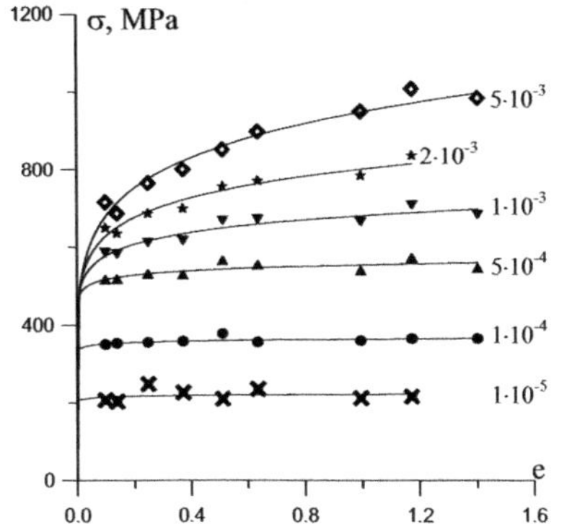

Figure 8b. Stress - pre-strain curves for different plasticity levels.

Conclusions

Cells or subgrain size in Armco-Fe obtained by different methods of severe plastic deformation (rolling, drawing, EC-pressure), can not be less than 100 nm. As the result, the yield point for such material is only 1000 MPa. Static and dynamic recovery are the main reasons limiting the minimal size of structural elements under deformation in high deformed materials.

The change of mechanical behavior near critical deformation e_c was observed in high deformed Fe-Armco after ECA pressure .The growth of the fracture toughness for specimens with cracks introduced into the plane perpendicular to the plane of deformation was shown, and the decreasing of the fracture characteristic for specimens with cracks introduced into the plane parallel to the plane of deformation was observed. Increasing of the deformation degree ($e>e_c$) promotes the change of the failure mechanism as: quasi-cleavage → quasi-cleavage with delamination.

Postulate of mechanics, that deformation stress being achieved in a material at repeated deformation corresponds exactly to deformation stress achieved in it at unloading moment at initial deformation, does hold well for materials which have not susceptibility to recovery process. But this postulate does not hold for materials which have inclination to recovery. Structural relaxation proceeding by recovery mechanism both during plastic deformation and during unloading promotes loss of strengthening in high deformed materials.

In the initial stage of repeated deformation the recovered cells structure interacts with moving dislocation by spatial way. In the micro-deformation level hardening stress is practically independent on previous deformation degrees in wide interval of deformation. Usual increase of strengthening with rise of deformation is observed only from yield point (e = 0,002).

Acnowledgements

The author wish to thank professor S. Firstov and Dr. M. Danylenko for fruitful discussion . I also thank Dr. D. Verbylo who essisted in the experiments.

References

1. Segal V.M., Reznikov V.I., Kopilov V.I., Pavlik D.A., Malyshev V.F. Processes of Plastic Structural Formation of Metals, Nauka i technika publishers, Minsk, (1994), 221p. (in Russian).
2. Beygelzimer Y.,Varyukhin V.,Orlov D, Efros B., Salimgareyev A. , Stolyarov V. Microstructural Evolution of Titanium under Twist Extruzion Ultrafine Grained Materials: Processing and structure Washington (2002), pp. 137 –142.
3. Tompson A.W. Substructure strengthening Mechanisms Met. Trsns. V 8A N7 (1977) pp.833-842.
4. Langford G., Cohen M., Strain Hardening of Iron by Severe Plastic Deformation. Trans. of the ASM, 62, (1966), pp. 623-637.
5. Clenn R.C., Langford G. and Keh A.S. Electron Microscope Observationof Wire-Drawn and rolled Steel in Two and Three Orthogonal Sections ASM Trans Quan 62, (1969), pp. 285-299
6. Langford G.,Cohen M. Microstructural Analysis by High - Voltage Electron Diffraction of Severly Drawn Iron Wires. Met. Trans.V. 6A, N4, (1975), pp. 901-909.
7. Rybin. V.V. Principles of formation microstructurs in the process of as developed plastic strain. Problems of Materials Sciience, N 1 (29) , (2002), pp. 11- 33 (in Russian).
8. Noskova N.I. Formation of mesoscopic shear bands in nanocrystalline materials. Problems of Materials Sciience, , N 1 (29), (2002), pp. 309 - 313 (in Russian).
9. Podrezov Yu.N. , Danilenko N.I Kopylov V.I., Firstov S.A Influence of structure on the mechanical properties of deformed BCC-metals Phys. of high pressure V.11 N2 (2001) pp.33-45
10. Bell J., Experimental Bases of the Mechanics of Deformed Bodies Part 1, London, (1984) 456 p.
11. Trefilov V.I., Moiseyev V.F., Pechcovsky A.P. Work Hardening and Fracture of the Polycryctal Metals, Ed. by Trefilov V.I., Naukova Dumka, Kiev, (1975), 248p. (in Russian).
12. Zehetbauer M.J. Features of Severe Plasstic Deformation as Compered to Conventional Deformation Model Microstructural Evolution of Titanium under Twist Extruzion Ultrafine Grained Materials: Procassing and structure Washington (2002), pp. 39 57.
13. MilmanYu.V, Miracle D.B. Korzhova N.P., Podrezov Yu.N. Mechanical behavior of Al_3Ti intermetallic and $L1_2$ phases on its basis Intermetallics 9 (2001),pp. 839-845.
14. Podrezov Yu.N. Minakov M.V. Strengthening of clean titanium at unisxial tension and rolling. Electron microscopy and strength of materials N9 (1998) pp 60-68. (in Russian).
15. Larikov L.N. Healing of defects in metals. Kiev, Naukova dumka, (1980) 280 p. (in Russia).

Structure and Properties of Ti Alloys Processed by ECAP

V.V. Stolyarov*[1], E.A. Prokofiev[1], R.Z. Valiev[1], T.C. Lowe[2], Y.T. Zhu[2]
[1]Institute of Physics of Advanced Materials, Ufa State Aviation Technical, Ufa, Russia, 450000
[2]Los Alamos National Laboratory, Los Alamos, NM 87545, USA

Abstract

In this paper, we present the microstructure and properties of commercially pure Ti and a TiNi shape-memory alloy processed by equal channel angular pressing (ECAP). Some of the samples were also subsequently processed by thermo and/or mechanical processing to further improve their mechanical and other properties, including strength, ductility, fatigue strength, impact toughness, tribological characteristics, corrosion resistance and functional properties.

Introduction

In recent years severe plastic deformation (SPD) has been established as an effective technique for producing bulk ultrafine grained (UFG) metals and alloys [1-3]. The most successful SPD technique has been the equal channel angular pressing (ECAP). These UFG materials processed by ECAP usually possess properties superior to their coarse-grained counterparts that are processed by conventional thermo/mechanical processing, and therefore are of great interest for use in many advanced applications. The UFG billets processed by ECAP usually have the geometry of a short, rectangular or round rod. Further processing is needed to further shape them into required semi-finished products, such as long sheets or rods. Therefore, post-ECAP thermo-mechanical processing and investigation of the structure and properties of UFG materials processed by these processing are also of great interest.

Commercially pure (CP) Ti and TiNi shape-memory alloys have excellent biocompatibility and are often used to make medical implants and devices [4,5]. Their properties can be significantly improved by processing them into UFG states, which makes it possible to improve the performance of medical implants and devices and to enable new medical applications. This paper presents the results of our recent studies on several ECAP-processed CP Ti and $i_{49.4}Ni_{50.6}$ shape-memory alloy.

Experimental Procedures

Three grades of CP Ti (VT1-0, Grade 2, Grade 4) and $Ti_{49.4}$ $Ni_{50.6}$ alloy in rod form with diameter of 40-60 mm were used in this study. The initial grain size was 8-20 μm for CP Ti and 25 μm for the TiNi alloy. CP Ti samples were processed for 8 ECAP passes at 450°C using a 90° die and the route B_c, in which the billet was rotated clockwise around its longitudinal axis for 90° between adjacent passes [6].

* Corresponding Author. vlst@mail.rb.ru

Y.T. Zhu and V. Varyukhin (eds.), Nanostructured Materials by High-Pressure Severe Plastic Deformation, 169-174.

The TiNi alloy was processed at 350 °C and 450 °C for 8 ECAP passes via route B_c using a 110° die set. Cold rolling was performed using round calibrated rollers for CP Ti (VT1-0) and flat rollers for CP Ti (Grade 2, Grade 4). Mechanical properties were investigated on standard samples with diameter of 5 mm and gage length of 25 mm for CP Ti and on samples with a cross-section of 0.25 x 1.0 mm and a gage length of 3 mm for TiNi alloy. Microstructures were studied using transmission electron microscopy (TEM).

Results and Discussion

Commercially pure (CP) Ti

Typical microstructures and selected area diffraction patterns (SAEDP) developed in a CP Ti (VT1-0) processed by ECAP and cold rolling for 80% strain (cross-section reduction) are shown in Fig.1. High density of dislocations was observed in the sample. Equiaxed nanocrystalline grain/subgrains are seen in cross section and elongated grain/subgrains are seen in longitudinal, indicating that the grains are elongated along the longitudinal direction.

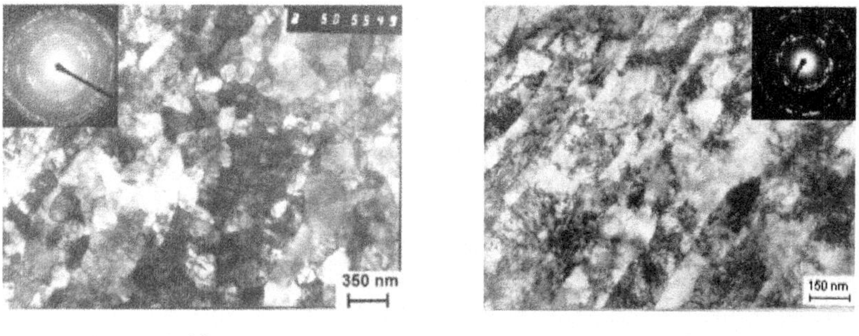

(a) (b)

Figure 1. Microstructure and SAEDP of a CP Ti (VT1-0) sample processed by ECAP for 8 passes and further cold rolling for 80% strain: (a) cross section, (b) longitudinal section [7].

Table 1 shows the mechanical properties of three grades (VT1-0, Grade 2, and Grade 4) CP Ti under various processing states [7, 8]. UFG billets in all processing states have significantly higher strength over their coarse-grained counterparts and at the same time have very good ductility. Such high strength and good ductility are very attractive for structural applications such as medical implants.
Number of other properties important for practical application was also investigated.

High fatigue strength up to 500 MPa was observed in UFG Ti under a testing condition of rotational bending [9]. This value meets the requirement of many medical implants and devices although still lower than the fatigue strength of Ti-6Al-4V alloy.

UFG Ti processed by ECAP, following cold rolling (80%) and final annealing at 300 °C for 1 hour is also found to have higher impact toughness, which is especially higher in low temperatures than that of coarse-grained Ti (Fig.2). The impact toughness of UFG Ti decreases with decreasing temperature, but has no brittle-ductile transition [10].

Table1. Mechanical properties of CP titanium in longitudinal section [7, 8]

Material	State	UTS, MPa	YS, MPa	EL, %
VT1-0	As received, ∅60 mm	550	430	42
	As received +CR80%	790	750	9
	ECAP (8), ∅60 mm	765	670	23
	ECAP (8)+CR80%	1040 1150*	940 1050*	12.5 12*
Grade 2	As received, ∅60 mm	490	355	30
	As received +CR77%	791	650	15
	ECAP (8), ∅60 mm	633	550	27
	ECAP (8)+CR77%	938	665	18
Grade 4	As received, ∅40 mm	>550	>480	>15
	As received +CR80%	1130	-	9
	ECAP (8), ∅40 mm	815	750	19
	ECAP (8)+CR83%	1135	1006	11

* - record values at CR 93%

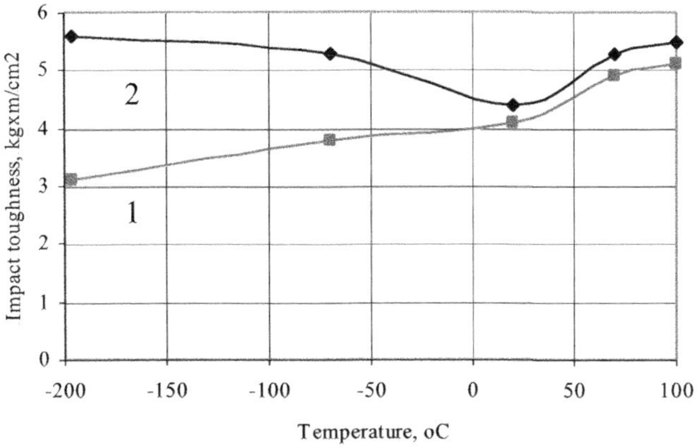

Figure 2. Temperature dependence of impact toughness of CP Ti (VT1-0) in the CG (1) and UFG (2) states.

UFG Ti is also found to have superior tribological property and corrosion resistance over their coarse-grained counter parts [11,12]. Experimental data indicate that UFG Ti has lower friction coefficient, which should in turn improve the wear resistance [11]. The superior corrosion resistance of UFG Ti is believed to result from rapid passivation and less impurity segregation on grain boundaries of UFG Ti [12].

$Ti_{49.4}Ni_{50.6}$ alloy

The microstructure after ECAP is relatively homogeneous. It consists mainly of grains of parental austenite B2 phase with a mean size of 310 nm with clear straight boundaries. The dislocation density is very low (Fig.3a). SAED pattern and TEM micrographs indicate that small amount of B19′ martensite phase also exists in some local areas (Fig. 3b). The microstructure on the longitudinal section is similar to that in the cross-section. SAED patterns indicate crystallographic texture of B2-phase. Particularly, there are practically no {200} reflections, which are to form near the {110} reflections.

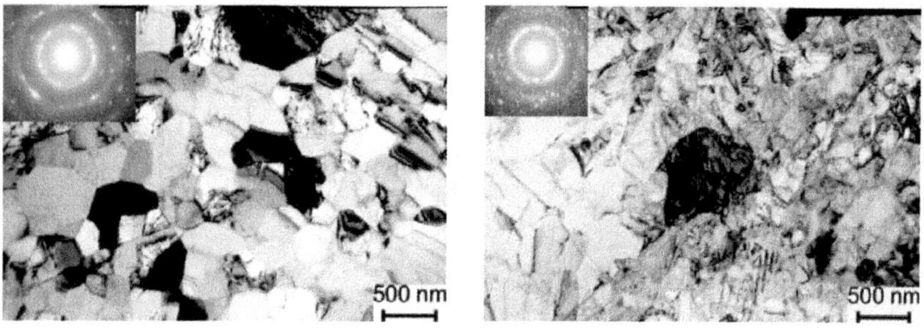

(a) (b)

Figure 3. Microstructure and SAEDP of ECAPed $Ti_{49.4}Ni_{50.6}$ alloy (8 passes at 450′C) in longitudinal section: (a) B2 austenite, (b) mixture of austenite and martensitic structures [11].

The mechanical properties, stress-strain curves and the photographs of tensile samples after fracture are presented in Table 2 and Fig. 4 [13]. As shown, ECAP significantly increased the strength, especially the yield strength, while retaining most of the ductility. Decrease in ECAP temperature from 450 °C to 350 °C and post-deformation annealing at 450 °C further increase the strength, but significantly decrease the elongation to failure, which however remains sufficient for practical applications. The strengthening by a post-deformation annealing is believed caused by the ageing of austenite, which is oversaturated with nickel as a result of quenching before ECAP. The annealing also significantly decreases σ_m, the stress to induce martensitic transformation. We also found that the strength is higher, and the ductility is lower in the cross-section than those in the longitudinal section.

Table2. Mechanical properties of ECAP $Ti_{49.4}Ni_{50.6}$ alloy, σ_m is the critical stress for inducing martensitic transformation, UTS is the ultimate strength, YS is the yield strength, and EL is the elongation to failure [13].

State	Sample Orientation	σ_m, MPa	UTS, MPa	YS, MPa	EL, %
Quenched (800°C, water)	longitudinal	140	1040	560	85
	cross	150	1050	580	76
ECAP (450°C, 8 passes),	longitudinal	220	1050	940	63
	cross	180	1210	1010	48
ECAP (350°C, 7 passes)	cross	230	1270	1190	28
ECAP (350°C) + annealing 450°C	cross	110	1440	1340	11

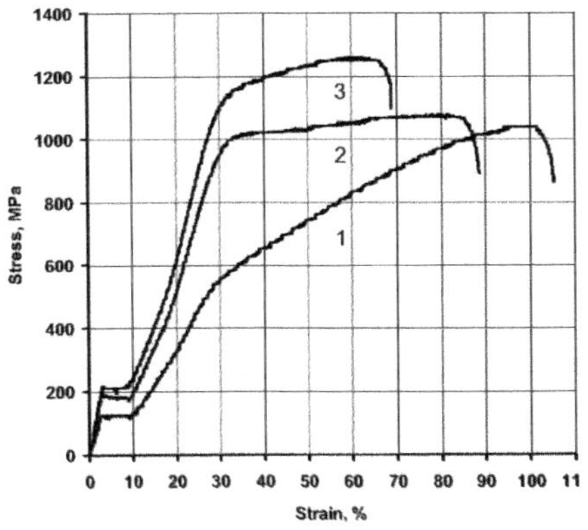

(a)

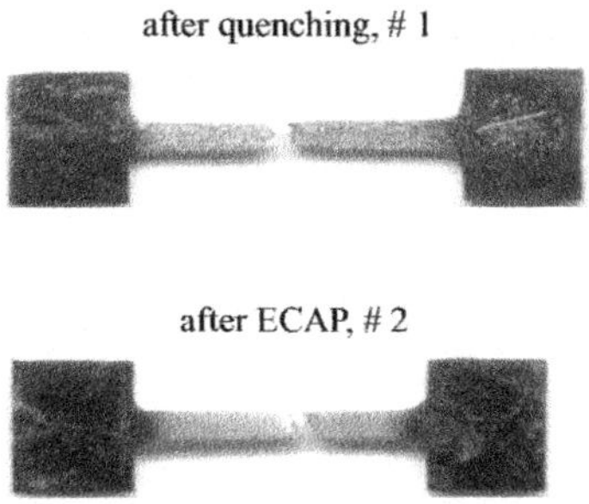

(b)

Figure 4. (a) Engineering stress-strain curves of $Ti_{49.4}Ni_{50.6}$ alloy in (1) quenched state and (2,3) after ECAP at 450°C: (1,2) – longitudinal; (3) cross sections. (b) Photographs of samples after tensile tests [13].

The UFG TiNi alloy processed by ECAP has lower work hardening than its coarse-grained in quenched state. The UFG TiNi alloy also has a uniform elongation that is surprisingly large in comparison with that of UFG Ti (Fig 4a).

The fatigue strength σ_{-1} was also found to improve from 385 MPa in coarse-grained TiNi alloy to 541 MPa in UFG state [14].

The ECAP processing at 450 °C for 8 passes decreased the martensite transformation temperatures, M_s, by 20-30 °C. Maximal fully reversible strain ε_r^{max} and maximal reactive stress σ_r^{max} increased to 9.2% and 890 MPa, respectively (at induced strain $\varepsilon_i \approx 5\%$)[13].

Preliminary results from tribology investigations of ECAPed TiNi alloy have shown that adhesive component of friction coefficient is smaller than that in its quenched state [15].

Conclusion

We have demonstrated that severe plastic deformation by ECAP can be successfully used to fabricate UFG CP Ti and TiNi alloy. This approach, particularly when coupled with subsequent conventional thermo-mechanical treatments, can result in properties that are unobtainable through normal thermo-mechanical processing. Examples from our recent investigations have been presented and discussed showing significantly improved mechanical properties and other properties such as fatigue strength, corrosion resistance, etc. in Ti UFG alloys processed by ECAP.

Acknowledgements

The authors gratefully acknowledge the support of US DOE via the International Science & Technology Center (ISTC), Moscow; INTAS program office, Brussels and Austrian Research Hertha Firnberg.

References

1. Valiev R.Z., Islamgaliev R.K., Alexandrov I.V., *Prog. Mat. Sci.* 45, 2, 2000, 102-189.
2. Zhu Y. T., Langdon T. G. et al eds, Proceedings of a symposium *"Ultrafine-Grained Materials II"* held during the 2002 TMS Annual Meeting TMS Publ., USA, 2002.
3. Lowe T.C., Valiev R.Z., eds, Proceedings of the NATO ARW on *"Investigations and Applications of Severe Plastic Deformation"*, NATO Sci. Series, Kluwer Publ., 80, 2000.
4. Brunette D.M., Tengvall P., Textor M., Thomsen P., *Titanium in Medicine*. Springer-Verlag, 2001, p.1019.
5. Valiev R.Z., *Nature Mater.* 3, 2004, 511-516.
6. Stolyarov V.V., Zhu Y.T., Alexandrov I.V., Lowe T.C., Valiev R.Z. *Mat. Sci. Eng.* A 299, 1-2, 2001, 59-67.
7. Stolyarov V.V., Zhu Y.T., Raab G.I., Zharikov A.I., Valiev R.Z. *Mat. Sci. Eng.* A, 2004, in press.
8. Stolyarov V.V., Zeipper L., et al, *Mat. Sci. Eng.* A, 2005, to be published.
9. Stolyarov V.V., Shestakova L.O., Zharikov A.I., Latysh V.V., Valiev R.Z., Zhu Y.T., Lowe T.C., Proceedings of 9th Inter. Conf. on Titanium, 1, 2000, 466-472.
10. Stolyarov V.V., Valiev R.Z., Zeipper L., Korb G., Proceedings of 10th Inter. Conf on Titanium held in Hamburg, Germany, 13-18 July 2003, eds. G. Lütjering and J. Albrecht, 2004, 1437-1444.
11. Stolyarov V.V., Shuster L.Sh., Migranov M.Sh., Valiev R.Z., Zhu Y.T. *Mat. Sci. Eng.* A 371, 2004, 313-317
12. Balyanov A., Kutnyakova J., Amirkhanova N.A., Stolyarov V.V., Valiev R.Z., Liao X.Z., Zhao Y.H. , Jiang Y.B., Xu H.F., Lowe T.C., Zhu Y.T., *Scr. Mater.* 51, 2004, 225-229
13. Quarter Report for ISTC project # 2398, June 2004
14. Annual Report for ISTC project # 2398, August 2003
15. Chertovskikh S.V., Shuster L.Sh., Stolyarov V.V., *paper collection*, USATU, Ufa, 2004, in Russian

Microstructure and Mechanical Properties of 6082 Aluminum Alloy Processed by Hydrostatic Extrusion

P. Widlicki[1*], H. Garbacz[1], M. Lewandowska[1], W. Pachla[2] and K.J. Kurzydlowski[1]

[1] Warsaw University of Technology, Faculty of Materials Science and Engineering, Woloska 141, 02-507 Warsaw, Poland

[2] High Pressure Research Center, Polish Academy of Science, Sokolowska 29/37, 01-142 Warsaw, Poland

Abstract

Hydrostatic extrusion process of 6082 aluminum alloy was investigated. The material in the initial state was fully recrystallized. The specimens were subjected to the hydrostatic extrusion in consecutive steps up to the cumulated true strain of 3,75. The extruded specimens were water cooled at the die exit. The microstructure of the alloy was examined using transmission electron microscopy (TEM) and image analysis. Microhardness measurements and tensile tests were performed to characterize the mechanical properties.

The extrusion process results in grain refinement, as evidenced by TEM observations. Hardness, microhardness measurements and tensile tests showed that also mechanical properties of the alloy significantly increase due to hydrostatic extrusion.

Introduction

Severe plastic deformation has become a popular method for fabrication of ultra-fine grained metals. The first developments and investigations of nanostructured materials processed using SPD methods were reported more than 10 years ago [1,2]. A rapid increase of different publications on this subject in the recent years has been observed. The widely used methods for such a processing are: equal channel angular pressing (ECAP) [3], cycling extrusion compression (CEC) [4] and high pressure torsion (HPT) [5]. Large plastic strain may also be achieved by a hydrostatic extrusion which results in ultrafine and homogenous structure in the entire volume of the material [6]. The aim of this work was to show the influence of the hydrostatic extrusion on the microstructure and mechanical properties of 6082 aluminum alloy.

Materials and methods

Examinations were carried out on 6082 aluminum alloy of chemical composition presented in Tab.1. Materials in the initial state and after succeeding stages of hydrostatic extrusion, with reduction coefficients (R) of 3,8, 4 and 2,78 were examined (Tab.2).

[*] Corresponding Author: pwidlicki@inmat.pw.edu.pl

Y.T. Zhu and V. Varyukhin (eds.), Nanostructured Materials by High-Pressure Severe Plastic Deformation, 175-180.

Table 1. Chemical composition of the 6082 alloy

Si	Fe	Cu	Mn	Mg	Cr	Zn	Inne	Al
0,7–1,3	max 0,5	max 0,1	0,4–1,0	0,6–1,2	max 0,25	max 0,2	max 0,15	balance

Table 2. Parameters of hydrostatic extrusion of the 6082 alloy

	Reduction coefficient R	**True strain φ**
Initial state	-	-
ϕ19,5 → ϕ10	3,8	1,34
ϕ10 → ϕ5	4	1,39
ϕ5 → ϕ3	2,78	1,02
Cumulated	**42,3**	**3,75**

Hardness measurements were performed using a Brinnel hardness tester with a ball diameter of 2,5mm.. Loads of 153,2 N and 306,5 N were applied for the initial and extruded material respectively.

The Vicker's microhardness tests were carried out using a Zwick microhardness tester with 200G loading for 15 seconds.

Examinations of mechanical properties were carried out at room temperature with a strain rate of $8,3*10^{-4}$ s^{-1}. Two specimens were deformed for each state. Mean values were calculated from each pair of the obtained values of the strength parameters.

Microstructure observations were carried out using Transmission Electron Microscope operated at 100kV. In order to determine the influence of the hydrostatic extrusion process on the grain size of the examined material, quantitative analysis of the microstructure was performed based on the obtained TEM images. The MicroMeter program [7] was used for the analysis.

Results and Discussion

Hardness and microhardness measurements

Results of hardness and microhardness measurements (Fig.1) show that the hydrostatic extrusion brought about a significant increase in both of those parameters. Increase of the hardness value from 36 to 73 HB after extrusion and increase of the microhardness from 50 to 99 HV0,2 was observed. It should be noted that the highest increase of the hardness and microhardness were observed for the first stage of extrusion (ϕ19,5→ϕ10). The increase of both of those parameters was much lower in the following stages.

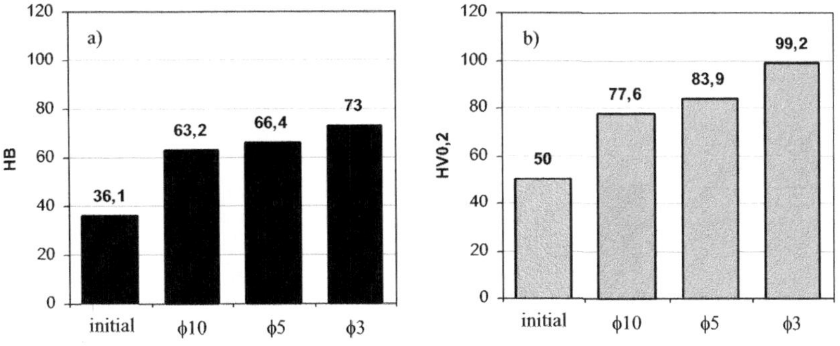

Figure 1. Influence of hydroextrusion on hardness (a) and microhardness (b) of the 6082 alloy.

Tensile tests

Results obtained in static tensile tests are presented in the form of σ - ε curves (Fig.2). Influence of the applied hydrostatic extrusion process on yield stress and tensile strength of the 6082 alloy was also analyzed (Fig.3).

Mechanical properties of the examined material increased with the increase of the cumulated true strain. A significant increase of tensile strength can be observed already after the first stage of hydroextrusion (ϕ19,5 → ϕ10). The succeeding hydroextrusion stages did not cause such radical strain hardening of the material.

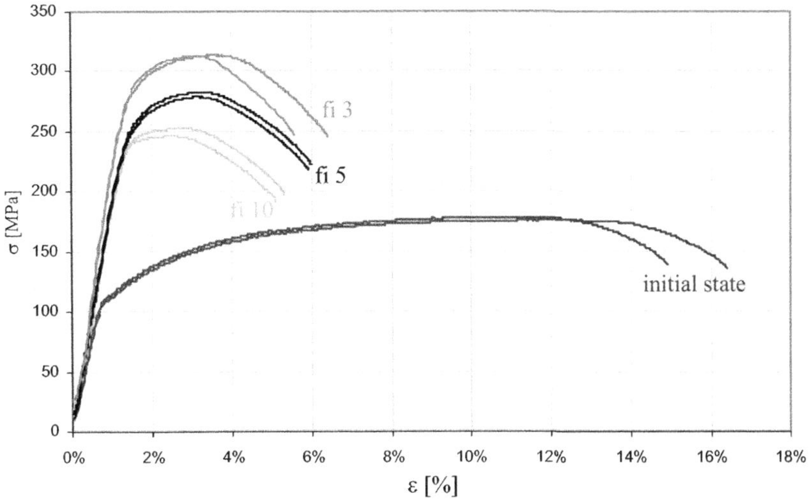

Figure 2. Curves σ-ε of the 6082 alloy in the initial state and after hydrostatic extrusion.

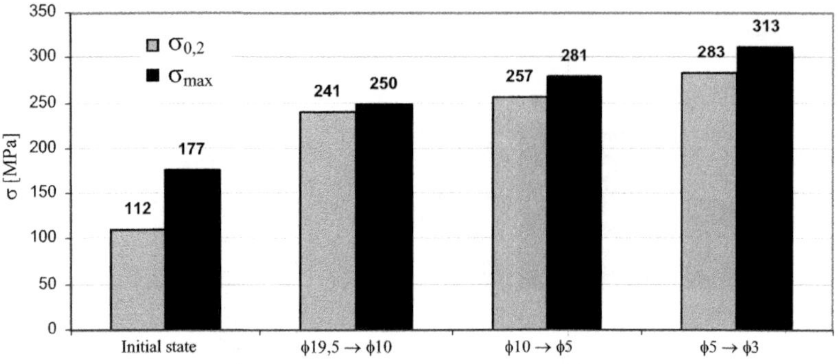

Figure 3. Influence of hydrostatic extrusion on strength ($\sigma_{0,2}$ and σ_{max}) of the 6082 alloy

Quantitative analysis of microstructure

Quantitative analysis of microstructure was performed based on the obtained TEM images. TEM images of the 6082 alloy are presented in Fig.4.

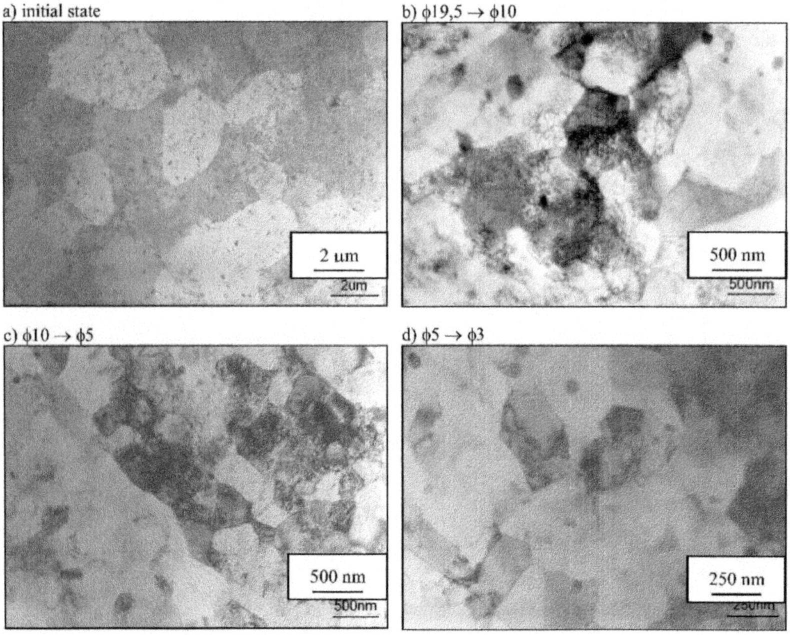

Figure 4. TEM images of microstructure of the 6082 alloy in the initial state (a) and after hydrostatic extrusion process (b, c and d)

Hydrostatic extrusion process brought about a significant refinement of the microstructure of the examined 6082 alloy (see Tab.3 and Fig.5). The highest decrease of the mean grain size (2,53μm → 428nm) was observed after the first stage of hydroextrusion (ϕ19,5 → ϕ10). The following stages of plastic deformation led to further slight grain refinement and recovery of grain interiors. Microstructure of the 6082 alloy after the last stage of hydrostatic extrusion is characterized by mean grain size $d_2 \approx$ 300nm and lack of dislocations inside the grains.

Table 3. Results of grain size distribution in the 6082 alloy

	Initial state	**ϕ19,5 → ϕ10**	**ϕ10 → ϕ5**	**ϕ5 → ϕ3**
d_2	2,53 μm	428 nm	341 nm	328 nm
SD	0,86	121	101	92
CV = SD/d_2	0,34	0,28	0,30	0,28

where: d_2 – mean equivalent diameter
SD – standard deviation
CV – variation coefficient

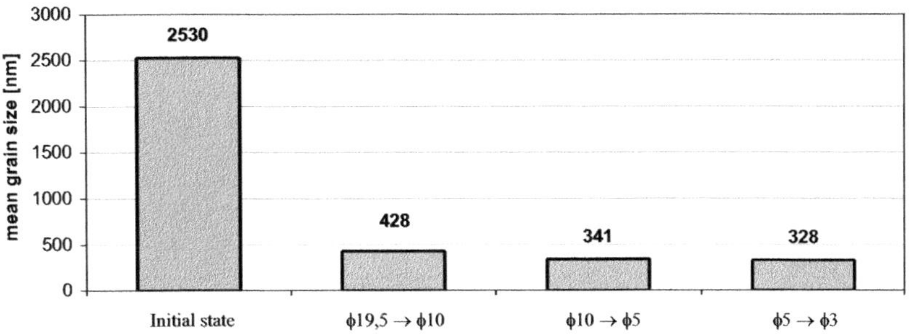

Figure 5. Influence of hydrostatic extrusion on grain size of the 6082 alloy

Conclusions

- Hydrostatic extrusion process caused a significant refinement of microstructure of the examined 6082 alloy. The process results in material with average grain size at about 300 nm.
- The 6082 alloy after hydrostatic extrusion process achieved much higher strength properties compared to the material in the initial state.

Acknowledgements

This research was financed by the State Committee for Scientific Research (KBN) as an ordered research assignment (grant PBZ-KBN-096/T08/2003).

References

1. R.Z. Valiev, O.A. Kaibyshev, R.I. Kuznetsov, R.Sh. Musalimov R.Sh, N.K Tsenev; DAN SSSR 301(4) (1988) 864
2. R.Z. Valiev, N.A. Krasilnikov, N.K. Tsenev; Mater Sci Eng A137 (1991) 35.

3. R.Z. Valiev, R.K. Islamgaliev, I.V. Alexandrov; Bulk nanostructured materials from severe plastic deformation; Progress in Mater. Science 45 (2000) 103-189
4. M. Richert, Q. Liu, N. Hansen; Mater. Sci. Eng A260 (1999) 275
5. A.P. Zhilayaev, B.K. Kim, G.U. Nurislamova, M.D. Baro, J.A. Szpunar, T.G. Langdon; Scripta mater. 46 (2002) 575
6. M. Lewandowska, H. Garbacz, W. Pachla, A. Mazur, K.J. Kurzydlowski; Solid state Phenomena 101 (2005) 65-69
7. T. Wejrzanowski, M.Sc. Thesis. Warsaw University of Technology (2000)

Grain Refinement and Viscous Fracture of Metals during Severe Plastic Deformation: Mathematical Simulation

Yan Beygelzimer

Donetsk Phys&Technology institute of Ukrainian National Academy of Sciences, 72 R.Luxembourg St., Donetsk,, UKRAINE, 83114

Abstract

We investigate structure formation of metals undergoing severe plastic deformation. We present two natural assumptions that reveal valuable information about this process. The assumptions are the self-similarity of grain refinement and micro-damage and the complementarity of these two processes. Based on these assumptions, we obtain a system of kinetic equations that explains the formation of structure and suggests new problems for experimental and theoretical investigations.

Introduction

Grain refinement under severe plastic deformations is currently studied by a large number of laboratories throughout the world, because ultra-fine grained structures have very appealing physical and mechanical properties. This surge of interest led to a number of publications that investigate grain refinement of particular metals and alloys under various conditions. (See, for example, [1].)

We already know that grain refinement is determined by a number of factors such as the accumulated strain, strain rate, the character of deformation, the level of hydrostatic pressure, and the temperature. From the point of view of mechanics and physics of solids, these factors represent different parameters of a loading process. It is convenient to view the loading process as a curve in the deformation space, as a trajectory of loading with scalar and vector parameters of the process shown for every point. Accumulated strain corresponds to the length of this curve. Such representation is called the image of a loading process.

The fact mentioned above can now be reformulated as stating that metal structure is determined by the image of the loading process. This formulation, by itself, is trivial – it is just a rephrasing using a new language. However, consider the following question: What are the factors determining the image of the loading process of a given material point of a specimen undergoing metal forming?

First, there are factors "external" to the deformable material, e.g., the method of deformation, parameters of deformation tools, position of the material point in the volume of a specimen, various mechanical and thermal affects on the specimen.

The image of the loading process also strongly depends on "internal factors", i.e., mechanical properties of the material determined by its structure. For example, some structures can lead to localization of deformation which drastically changes the stress-strained state of the specimen.

Therefore, the original statement has to be augmented: Metal structure is indeed determined by the image of the corresponding loading process, but the loading process itself may depends on this structure.

This inter-dependency of macro- and micro- characteristics significantly complicates the description of SPD processes and makes it hard to control them.

* Corresponding Author. tean@an.dn.ua

Y.T. Zhu and V. Varyukhin (eds.), Nanostructured Materials by High-Pressure Severe Plastic Deformation, 181-185.

We propose an approach to this problem based on a continuum model (analogous to models in plasticity theory) with internal parameters that reflect the structure of the material. Internal parameters allow us to account for the interdependency between the stress-strain state and the structure.

The model describes two interrelated processes simultaneously happening in metals under plastic deformation, namely grain refinement and fracture. The model allows one to formulate the problem of computing the stress-strained state of a specimen undergoing an SPD process, accounting for the interdependency of this state with the structure and properties of the deformable material.

A number of assumptions were used. This made the developed instrument (model) much less precise than, for example, electronic or optical microscopes. On the other hand, however, this allowed us to make several general conclusions that may help to systematize the experiment.

In this talk I will present only main assumptions and consequences of the model. For more details see article [2].

Main assumptions

The model is based on the following assumptions:

1. There are regions of polycrystal where dislocations get plugged during plastic deformation. Such regions cause bending of the crystalline lattice. We called this regions as the accumulative zones (AZ). The model postulates that AZ emerges in places of hurdles (e.g., grain boundaries) that exist in polycrystal before deformation, and also in high-angle boundaries that emerge during deformation.
2. There are various relaxation mechanisms of the AZ. When talking about large deformations at low temperature, we distinguish two main mechanisms: emergence of high angle boundaries and emergence of voids. The first mechanism leads to grain refinement, the second mechanism leads to fracture.
3. When the fragments become sufficiently small, they stop dividing. This is due to the fact that elastic stresses relax by fragment sliding along high-angle boundaries. In other words there is a limiting fragment size d_c.
4. When the fragment size d satisfies inequality $d_c \ll d \ll L_0$ (L_0 is the diameter of the original grain), the mechanism of fragment division is unchanging. That imply the self-similar fashion of grain refinement when $d_c \ll d \ll L_0$. In other words the net of the high-angle boundaries obtained under different strain statistically differ only by their scale.

Kinetic equations

On the based of this assumptions we have got following system of the kinetic equations:

$$\begin{cases} \dfrac{d\overline{N}}{d\gamma} = \left(C_1 + C_2 C_5 \overline{N}_b\right) F\left(\overline{S}\right) - \left(C_3 + C_4\right)\overline{N} \\ \dfrac{d\overline{N}_b}{d\gamma} = C_4 \overline{N} - C_5 \overline{N}_b \\ \dfrac{d\overline{S}}{d\gamma} = \dfrac{C_5 \overline{N}_b}{\overline{S} + \overline{S}_0} \\ \dfrac{d\Theta}{d\gamma} = C_3 \overline{N}^{\frac{3}{2}} d_c^{-3} v - C_6 \Theta \end{cases} \tag{1}$$

Here following table of symbols are introduced: γ - accumulative strain; v - volume of a void; $\overline{N} = N d_c^2$, $\overline{N}_b = N_b d_c^2$, $\overline{S} = S d_c$, $\overline{S}_0 = \dfrac{d_c}{L_0}$; N - number of AZ per unit of the cross section area; S - total length of the high-angle boundaries per unit of the cross section area; N_b - number of embryo of the high-angle boundary per unit of the cross section area; Θ - porosity; p – hydrostatic pressure; K – shear yield stress; $C_1 - C_6$ – parameters defined by testing experiment with the material (see [2]).

The initial conditions are:

$\overline{N} = \overline{N}_b = 0$, $\overline{S} = \overline{S}_0$ when $\gamma = 0$

The resulting system of kinetic equations represents the evolution of the structure of the representative volume of polycrystal during quasimonotone loading. In order to describe non-monotone loadings, we divide the evolution path on quasi-monotone segments and to join only the values of Θ, S, and N_b when going from one segment to the next. The number N of AZ is set to 0 in the beginning of every segment. This models the relaxation of AZ under a change of the loading direction.

Constitutive Equations

The constitutive equations allow one to obtain the relation between stress and strain tensors. For the developed model they are:

$$\dot{e}_{ij} = \lambda \frac{\partial f}{\partial \sigma_{ij}}; \tag{2}$$

Here the following conditions are fulfilled:

$$\text{at } f = 0, \ \frac{df}{dt} = 0, \ \frac{\partial f}{\partial \sigma_{ij}} d\sigma_{ij} \neq 0, \quad \lambda > 0 \tag{3}$$

$$\text{at } f = 0, \ \frac{df}{dt} = 0, \ \frac{\partial f}{\partial \sigma_{ij}} d\sigma_{ij} = 0, \quad \lambda = 0 \tag{4}$$

$$\text{at } f = 0, \qquad \frac{df}{dt} < 0, \qquad \lambda = 0 \tag{5}$$

$$\text{at } f < 0, \qquad \lambda = 0 . \tag{6}$$

Here σ_{ij} and $\dot{e}_{ij}$ are the stresses and strain rate tensors respectively; f is the loading function,

$$f = \frac{\sigma^2}{\psi(\Theta)} + \frac{\tau^2}{\varphi(\Theta)} - (1-\Theta)\left(K(S) - C_3 \overline{N}^{\frac{3}{2}} d_c^{-3} v\sigma\right)^2 \quad ;$$

λ is the Lagrange factor;

$$\tau = \sqrt{\left(\left(\sigma_{ik} - \frac{1}{3}\sigma\delta_{ik}\right)\left(\sigma_{ik} - \frac{1}{3}\sigma\delta_{ik}\right)\right)}, \quad \sigma = \frac{1}{3}\sigma_{ik}\delta_{ik}$$

$$\psi(\Theta) = \frac{(1-\Theta)^{2n-1}}{6a\Theta^m}, \quad \varphi(\Theta) = (1-\Theta)^{2n-1};$$

m, n – parameters of the materials (see [4])

Relationship (2) is the mathematical expression of the gradient condition [3]. Conditions (3)-(6) show that the plastic deformation takes place only when the stresses satisfy the yielding condition $f = 0$ at the time moment under consideration and the closest one to it and, the stress increment vector does not lie on the loading surface.

Condition (3) is assumed by us in an extended form to take into account the possible loss of the material stability (for stable materials in condition (3) ($\partial f/\partial\sigma_{ik}$) $d\sigma_{ik} > 0$, [3]).

If one substitutes the Mises equations for equations (1)-(6) in any FEM package, one could directly compute the stress-state state of a metal and its dependence on the structure

Yield Surface

In the framework of plasticity theory, every material has a yield surface in the stress space. Thus for example, yield surface corresponding to the Mises material is a cylinder, while the surface corresponding to the Tresca – Saint-Venant material is a hexagon prism. The material proposed here has a thick yield surface with a fractal structure described in [5].

Consequences of the model

- A material does not fracture or hardens in the range of very large plastic deformations under pressure. In other words materials become ideally plastic under sufficiently large pressure and strain.
- During the self-similar stage, the distribution of fragment sizes is self-similar. This implies that the distributions for different deformation stages must be the same

in the normalized coordinates.

- Self-similarity implies lognormal distribution for the fragments sizes (according to A. Kolmogorov, 1941)
- During the self-similar stage of grain refinement, the fragment boundary net in the cross-section of the specimen represents a fractal set with dimension η, $1<\eta<2$.
- During the self-similar stage of grain refinement, the yield stress related with average fragment size by the Hall-Petch law, were exponent related with fractal dimension of the fragment boundary net:

$$K \sim \frac{1}{\bar{d}^{\eta-1}}$$

- When sufficiently many indivisible fragments of size *dc* appear, we have

$$K \sim \frac{1}{\bar{d}}$$

- Grain refinement intensity grows and fracture decrease with the increase of pressure in the center of deformation.
- To achieve intense grain refinement, one needs to choose deformation scheme with a small value of ductility at p=0 and perform deformation under pressure.
- Quasi-monotone deformations provide higher grain refinement intensity than cyclic deformations
- In order to increase the intensity of grain refinement under cyclic deformation, one has to increase the amplitude of deformation.
- Cyclic deformations provide higher ductility than quasi-monotone deformations
- A metals becomes more ductile after it undergoes a large deformation under pressure.
- Strain localization is a manifestation of the interdependency between macro- and micro- levels. One of the possible causes of strain localization during simple shear is an avalanche-like increase in porosity corresponding to a small metal strengthening. This is accompanied by the appearance of a voids layer along the shift plane. This decreases metal ductility in the direction perpendicular to this plane, and in certain cases can fracture the specimen along the plane. The analysis of the model shows that a sufficiently high pressure in the center of deformation prevents this undesirable effect.

References

1. Valiev R.Z., Islamgaliev I.V., Alexandrov I.V., Progr. Mater. Sci., **45**, (2000), 103-150
2. Beygelzimer Y., Mechanics of Materials , (2004) (in press)
3. Hill R., *The Mathematical Theory of Plasticity*. Clarendon Press, Oxford, New York, 1998, p.50
4. Beygelzimer Y., Efros B., Varyukhin V., Khohlov A., Engineering Fracture Mechanics, **48** (5), (1994), 629-640.
5. Beygelzimer Y., Spuskanyuk A.., *Philosophical Magazine* A, **79**, (10), (1999), 2437-2459

Structure and Properties of Ultrafine Grained Nickel after Severe Plastic Deformation

B.M. Efros[1*], V.A. Ivchenko[2], N.B. Efros[1], S.G. Synkov[1], L.V. Popova[2], T.P. Zaika[1] and L.V. Lolade[1]

[1]Phys.&Tech. Institute of the NASc of Ukraine, Donetsk, 83114, Ukraine

[2]Institute of Electrophysics of the RAN, Ekaterinburg, 620016, Russia

Abstract

Method of packet hydroextrusion was used to investigate the effect of severe plastic deformation on defect structure and strengthening of polycrystalline nickel. The grain-boundary component of the strengthening is shown to be due to the increase in the defect density. The formation of sub blocks in the structure of ultrafine grained nickel subjected to severe plastic deformation has been detected by the field ion microscopy method.

Introduction

Grain size reduction is one of the most effective methods of improving the strength characteristics, as it is not accompanied by embrittlement occurring with other methods of strengthening. The higher the grain dispersive in the crystal lattice, and thus the larger the relative extent of their boundaries, the higher should be the strengthening. At present, there are a lot of methods for grain size reduction [1]. Severe plastic deformation (SPD) is one of the most prospective methods of production of the ultrafine grained (UFG) materials [2].

This paper deals with the analysis of SPD effect on structure and properties of FCC metals with nickel as an example. It is aimed at investigation of the influences of grain size d on strain hardening depending on dislocation density ρ_D estimated by resistance method, as well as of the action of SPD on the formation of defect structure and mechanical properties of UFG nickel.

Experimental Procedures

The investigated was polycrystalline nickel (99.97 %) quality. With this material it is possible to determine the density of dislocations more exactly by using the resistance procedure in the case of deformations to a high degree [3].

The method of packet hydroextrusion (PHE) was applied for having SPD under pressure [4]. To obtain samples of different grain size ($d \approx 50 - 5$ μm), the nickel rods with the initial diameter of 8 mm subjected to preliminary PHE were annealed in the temperature range 450 to 1000°C.

The structure and properties of samples were studied by methods of optical (OM), transmission electron (TEM) and field ion microscopy (FIM), methods, as well as by using specific resistance (ρ) and HV microhardness. Structure parameters were quantitatively analyzed by the standard secant method.

* Corresponding Author. efros@hpress.dipt.donetsk.ua

Y.T. Zhu and V. Varyukhin (eds.), Nanostructured Materials by High-Pressure Severe Plastic Deformation, 187-192.

Results and Discussion

Investigations of the influence of grain size on structure and properties of nickel were done by PHE method (one-stage route regime) within the range of relative deformation degree $\varepsilon = 0 - \approx 60$ %.

For formation the UFG structure in nickel with the initial grain size $d \approx 8$ μm the PHE method was also used (multi-stage route regime) with the maximum logarithmic deformation $e \approx 12$.

One-Stage Route Regime of PHE of Nickel

Changes in the degree of strain hardening ΔHV for nickel samples depending on ε value during the PHE (one-stage route regime) for different initial grain size d are shown in Fig. 1a. With degree ε increase the cold-work hardening value ΔHV increases in a nonlinear way. Moreover, it increases noticeably with grain size reduction, especially when ε is high.

Let us consider how the grain size affects the degree of strain hardening from the point of view of dislocation density change $\Delta\rho_D$ by using the results of ρ specific resistance measurement. Fig. 1b illustrates the increase of Δρ during the PHE of nickel samples depending on the degree of deformation ε (the initial size d changes from 5 to 53 μm). It should be noted that with such ε the samples having finer grains have a higher value of Δρ ($\Delta\rho = \Delta\rho_d + \Delta\rho_P$), where $\Delta\rho_d$ is contribution from dislocations and $\Delta\rho_P$ is contribution from point defects. So, to estimate $\Delta\rho_D$, $\Delta\rho_d$ should be know. The section on the curves of isochronous annealing, where $\Delta\rho/T \approx 0$ ($T \approx 335°$ C) helps in determining the value of $\Delta\rho_d$ unambiguously and in converting $\Delta\rho_d$ to ρ_D by using coefficient η [3].

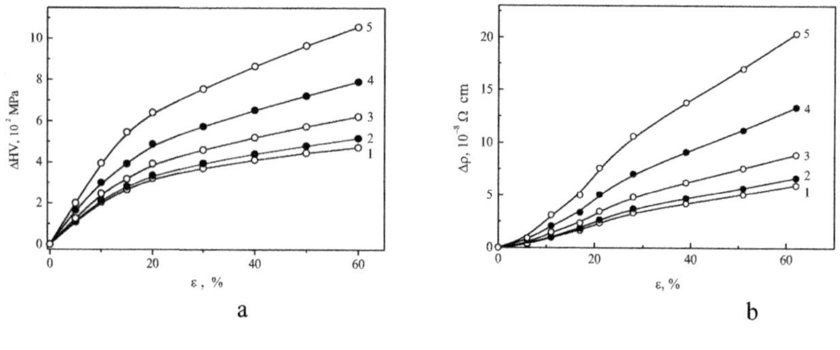

Figure 1. Change in ΔHV (a) and Δρ (b) versus deformation degree ε by the method PHE of nickel samples with different initial grain size d, μm: 1 – 53; 2 – 31; 3 – 18; 4 – 8; 5 - 5

The analysis of $\Delta\rho_d$ experimental values change depending on value of grain size d for the deformed nickel samples within the framework of equation

$$\Delta\rho_d = A + Bd^{-1}, \tag{1}$$

where A, B and n – constants.

The first in the right-hand part of esq. (1) stands for the density of dislocations $\Delta\rho_D$ accumulated in the grain, the second term is due to the available grain boundary. With grain size decreasing the $\Delta\rho_D$ in the specimen is increasing.

In Fig. 2 there are dependences of HV on $\sqrt{\Delta\rho_d}$ value for Ni samples with different initial grain size d after the PHE. The linear dependence of HV on $\sqrt{\rho_d}$ is maintained to a sufficient accuracy independent of grain size and degree of deformation. Thus, the value of deformation hardening can be described within the framework of changes in specific resistance $\Delta\rho_d$ due to the contribution from dislocation ρ_D as

$$HV = HV_0 + \alpha_d Gb\sqrt{\Delta\rho_d}, \quad (2)$$

where HV_0, α_d is constants, G is shear modulus, b is Burgers vector. By using coefficient η [3] the following relationship is obtained

$$HV = HV_0 + \alpha\, Gb\, \sqrt{\Delta\rho_D}, \quad (3)$$

here $\alpha = \alpha_d/\sqrt{\eta}$.

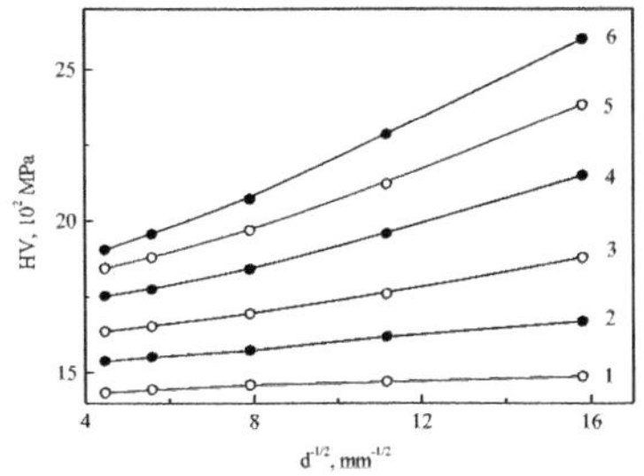

Figure 2. Change of HV versus d $^{-1/2}$ for nickel samples after HPE for different degree ε, %: 1 – 0.2; 2 - 6; 3 – 11; 4 – 21; 5 – 39; 6 - 62

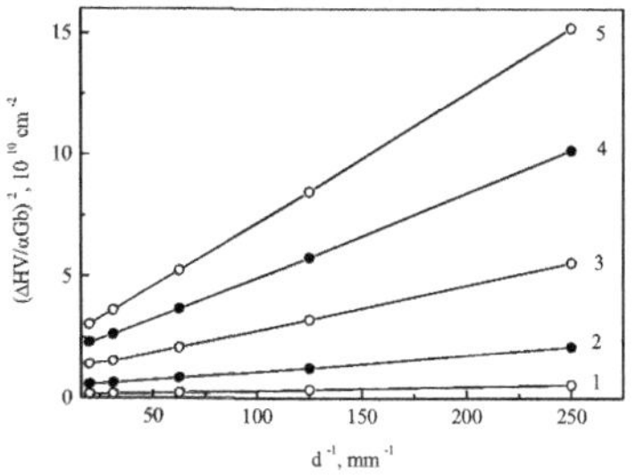

Figure 3. Change of $(\Delta HV/\alpha Gb)^2$ versus d $^{-1}$ for nickel samples after HPE for different degree ε, %: 1 - 6; 2 – 11; 3 – 21; 4 – 39; 5 - 62

So, the grain-boundary dependence of HV can be described by the following two dependences: the dislocation density $\Delta\rho_D$ accumulated in polycrystals during the deformation is inversely proportional to grain size d; hardness HV grows linearly with $\sqrt{\Delta\rho_D}$ increase. We have general expression in the form

$$HV = HV_0 + \alpha\, Gb\, \sqrt{[\, A_1(\varepsilon) + (B_2\varepsilon/d)]}, \quad (4)$$

Depending on the relation between terms $A_1(\varepsilon)$ and $B_2\varepsilon/d$, the grain boundary dependence of HV can be described by the following functions

$$HV = HV_0 + \alpha\, Gb\, [\sqrt{B_2(\varepsilon)}]d^{-1/2} \text{ for } A_1(\varepsilon) << B_2\varepsilon/d \quad (5)$$

$$HV = HV_0 + \alpha\, Gb\, [\sqrt{(A_1(\varepsilon)}] + (\alpha GbB_2\varepsilon)(2[\sqrt{A_1(\varepsilon)}])^{-1}\, (d^{-1}) \text{ for } A_1(\varepsilon) >> B_2\varepsilon/d \quad (6)$$

Fig. 3 illustrates the HV- d $^{-1/2}$ dependence (the Hall-Petch relationship). It is seen that for low degrees of nickel samples deformed by PHE with $\varepsilon \leq 5\%$ there is the linear dependence between HV and $d^{-1/2}$ which is transformed with deformation increase into a concave curve much deviating from the Hall-Petch relationship.

We rewrite relationship (4) as

$$[(\Delta HV)/\,\alpha Gb]^2 = A_1(\varepsilon) + B_2\varepsilon/d^{-1} \quad (7)$$

Using $\Delta HV(\varepsilon)$ (see Fig. 1 and $\alpha = 1.35$ it is possible to calculate the value of the left-hand part of equation (7), depending on inverse value of grain size for different degrees of deformation ε (Fig. 3). It is seen that in the investigated range of $\Delta\varepsilon$ can be described by equation (7).

So, the first term in the right-hand part of equation (7), when is in dependent d, is the value of dislocation density ρ_D accumulated in grain body during the deformation, whereas term $B_2\varepsilon/d^{-1}$ is the grain-boundary component of strain hardening for polycrystalline nickel. The obtained results evidence that the grain boundary component of hardening is conditioned by increased accumulation of dislocations in nickel samples during the PHE. The increased density of dislocations for Ni samples having finer grain results in decreasing the size of dislocation cells.

Multi-Stage Route Regime of PHE of Nickel

It is know that the UFG materials produced by various methods of SPD possess the unique physical and mechanical properties the explanation of which needs some information about the atomic structure of the investigated objects. The PHE method has been used for the formation of UFG structure in nickel (i.e. the regime of multi-stage rout) at each stage of PHE the pressure did not exceed 1.2 GPa, and the single relative deformation ε - 80 %. In the limiting case, the maximal accumulated logarithmic deformation reached $e = 12$.

During SPD by HPE in the grains a considerable density of dislocation is accumulated, and then the cellular structure is formed, next follows the fragmented (subgrain) structure, with the increase in the degree of deformation. The dislocation density (within the boundaries of fragments) and the angle of disorientation (at boundaries of the fragments) increase in proportion to deformation degree. As a result, the boundaries are transformed into new large-angle ones. The character of changes in grain (fragment) size d, values of microhardness HV and specific resistance ρ of nickel samples as a function of logarithmic deformation e is illustrated in Fig. 4.

The analysis of the obtained results enables us to assume that there are three stages of dislocation structure evolution by OM and TEM methods in nickel samples under SPD by of PHE, which is in agreement with the available data [5,6]. The first stage corresponds to disoriented cellular structure ($e \leq 3$). With e increase the generated dislocations can no longer be built into the boundaries, they gradually fill the cell volume, thus forming dislocation aggregates and tangles. When the dislocation density ρ_D approximately reaches $\approx 10^{16}$ m^{-2}, there occurs the transition to the second stage. This stage corresponds to the formation of transient structure ($e \approx 3 - 5$); the cells become finer and more disoriented. The cell boundaries rearrange to the boundaries of fragments. Evidently at this moment the mechanisms of rotational plasticity join the process. It has been found that at this stage the deformation-substructure type changes simultaneously with the velocity of grain (subgrain) size diminution. This change is also accompanied by changes in texture of the specimens. The third stage of structure evolution in the course of deformation is characterized by the retardation of the velocity of subgrain size diminution. Such UFG-structures are dissipative. [5,6]. The character of changes in hardness, specific resistance (an indirect characteristics of the density of defects) and size of structure elements corresponds to the above scheme of nickel structure evolution in the process of PHE. HV value for nickel in the UFG state hat 2-3 times (as compared to the initial state) during the logarithmic deformations $e > 8$ and was in the range $\approx 4 - 4.5$ GPa (see Fig. 4). Thermal stability of the UFG structure of nickel samples formed under the PHE is illustrated in Fig. 5.

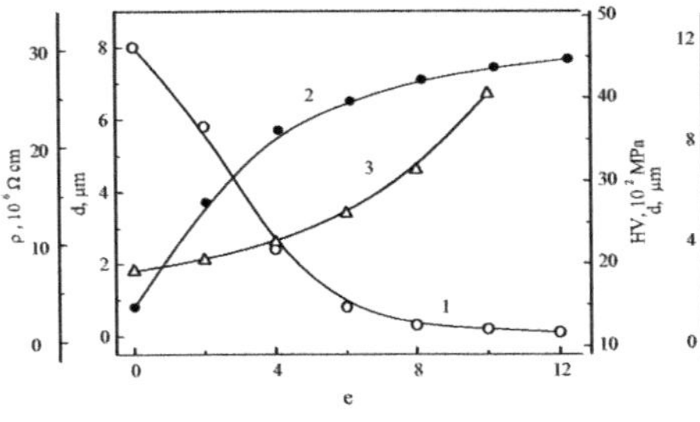

Figure 4. Changes in grain size d (1), HV microhardness (2) and specific resistance (3) versus logarithmic deformation e for nickel samples after PHE

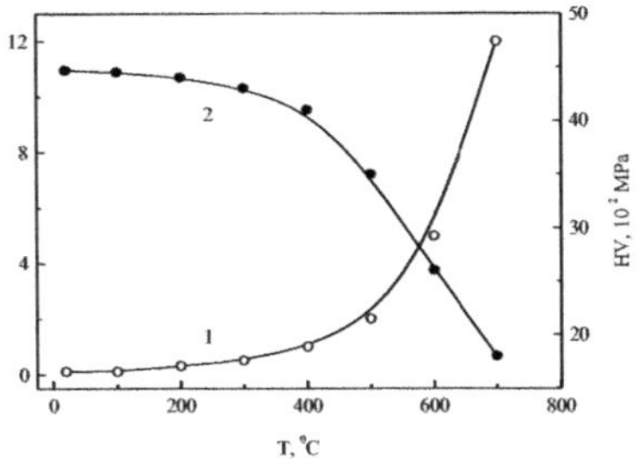

Figure 5. Changes in grain size d (1) and HV microhardness (2) versus tempering temperature T for nickel samples after PHE with logarithmic deformation e = 12

FIM method has been applied to study the structure of defects in UFG nickel at the atomic level. FIM potentials make it possible to investigate real structure of a crystal lattice at the level of separate atoms, to work with atomically pure surface at cryogenic temperatures, and at the same time to analyze the atomic structure of an object by a controllable sequential removal of surface atoms with the help of electric field. The samples to be investigated were in the form of needle-like emitters with the radius of curvature at the apex of 30-50 nm prepared from materials treated by SPD [7].

In Fig. 6 there are ion images of UFG nickel sample surfaces after PHE with logarithmic deformation e = 12.

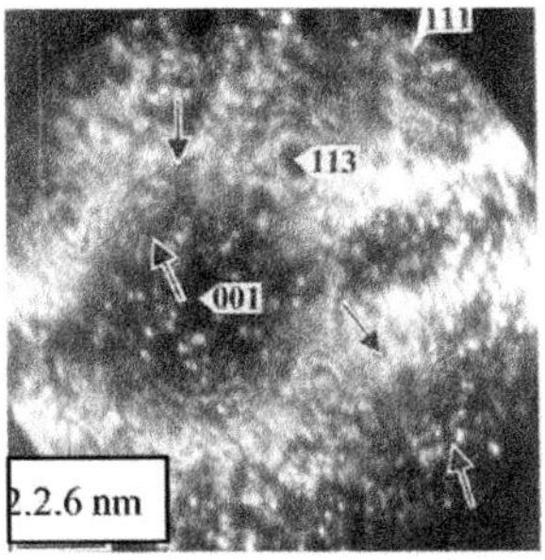

a

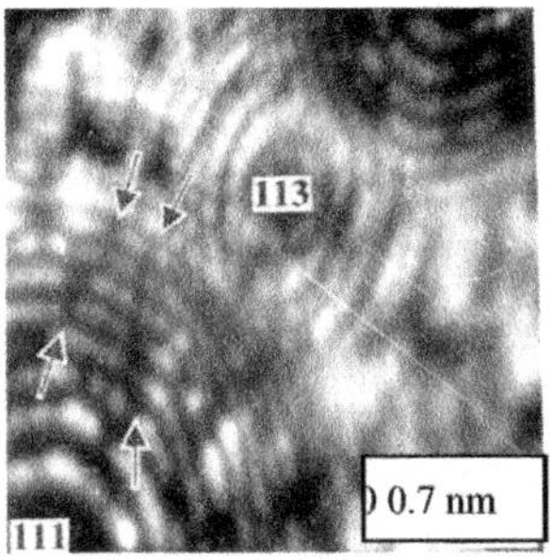

b

Figure 6. Ion field images of nickel atomically smooth surface after removal of: a – 73 atomic layers (26 nm), V = 20 kV; b – 662 atomic layers (240 nm), V = 12.5 kV. The arrows show the boundaries of ultra dispersive blocks, 1 – 10 nm in size

Their analysis of ion image has shown that there were boundaries of ultra dispersive subgrains constituting the microstructure of separate grains. The size of the revealed blocks was estimated both on the surface of ion micro image of UFG nickel samples and during the controllable removal of one atomic layer after another, it constituted from 1 to 10 nm. The subgrains were found to be perfect crystallites practically not disoriented relative each other. In the process of studying the atomic structure of UFG nickel samples at the boundaries of subgrain separation there were the emergent dislocations. The width of boundary region was a distance comparable with the interatomic distance. The obtained results are identical, in the first approximation, to the results of atomic structure investigation for UFG tungsten samples (e = 7) after SPD by high pressure torsion [7].

Conclusion

Investigations have been done of the influence of grains size on the value of nickel hardening during the SPD by PHE method (one-stage route regime). It is shown that the grain-boundary component of the hardening is conditioned by the increase of dislocation density during the deformation. A relationship between the value of strain hardening and the grain size has been derived. It coincides with the Hall - Petch relationship for low values of the degrees of deformation.

The results of investigations of structural and properties changes in nickel under SPD by PHE method (multi-stage route regime) shows with the logarithmic deformation $e \geq 8$, the UFG structure is formed.

It was for the first time when the FIM method gave evidence of a high density of defects in UFG structure of nickel. In particular, an ultra-disperse (1-10 nm) subgrain structure has been detected in the volume of grains. The boundaries of the observed subgrains have been found to be of dislocation nature.

References

1. Gleiter H., *Progr. Mat.* 33 (1989), 223-330.
2. Valiyev R.Z. and Aleksandrov I.V., *Nanostructural materials produced by severe plastic deformation*. Moscow: Logos, 2000, p. 271.
3. Efros B.M., Synkov S.G., and Ivchenko V.A., *High-Pressure Physics and Technology* 12, no 2 (2002), 27-37.
4. Synkov S.G., Synkov V.G., and Sapronov A.N., *High-Pressure Physics and Technology* 6, no 2 (1996), 141-148.
5. Teplov V.A., Pilyugin V.P., and Talutz G.G., *Metals*, 1992, no 2: 109-116.
6. Beygelzimer Y., Varyukhin V., and Efros B., *Physical Mechanics of Hydrostatic Processing of Materials*. Donetsk: DonFTI Publ., 2000, p. 198.
7. Ivchenko V.A., Efros B.M., Popova E.V., and Efros N.B., *High-Pressure Physics and Technology* 13, no 3 (2003), 109-115.

Deformation Mechanisms Inducing Microstructure Refinement in Commercially Pure Aluminium Processed via ECAP: Comparison to Cold-Rolling and Hot-Torsion

M. Cabibbo* and E. Evangelista

INFM / Dipartimento di Meccanica, Università Politecnica delle Marche, Via Brecce Bianche, 60131-Ancona, Italy.

Abstract

The subgrain and grain evolution with the strain of an equal channel angular pressed commercially pure aluminium alloy has been investigated by means of transmission electron microscopy techniques. The deformation was carried out through route C. The process was capable of producing a nano-scale grained microstructure with very fine subgrains in the grain interior, after a true strain of 8. The fraction of high-angle grain boundaries, reached after 8 passes, was of 72%. The microstructure evolution was compared to the one induced by cold rolling. The substructure parameter was practically the same, although, geometrically necessary boundaries and incidental dislocation boundaries, in the material subjected to equal channel angular pressing, showed a lower increasing slope with the strain, basically due to a different deformation mechanism. The equal channel angular pressing microstructure evolution was also compared to the one occurring during hot torsion.

Introduction

Different techniques for producing ultrafine-grained (UFG) materials for structural applications have been introduced and patented, especially in the last decades [1,2]. The advantages of fabricating materials with sub-micron size-grained microstructure as structural components lie in their increased mechanical properties such strength, hardness, ductility, fatigue response and low-temperature superplasticity [3-7]. Among the techniques involving severe plastic deformations to induce a very-fine microstructure, equal channel angular pressing (ECAP) constitutes one of the most promising such techniques. Although ECAP has basically the same potential and microstructure-refining efficiency as cold-rolling (CR), ECAP materials have the crucial advantage of maintaining their shape whatever the number of passes to which the material is subjected to acquire the needed UFG microstructure. This method was introduced and developed in the former Soviet Union by Segal et al. [2].The present paper is devoted to a thorough characterization of microstructure and grain boundary statistics in pure nickel samples fabricated by means of ECAP, HPT and their combination. The grain boundary (GB) surface energy of high angle boundaries was also evaluated.

Systematic studies of the role of ECAP shear on subgrain and grain evolution have been performed only in recent years [4-8]. Several investigations [9,10] have shown that, during deformation, grains in polycrystals subdivide into many small crystallites, each having a crystal orientation rotated to the neighbouring and to the original grains.

Thus, during straining, dislocations generally arrange into a mosaic-like pattern

* Corresponding Author. marcello.cabibbo@univpm.it

Y.T. Zhu and V. Varyukhin (eds.), Nanostructured Materials by High-Pressure Severe Plastic Deformation, 193-198.

in response to both self-stress and to the applied stress. The mosaic-like configuration is basically composed of dislocations arranged in nearly two-dimensional boundaries surrounding regions which are almost dislocation-free. Parameters such as dislocation density and dislocation boundary spacing (pertaining to both subgrains and grains) are generally used for flow stress and grain-subgrain refining mechanism determination [11-15]. Under severe plastic deformation (SPD), dislocation boundaries evolve into a regular pattern of grain subdivisions belonging to two scales [13-15]: large-scale long, continuous dislocation boundaries, called geometrically necessary boundaries (GNBs), and small-scale incidental dislocation boundaries (IDBs) [13]. The boundary spacing and misorientation angle distribution of GNBs and IDBs evolve differently as a function of strain; in particular, they exhibit different morphologies at small to medium strains, but similar at high strains [13-16]. Studies of misorientation angles and of the early subgrain formation process are actually less investigated (e.g. [17,18]).

The present paper reports some results on TEM studies of the early steps of subgrains formation within the original grain networks in severely deformed AA1200 commercially pure aluminium. The deformation mechanism induced by ECAP is also compared with the microstructure evolution generated by cold rolling and hot-torsion.

Experimental Procedures

The chemical composition of the AA1200 commercially pure aluminium is as follows: 0.7Si, 0.3Fe, 0.1Zn, 0.05(Cu+Mn) (wt.%). The material was cast and machined by Hydro Aluminium (Norway) as rod-shaped bars 10 mm in diameter and 10 cm in length.

ECA pressing was carried out at room temperature using a solid die described elsewhere [17,18] and inducing a true strain of $\varepsilon \approx 1$ ($\varepsilon = 1.055$) at each pass [1-3,5]. Deformation route C, which involve a 180° billet rotation at every pass, was used and ECAP was carried out for a maximum of 8 passes.

Microstructure inspections were carried out along the ED-TD plane (i.e. the extrusion-transverse direction plane). Specimen preparation for either optical microscopy and TEM and inspection details are described elsewhere [17,18].

Results and Discussion

The shear bands deformation configuration, introduced by the ECAP through route C, is shown in Fig. 1 which constitutes a compound polarised optical microscope image of a specimen already subjected to 1 pass and in the process of being pressed the second time.

Microstructure evolution during ECAP

Fig. 2 (a) to (e) shows some TEM images documenting the microstructural evolution from the 1st to the 8th pass. After 1 ECAP pass the microstructure exhibited most of the GNBs (basically high angle boundaries, HABs), originated from the induced shear bands, oriented along the deformation shear direction, i.e. at an angle of ~45° to the pressing direction.

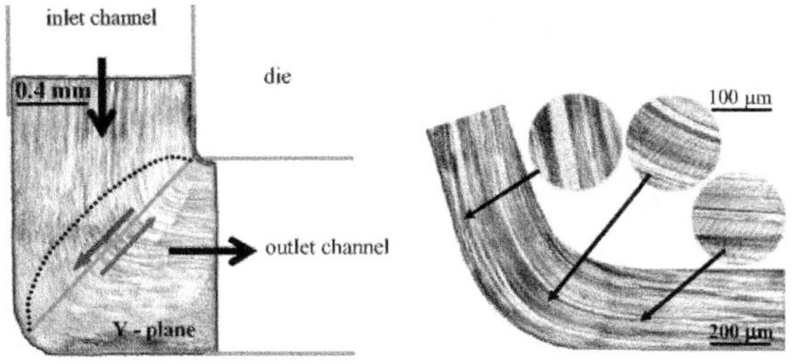

Figure 1. POM image of a specimen, already subjected to 1 pass, and introduced in the ECAP die for the second pass using route C (a 180° rotation along the specimen longitudinal axis). In the left-handed image, the line indicates the zone of the maximum shear deformation, the arrows indicate the shear direction for the odd (2N+1) and even (2N) number of passes.

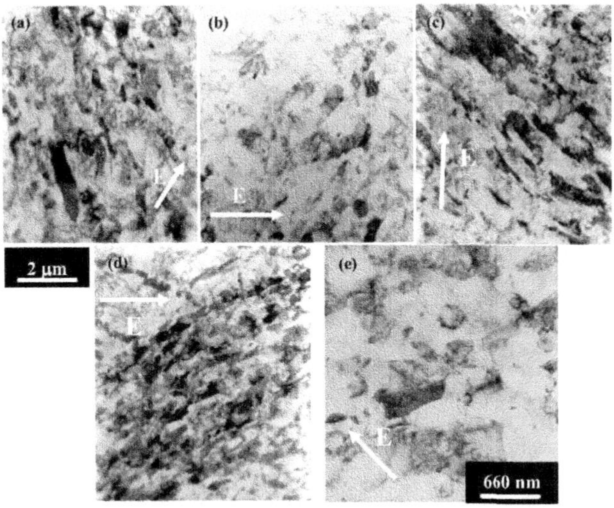

Figure 2. TEM image of the material subjected to 1 (a), 2 (b), 5 (c), 7 (d) and 8 passes (e). E indicates the extruded direction (ECAP pressing direction).

The spacing between bands, i.e. HABs, ranged 600-800 nm. After 2 passes an almost equiaxial grain and subgrain structure was formed. A high density of dislocation tangles was also observed in the interior of either the regions surrounded by GNBs and original grains, a microstructural characteristic exhibited after each even pass. After the 5th pass, the microstructure was characterized by a considerable size reduction of grains (regions surrounded by GNBs and having HABs character) down to about 500 nm. Thus, a sub-micrometer-grained microstructure was achieved already after 5 passes via route C. The 8th pass reduced mean grain and subgrain size to a lesser extent, while the HAB fraction rose to ~72% of the total amount of boundaries.

During plastic deformation, grains tend to subdivide into high angle blocks with different combinations of slip systems and whose lateral spacing is quite similar to the shear bands, that controls the deformation process. Eventually, boundaries generated between these blocks. As deformation progresses, the blocks are induced to rotate to different end orientations, the boundaries between the blocks then evolve into HABs. Such an out-of-sequence strain reversal, along the same shear plane, makes it more difficult for the high angle blocks (which are essentially the new fine grains) to rotate back to the initial state. This also contributes to induce an almost dislocation-free grain interior microstructure at every odd passes, the dislocations being arranged geometrically to form cell boundaries as well as to thicken grain boundaries. As a counterpart, a large number of tangles dislocation forms at every following even pass, as illustrated in Fig. 3. Thus, each even pass constitutes an incubation-like step toward the low-angle and high-angle boundary refinement process induced by the next ECA odd pressing.

a) 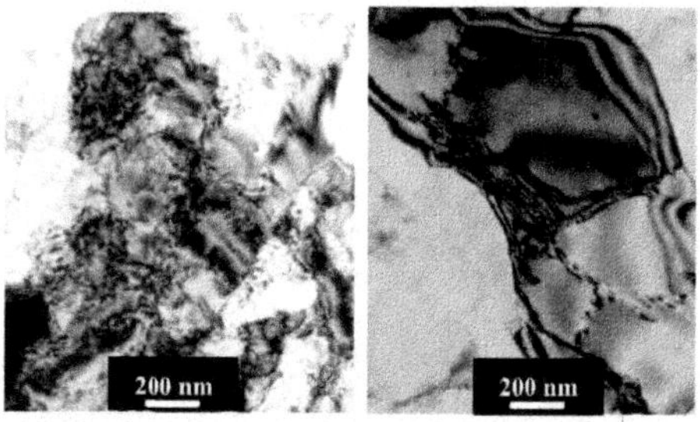 b)

Figure 3. Typical microstructure of a specimen subjected to an even number of passes (4) (diffuse presence of tangle dislocations in the subgrains and grains interior) (a), and to an odd number of passes (5) (dislocation arranged to form low angle boundaries) (b).

Microstructure refinement process: ECAP vs. Cold-Rolling

Fig. 4 (a) is a plot of the high and low angle boundary fraction (GNBs and IDBs) as a function of the strain. The high angle fraction reached a value of 72% after the maximum strain of 8, and a saturation limit seemed to be reached. Note that the data point have a waving increasing trend due to the mechanism of formation of high density of tangle dislocations in both subgrains and grains interior at every even passes. As already discussed, this phenomenon is basically due to the changing of the shear direction from the previous odd pass. Fig. 4 (a) also reports the substructure evolution which has been compared to the one of same alloy subjected to cold-rolling (data after Hansen (1969) and Shun (1974)[19]). The substructure parameter is defined as: $\Delta = 1/(1/\delta + 1/d)$, where d represents the mean grain size, and δ the subgrain size. The data were plotted in normalized form to the 1st ECAP pass for the ECA-pressed materials, and to an equivalent strain of $\varepsilon = 1.5$ for the CR material (corresponding to a mean spacing of 1.0 μm) [19]. The evolution in the common range of strain for both ECAP and cold-rolling practically overlapped. This denotes a quite similar microstructure refinement mechanism in both cases. After already a deformation of 1 the substructure reached a value below 1, which eventually slowed down to an asymptotic-like limit of 0.6 - 0.7.

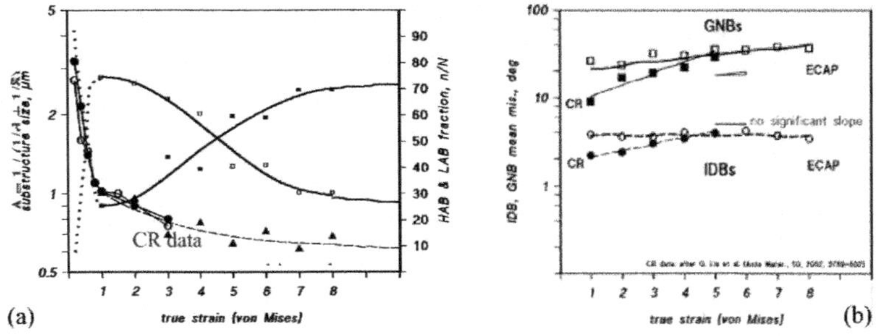

Figure 4. HABs (GNBs and original grain boundaries) and LABs (essentially IDBs) as a function of strain, and comparison of the substructure parameter (Δ) as a function of strain for either ECAP and CR (the latter are data by Hansen (1969) and Suhn (1974)) (a). Comparison of GNBs and IDBs mean misorientation as a function of strain in the case of ECAP and CR of the same alloy (CR data by Hansen et al. [16]) (b).

Fig. 4 (b) shows the misorientation plot of either GNBs (basically high angle boundaries, constituted by newly formed fine grain and original elongated grain boundaries) and IDBs (low angle boundaries) as a function of strain. IDBs have a trend practically flat, whilst GNBs revealed a constant positive trend whose slope was calculated to be 0.44. Hansen et al. [16] for CR of the same 1000 aluminium alloy series reported values for GNB and IDB of 0.7 and 0.4, respectively. This difference between ECAP and CR may be justified as follows: by increasing the strain (i.e. the number of passes), more and more shear bands are introduced in the microstructure. These shear bands propagate within the already elongated original grains and into the previously formed grains and subgrains, thus most of the IDBs are replaced by GNBs. The former remain practically constant in misorientation with increasing strain. This deformation process accounts for the lower increasing of GNBs misorientation compared to plain strain induced by CR. The process of transformation of IDBs into GNBs is rather slower in CR due to the shear planes rotating toward the rolling direction as the deformation proceeds. As a matter of facts, GNBs behave as strong barriers to the slip due to their high dislocation concentration, while, the lower angle IDBs are penetrable by mobile dislocations. This way, the flow may take place within the channels formed by the GNBs with a slip length related to the distance between them.

Subgrains formation and evolution with strain: ECAP vs. Hot Deformation

When Al is deformed above 400°C (in the range 0.4 to $0.6T_m$) it develops subgrains that maintain a constant size. As the grains elongate through strains up to over 8, low-angle boundaries rearrange to maintain the subgrains equiaxed, whilst, the original boundaries are not clearly visible because they have become serrated. The serrations in neighbouring grain boundaries may touch as grains become thin causing them to pinch off into shorter grains. The high-angle GNB blocks observed in CR and ECAP (blocks typically wide 5μm and containing subgrains wide 0.5μm) (see Fig. 2 (d)) are generally not observed in hot deformation, where subgrain boundary misorientation saturates at about 4° [20,21]. The substructure evolution is intermediate between cold and hot working [20,21].

Indeed, the microstructural evolution consistently differentiates from the one in

ECAP where subgrains continuously keep rearranging and forming from tangles dislocations. Moreover, subgrains, formed in ECAP, tend to align and elongate, as well as the grains, along the pressing direction.

Conclusion

The grain and subgrain evolution with strain of an 1200 aluminium alloy subjected to ECAP, via route C, was investigated by means of TEM techniques.

A sub-micrometer-grained microstructure was achieved after 5 passes and the maximum fraction of HAB reached the value of ~72% after a strain of 8.

The evolution of the substructure parameter for both ECAP and cold-rolling overlapped. Yet, GNBs and IDBs misorientation evolution with strain was considerably different denoting a different mechanism of formation and evolution of low and high angle boundaries between ECAP and CR. Concerning the high-temperature deformation, as true strain increases the subgrain spacing remains practically unchanged maintaining an equiaxed morphology.

Acknowledgements
This study was supported by the Italian Ministry of Research (MIUR).

References

1. Zhu Y.T., Lowe T.C., *Mater. Sci. Eng.* A291 (2000), 46.
2. Segal V.M., Reznikov V.I., Drobyshevskiy A.E., Kopylov V.I., *Russian Metallurgy*, 1 (1981), 99.
3. Segal V.M., *Mater. Sci. Eng.* A197 (1995), 157.
4. Berbon P.B., Tsenev N.K., Valiev R.Z., Furukawa M., Horita Z., Nemoto M., Langdon T.G., *Metall. Mater. Trans.* A29 (1998), 2237.
5. Golina A., Prangnell P.B., Markushev M.V., *Acta Mater.* 48 (2000), 1115.
6. Chang J.Y., Yoon J.S., Kim G.H., *Scripta Mater.* 45 (2001), 347.
7. Shan A., Moon I.G., Park J.W., *J. Mater. Proc. Technology*, 122 (2002), 255.
8. Mishin O.V., Juul Jensen D., Hansen N., *Mater. Sci. Eng.*, 342 (2003), 320.
9. Wert J.A., Liu Q., Hansen N., *Acta Mater.*, 45 (1997), 2565.
10. Skalli A., Fortunier R., Fillit R., Driver J.H., *Acta Metall.*, 33 (1985), 997.
11. Hughes D.A., *Mater. Sci. Eng.*, A319-321 (2001), 46.
12. Hughes D.A., Liu Q., Chrzan D.C., Hansen N., *Acta Mater.*, 45 (1997), 105.
13. Godfrey A., Hughes D.A., *Mater. Charact.*, 48 (2002), 89.
14. Sun P.L., Kao P.W., Chang C.P., *Mater. Sci. Eng.*, A283 (2000), 82.
15. Hansen N., Juul Jensen D., *Acta Metall.*, 40 (1992), 3265.
16. Liu Q., Huang X., Lloyd D.J., Hansen N., *Acta Mater.* 50 (2002), 3789.
17. Cabibbo M., Evangelista E., ICAA-9, ed. IMEA, Brisbane, Australia, 2004, 190.
18. Cabibbo M., Evangelista E., Latini V., *J. Mater. Sci.*, 39 (2004), 1.
19. Nes E., Dons A.L., Ryum N., ICSMA 6, ed. R.C. Gifkins, Pergamon Press, 1982, 425.
20. McQueen H.J., Blum W., *Mat. Sci. Eng.*, A290 (2000), 95.
21. Nes E., *Prog. Mat. Sci.*, 41 (1998), 129.

IV. MECHANICAL PROPERTIES

The Deformation Characteristics of Pure Aluminum Processed by Equal-Channel Angular Pressing

C. Xu[1], M. Furukawa[2], Z. Horita[3] and T.G. Langdon[1,*]

[1]Departments of Aerospace & Mechanical Engineering and Materials Science University of Southern California, Los Angeles, CA 90089-1453, USA

[2]Department of Technology, Fukuoka University of Education, Munakata Fukuoka 811-4192, Japan

[3]Department of Materials Science and Engineering, Faculty of Engineering Kyushu University, Fukuoka 812-8581, Japan

Abstract

It is now well-established that the process of equal-channel angular pressing (ECAP) leads to substantial grain refinement in polycrystalline metals and to the production of materials having unusual and useful properties including high strength and a capability for superplastic forming. This paper examines the fundamental deformation characteristics associated with the application of ECAP to samples of pure aluminum. Results are presented describing the characteristics associated with the pressing of aluminum single crystals and the development of a homogeneous microstructure in polycrystalline aluminum.

Introduction

Several procedures are now available for the processing of metals through the application of severe plastic deformation (SPD) [1] although the most attractive appear to be Equal-Channel Angular Pressing (ECAP) [2-4] and High-Pressure Torsion (HPT) [5,6]. In ECAP the sample, in the form of a rod or bar, is pressed through a die under a high pressure whereas in HPT the sample, in the form of a disk, is subjected to a high pressure and concomitant torsional straining. In practice, processing by ECAP has the advantage that it produces reasonably large bulk samples and it can be scaled fairly easily to larger sizes [7]. Accordingly, significant attention has been devoted to this processing method.

An attractive feature of SPD processing is the ability to produce an ultrafine grain size in bulk metals [8,9]. Typically, the grain sizes produced by SPD are in the submicrometer range but under some conditions, especially through the use of HPT, the grains may be reduced to the nanometer range. These small grain sizes lead to a high strength at ambient temperatures through the Hall-Petch relationship [10,11] and they provide a potential for achieving superplasticity, and a superplastic forming capability, at high temperatures [12,13]. Examples of this high strength are readily available for a range of aluminum-based alloys [14] and there are examples of the occurrence of high strain rate superplasticity in Al alloys after ECAP [15,16].

This paper examines the fundamental deformation characteristics of pure aluminum when processing by ECAP. Examples are presented for the pressing of single crystals and the homogeneity associated with the ECAP processing of polycrystalline pure aluminum.

---

* Corresponding Author. langdon@usc.edu

Y.T. Zhu and V. Varyukhin (eds.), Nanostructured Materials by High-Pressure Severe Plastic Deformation, 201-208.

Experiments on Single Crystals

An earlier report provides a detailed description of the experiments conducted on an aluminum single crystal [17]. Briefly, a single crystal of high-purity (99.999%) aluminum was grown from the melt in the form of a rod with a diameter of ~20 mm and a length of ~70 mm. The crystallographic orientation was determined using the Laue back-reflection method and a sample was cut for ECAP using electro-discharge machining. This sample was square in cross-section with dimensions of 4×4 mm^2 and a length of ~30 mm. Prior to ECAP, the sample faces were polished to a mirror-like finish.

Figure 1 shows a section through the ECAP die and illustrates the orientation of the single crystal. The ECAP die contains a single channel which is bent through an angle of $\Phi = 90°$ near the center of the die. In addition, although not shown in Fig. 1, there was an outer arc of curvature at the point of intersection of the two parts of the channels given by $\Psi \approx 30°$. Three orthogonal axes are depicted in Fig. 1 where X is the transverse plane perpendicular to the direction of flow, Y is the flow plane parallel to the side face and Z is the longitudinal plane parallel to the top face at the point of exit from the die, respectively. The two solid arrows in Fig. 1 denote the direction of pressing. The theoretical shear plane depicted within the crystal lies at 45° to the X direction at the point of intersection of the two parts of the channel. Since it is anticipated that slip will occur on the {111}<110> system, the crystal is oriented so that the {111} slip plane lies parallel to the theoretical shear plane, the [110] slip direction lies parallel to the direction of shear and the [112] direction lies along the Y axis. For simplicity, it is convenient to designate this condition, where the slip direction and the shearing direction are coincident, as the 0° orientation.

The magnitude of the resolved shear stress may be deduced by using Fig. 2 which illustrates the configurations of both the slip and shear directions and the slip and shear planes. Since the specimen is pressed in a vertical sense, the shear stress, τ, operating on the slip plane in the slip direction may be estimated from the expression $(F/A) \cos \theta \cos \lambda$, where the angles θ and λ are defined in Fig. 2 and F and A are the force in the shear direction and the area of the shear plane, respectively.

The single crystal was pressed through the split die illustrated in Fig. 1 at room temperature using MoS_2 as a lubricant. It can be shown from first principles that, for an ECAP die having $\Phi = 90°$ and $\Psi \approx 30°$, the strain imposed in each passage through the die is ~1 [18]. In addition, model experiments [19] and experiments on polycrystalline samples [20] have shown that no significant strain inhomogeneities are introduced by using a die where $\Psi \approx 20° - 30°$ because of the development of a "dead zone" at the outer corner where the sample is no longer in contact with the die wall [21,22].

The single crystal was pressed to the approximate mid-point of the die and then removed from the die for inspection. The sample was sectioned longitudinally on the Y plane into two approximately equal pieces and the exposed surface was carefully polished and etched for inspection by optical microscopy. The result is shown in the upper portion of Fig. 3 where this section represents the Y plane, the unsheared portion of the crystal lies vertically on the left and the sheared portion lies horizontally on the right. Also visible in Fig. 3 at the lower left is the outer arc of curvature within the die where $\Psi \approx 30°$. The four lower optical photomicrographs in Fig. 3 were taken at the various points labeled 1 to 4 marked on the surface of the crystal in the upper portion.

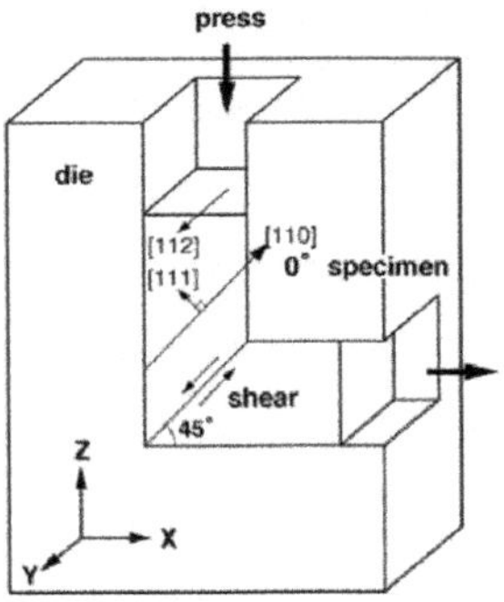

Fig. 1 Schematic illustration of cross-section through ECAP die.

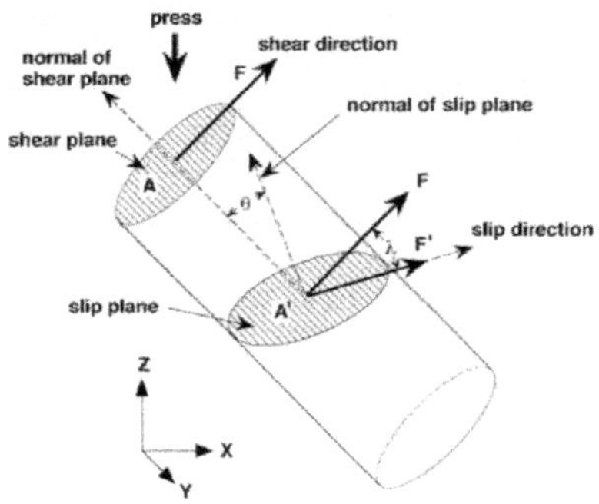

Fig. 2 Schematic illustration of method for estimating shear stress τ.

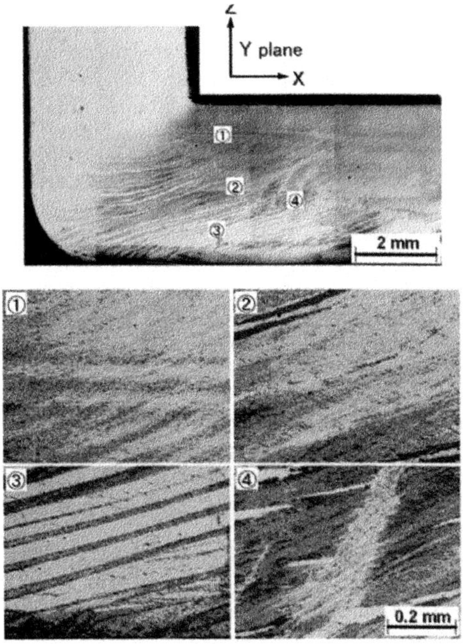

Fig. 3 Cross-section of single crystal after pressing approximately half-way through the die (upper): four optical photomicrographs from the points labeled 1 to 4 (lower).

Close inspection of the photomicrographs shown in the lower portion of Fig. 3 reveals several significant trends. At position 1, which is close to the shear plane, there are long horizontal striations or deformation bands at low magnification and additional thin striations at angles of 30° and 60° with respect to the horizontal X direction; at position 2 near the center of the deformed crystal, there are prominent striations at 20° and very thin striations at 50°; at position 3 near the bottom surface, there are very sharp striations at 15°; and at position 4 within the deformed region in the exit channel of the die, there are prominent striations or shear bands at 65° and additional striations at 20°. These results, and other extensive observations from these experiments on single crystals, show that deformation bands develop very close to the theoretical shearing plane in the bulk of the single crystal but different shearing characteristics occur near the bottom surface of the crystal where these differing characteristics are probably due to the influence of frictional effects in the vicinity of the die wall. The results in the bulk of the crystal are therefore consistent with the crystallographic expectations and they suggest that the intense plastic deformation introduced in ECAP leads, in the first pass through the die, to the presence of glissile dislocations on the primary slip plane. This information will be used later to develop a model for grain refinement in ECAP.

Experiments on Polycrystals

Experiments were conducted using high purity (99.99%) aluminum supplied in the form of an ingot: earlier reports described several fundamental investigations using this material [23,24]. In the present experiments, the material was swaged into rods having a diameter of 10 mm and cut to lengths of ~70 mm. Each rod was annealed for 1 hour at 773 K to give an initial grain size of ~1 mm. These rods were processed by ECAP using a die having internal angles of $\Phi = 90°$ and $\Psi = 20°$. Each rod was pressed at room temperature (298 K) up to a total of 8 passes using processing route B_C in which the rod is rotated by 90° in the same sense about the longitudinal axis between each separate pass [25]. Following ECAP, the billets were sectioned perpendicular to the longitudinal axes and then mounted and polished to a mirror-like finish. Microhardness measurements were taken on the polished surfaces using a Vickers indenter, with the values of the microhardness, Hv, recorded in a regular grid pattern with spacings between each measurement of ~0.5 mm. These values were then plotted in the form of contour maps to depict the variations of the local hardness over the cross-sectional planes [20].

An example is shown in Fig. 4 where the contour lines designate the perimeters of areas having similar hardness values: the map at upper left depicts the as-received and unpressed condition and the remaining maps from upper right denote the values after pressing through 1, 2, 3, 4 and 8 passes, respectively. These plots confirm the presence of a homogeneous microstructure in the as-received condition but the microhardness increases significantly after a single pass and there is inhomogeneity because of the presence of a broad region of lower hardness in the vicinity of the lower surface. This inhomogeneity is essentially removed on subsequent passes through the die. In Fig. 5, (a) and (b) show the presence of elongated and equiaxed grains near the center of the specimen and (c) and (d) show similar grains near the lower surface. However, the equiaxed grains occupied ~50% of the total area near the center but only ~10% near the bottom surface. Thus, the inhomogeneity occurring near the bottom surface of the channel is consistent with the observations in Fig. 3.

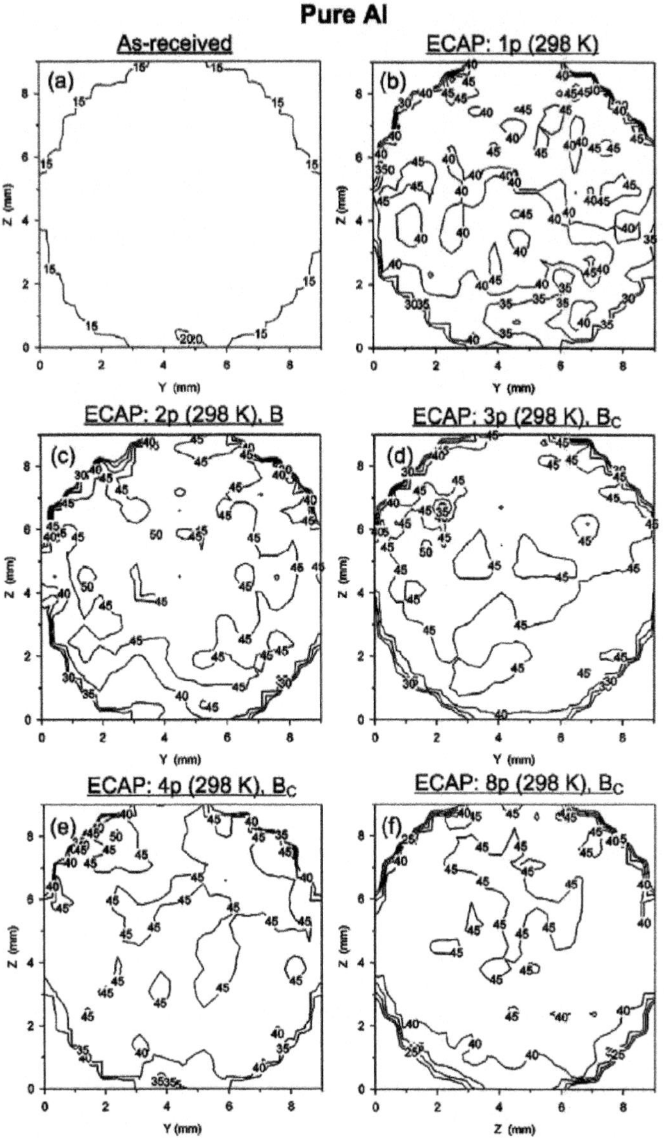

Fig. 4 Contour maps showing the Vickers microhardness through the cross-sections of pure aluminum billets in the unpressed condition and after pressing through 1, 2, 3, 4 and 8 passes at room temperature using route B_C.

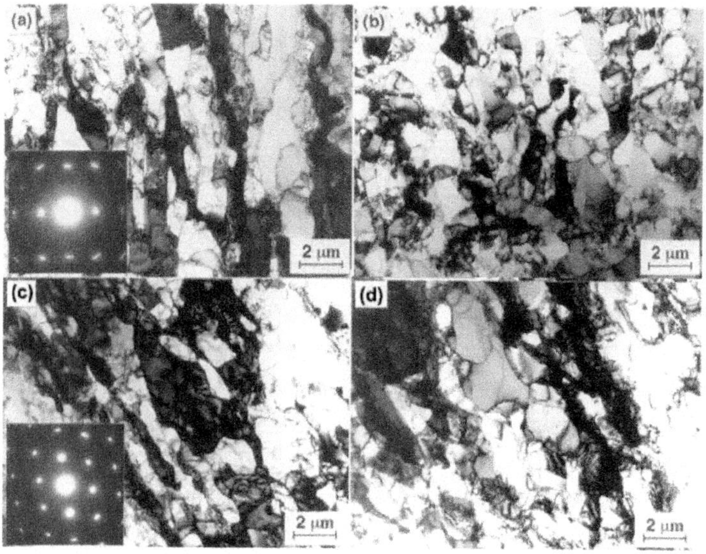

Fig. 5 Representative microstructures of pure aluminum after ECAP through one pass at room temperature showing regions of (a) elongated grains and (b) equiaxed grains in the central part of the sample and (c) elongated grains and (d) equiaxed grains near the bottom surface.

The increase in hardness on the first pass through the die is shown in Fig. 6 where the average value of Hv is plotted against the number of passes in ECAP. Thus, the hardness increases by a factor of approximately 2× in the first pass but thereafter there is no significant change in the value of Hv with additional passes through the die.

The results from these and other similar experiments may be used to develop a model for the process of grain refinement in ECAP. This is illustrated schematically in Fig. 7 where the microstructures are shown on the X plane after totals of 1, 2 and 4 passes: the orientation with respect to the three orthogonal X, Y and Z planes is indicated by the small inset between the top illustrations using the orientations already defined in Fig. 1 and the parameter ξ represents the fraction of high-angle boundaries measured in pure aluminum using orientation imaging microscopy [26]. It is apparent that a banded structure, having a width of d, is formed along the Y direction in the first pass and, because of the rotation associated with route B_C, this banded array is rotated by ~27° with respect to the Y axis and a new array is formed parallel to the Y direction in the second pass: this and other angles of rotation have been estimated for various processing routes in ECAP [25,27]. Finally, the microstructure evolves after 4 passes into an array of essentially equiaxed grains with an average grain size, d, equal to the width of the initial banded structure formed in the first pass.

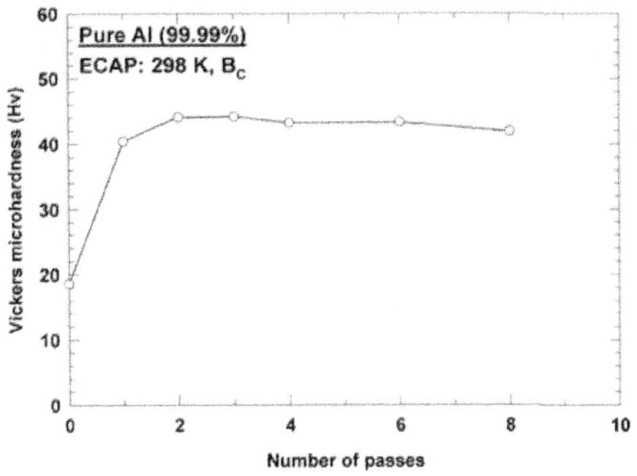

Fig. 6 Average Vickers microhardness, Hv, versus number of passes in ECAP for pure aluminum processed at room temperature using route B_C.

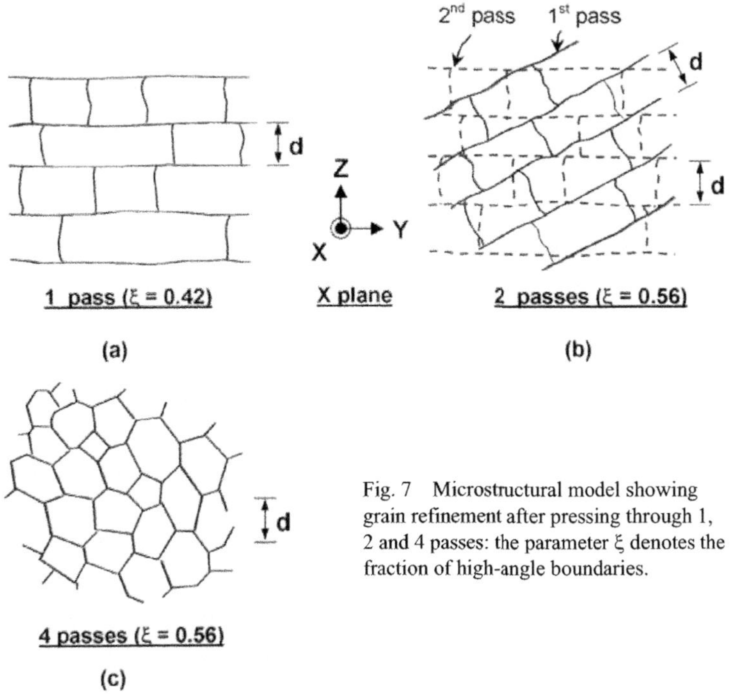

Fig. 7 Microstructural model showing grain refinement after pressing through 1, 2 and 4 passes: the parameter ξ denotes the fraction of high-angle boundaries.

Summary and conclusions

1. Detailed experiments have been conducted on single crystal and polycrystalline aluminum to evaluate the microstructural characteristics after processing by equal-channel angular pressing.
2. These two sets of data are mutually consistent and they permit the development of a microstructural model describing the development of grain refinement in ECAP.

Acknowledgements

This work was supported in part by a Grant-in-Aid for Scientific Research from the Ministry of Education, Science, Sports and Culture of Japan, in part by the Light Metals Educational Foundation of Japan and in part by the National Science Foundation of the United States under Grant No. DMR-0243331.

References

1. Valiev R.Z., Islamgaliev R.K., Alexandrov I.V., *Prog. Mater. Sci.* 45 (2000), 103-189.
2. Segal V.M., *Mater. Sci. Eng.* A197 (1995), 157-164.
3. Segal V.M., *Mater. Sci. Eng.* A271 (1999), 322-333.
4. Furukawa M., Horita Z., Nemoto M., Langdon T.G., *J. Mater. Sci.* 36 (2001), 2835-2843.
5. Zhilyaev A.P., Nurislamova G.V., Kim B.-K., Baró M.D., Szpunar J.A., Langdon T.G., *Acta Mater.* 51 (2003), 753-765.
6. Vorhauer A., Pippan R., *Scripta Mater.* 51 (2004), 921-925.
7. Horita Z., Fujinami T., Langdon T.G., *Mater. Sci. Eng.* A318 (2001), 34-41.
8. Valiev R.Z., Krasilnikov N.A., Tsenev N.K., *Mater. Sci. Eng.* A137 (1991), 35-40.
9. Valiev R.Z., Korznikov A.V., Mulyukov R.R., *Mater. Sci. Eng.* A168 (1993), 141-148.
10. Hall E.O., *Proc. Roy. Soc. B* 64 (1951), 747-753.
11. Petch N.J., *J. Iron Steel Inst.* 174 (1953), 25-28.
12. Xu C., Furukawa M., Horita Z., Langdon T.G. *Adv. Eng. Mater.* 5 (2003), 359-364.
13. Xu C., Furukawa M., Horita Z., Langdon T.G., *J. Alloys Compounds* 378 (2004), 27-34.
14. Horita Z., Fujinami T., Nemoto M., Langdon T.G., *Metall. Mater. Trans. A* 31A (2000), 691-701.
15. Valiev R.Z., Salimonenko D.A., Tsenev N.K., Berbon P.B., Langdon T.G., *Scripta Mater.* 37 (1997), 1945-1950.
16. Horita Z., Furukawa M., Nemoto M., Barnes A.J., Langdon T.G., *Acta Mater.* 48 (2000), 3633-3640.
17. Fukuda Y., Oh-ishi K., Furukawa M., Horita Z., Langdon T.G., *Acta Mater.* 52 (2004), 1387-1395.
18. Iwahashi Y., Wang J., Horita Z., Nemoto M., Langdon T.G., *Scripta Mater.* 35 (1996) 143-146.
19. Wu Y., Baker I., *Scripta Mater.* 37 (1997), 437-442.
20. Xu C., Langdon T.G., *Scripta Mater.* 48 (2003), 1-4.
21. Kim H.S., Seo M.H., Hong S.I., *Mater. Sci. Eng.* A291 (2000), 86-90.
22. Kim H.S., *Mater. Sci. Eng.* A315 (2001), 122-128.
23. Iwahashi Y., Horita Z., Nemoto M., Langdon T.G., *Acta Mater.* 45 (1997), 4733-4741.
24. Iwahashi Y., Horita Z., Nemoto M., Langdon T.G., *Acta Mater.* 46 (1998), 3317-3331.
25. Furukawa M., Iwahashi Y., Horita Z., Nemoto M., Langdon T.G. *Mater. Sci. Eng.* A257 (1998), 328-332.
26. Terhune S.D., Swisher D.L., Oh-ishi K., Horita Z., Langdon T.G., McNelley T.R., *Metall. Mater. Trans. A* 33A (2002), 2173-2184.
27. Furukawa M., Horita Z., Langdon T.G., *Mater. Sci. Eng.* A332 (2002), 97-109.

Microstructural Aspects of Cyclic Deformation and Fatigue of Ultrafine-Grained Metals

Haël Mughrabi*, Heinz Werner Höppel and Martin Kautz
Institut für Werkstoffwissenschaften, Lehrstuhl 1, Universität Erlangen-Nürnberg, Martensstr. 5, D-91058 Erlangen, Germany

Abstract

In recent years, the cyclic deformation and fatigue behaviours of different model and commercial ultrafine-grained (UFG) metals prepared by severe plastic deformation (SPD) techniques have been studied. The present contribution is based largely on the authors' recent work on fatigued UFG high-purity copper, commercial purity Al, the age-hardenable aluminium alloy AA6061 and α-brass, all prepared by ECAP (equal channel angular pressing). Emphasis is laid on the microstructural mechanisms governing cyclic microplasticity, cyclic deformation, the cyclic stress-strain (CSS) response, as displayed in the CSS-curve, fatigue damage and fatigue life. The roles of fatigue strength, fatigue ductility and microstructural stability of UFG materials in controlling fatigue life in high-cycle and in low-cycle fatigue will be discussed.

Introduction

In the last few years, it has been demonstrated impressively that ultrafine-grained (UFG) metals with grain sizes in the submicrometer range, which can be produced by a number of different techniques of Severe Plastic Deformation (SPD), exhibit remarkable mechanical properties. Among the SPD techniques, equal channel angular pressing (ECAP) is of particular interest, since it allows the production of bulk specimens of macroscopic dimensions, compare [1,2,3]. Several recent studies have considered and modeled in detail the ECAP forming process in terms of dislocation processes, continuum mechanics and by finite element techniques, as described e.g. in [4,5,6]. None of these studies, however, attempt or are able to predict the evolution of the grain refinement, compare [5]. In this context, it is worthwhile to recall that Buck and Essmann, already in 1964 [7], made the interesting TEM (transmission electron microscopy) observation that, after low-temperature tensile deformation of a copper single crystal, large-angle grain boundaries (GBs) had formed in addition to the other deformation-induced dislocation substructures, such as dislocation tangles and/or cells/subgrains that are commonly observed. It is proposed that the phenomenon found by these authors is actually the basic process underlying grain refinement, namely the generation of substantial misorientations through the formation of GBs. With respect to the dependence of the unidirectional and cyclic mechanical properties on grain (or subgrain/cell) size, grains bounded by large-angle GBs and dislocation cells/subgrains should be distinguished more clearly than in earlier studies.

As a result of their fine microstructure, UFG materials produced by ECAP can possess a very high static strength, paired with a quite remarkable ductility [8]. In addition, under service conditions involving cyclic loading, good fatigue properties

* Corresponding Author. mughrabi@ww.uni-erlangen.de

Y.T. Zhu and V. Varyukhin (eds.), Nanostructured Materials by High-Pressure Severe Plastic Deformation, 209-216.

are desirable in the LCF (low-cycle fatigue) and in the HCF (high-cycle fatigue) regimes. Earlier work, initially mainly on fatigued UFG copper [9,10,11,12,13], showed that while the fatigue lives of UFG material are enhanced in the HCF range, compared to the behaviour of conventional grain size (CG) material, fatigue lives in the LCF range are reduced. This behaviour can be rationalized in terms of the enhanced fatigue strength, responsible for the HCF performance, as expressed by the fatigue strength coefficient σ_f', and the reduced fatigue ductility, governing the LCF life, expressed by the fatigue ductility coefficient ε_f', as explained in [13], see [14] for fundamentals of fatigue relationships. While many UFG materials show a qualitatively similar behaviour under cyclic loading conditions, the details differ from case to case, as summarized by Vinogradov and Hashimoto [15] and in a recent brief review [16] and in earlier publications of the authors [17,18,19,20].

Experimental Details

In the authors' research group, fatigue studies were performed on the following different UFG materials that had been produced by the ECAP technique:

- Cu, purity 99.99%, grain size ca. 300 nm [12,17,19]
- α-brass, grain size ca. 750 nm [18]
- Al, commercial purity, grain size ca. 800 nm [17,18,20]
- the age-hardenable aluminium alloy AA6061, overaged condition, grain size ca. 450 nm [20].

The Cu and α-brass UFG materials were produced in Prof. R.Z. Valiev's laboratory in Ufa, Russia, while the Al and the alloy AA6061 specimens were produced in Prof. T.G. Langdon's laboratory at the University of Southern California in Los Angeles. Further details concerning the specimen preparation by ECAP, etc., can be found in the cited references. All fatigue tests were performed in symmetric push-pull on a servohydraulic MTS 810 system in either total strain control with constant plastic strain amplitude $\Delta\varepsilon_{pl}/2$ or in stress control at constant stress amplitude $\Delta\sigma/2$. Subsequently, fatigue-induced microstructural changes and damage were studied by TEM and SEM (scanning electron microscopy).

Results and Discussion

Inelastic reverse/forward strains upon unloading/reloading in the microyield regime

Since the plastic strain amplitudes in fatigue are typically low, lying in the "microyield" range of ca. 10^{-5} to 10^{-2}, it appeared interesting to investigate first the tensile microyielding loading/unloading behaviour on UFG Cu [13,16,17,18,19] and on UFG Al [16,17,18,20] specimens and, for comparison, also on the corresponding CG specimens with grain sizes of ca. 40-45 µm. Here, some typical results will be summarized and interpreted in terms of a recently developed composite model [21].

In Figs. 1a and 1b, the microyielding loading and unloading behaviours of UFG and CG Al are compared in plots of axial stress σ vs. plastic strain ε_{pl}. It is apparent that the UFG material exhibits much more pronounced inelastic reverse (forward) yielding during unloading (loading) than the CG material. This effect is attributed to the development of larger deformation-induced internal back (or forward) stresses in the UFG material compared to the CG material. A very simple composite model in which the UFG material is assumed to consist of soft and hard regions has been de-

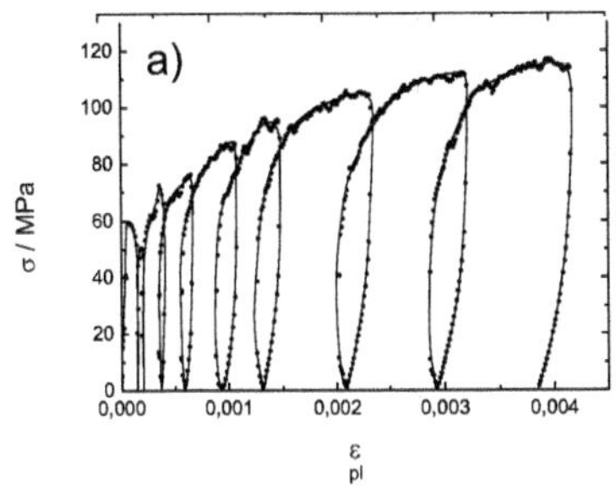

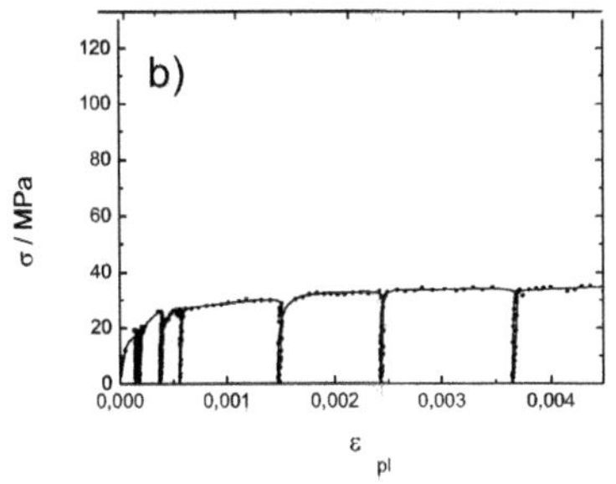

Figure 1. Unloading and reloading tensile stress-strain curves in the microyield range. a) UFG Al. b) CG Al. From [16,20].

veloped [21] in order to analyze the microyielding behaviour. From this analysis, it could be concluded that the reason for the higher internal stresses lies in the fact that the hard regions are not only much harder than the soft regions but that, in addition, they occupy a significantly larger volume fraction than the soft regions. While the latter result seemed somewhat surprising, it finds a natural explanation, when one considers that plastic yielding probably begins by emission of dislocations from the GBs. Subsequently, these dislocations can move easily into the grain interiors which occupy a larger volume fraction and which are initially harder, being virtually free of dislocations before the onset of yielding. This interpretation of the results obtained in the analysis of the composite model was discussed in detail in ref. [21].

Effects of grain size and purity on cyclic deformation and stress-strain response

In strain-controlled cyclic deformation studies on CG materials, the cyclic stress amplitude frequently saturates after a larger number of cycles, and the cyclic saturation stress does not depend very markedly on grain size D, compared to the much stronger grain size dependence observed in unidirectional deformation. Thus, a plot of the cyclic saturation stresses versus the plastic strain amplitude, i.e. the so-called cyclic stress-strain curve (CSSC), is only weakly dependent on grain size D, as shown, e.g., for CG Cu [22] or CG Ni [23]. On the other hand, the CSSC of UFG material lies well above that of CG material, as has been shown, for example, for UFG Cu [13] and Ni [24]. In Fig. 2, CSSC data are shown for UFG Cu in different conditions [25] and compared with data of CG Cu [22].

Comparing in Fig. 2 the CSS data of UFG Cu in the as-ECAP-processed state, obtained in stress- and strain-controlled tests, respectively, an interesting feature becomes apparent: It should be noted that the CSSC obtained (in the LCF regime) in the strain-controlled tests lies well below that found (mainly in the HCF range) in the stress-controlled tests. The reason is that true cyclic saturation was not attained in the LCF strain-controlled tests, because significant cyclic softening occurred, caused by marked dynamic grain coarsening (at the rather low homologous temperature of ca, 0.2!) and macroscopic shear banding [12,13,16,17,26]. On the other hand, in the HCF stress-controlled tests, pronounced cyclic saturation occurred, indicating that the UFG microstructure was rather stable during cyclic loading. This was also confirmed by TEM observations showing that grain coarsening was marginal,

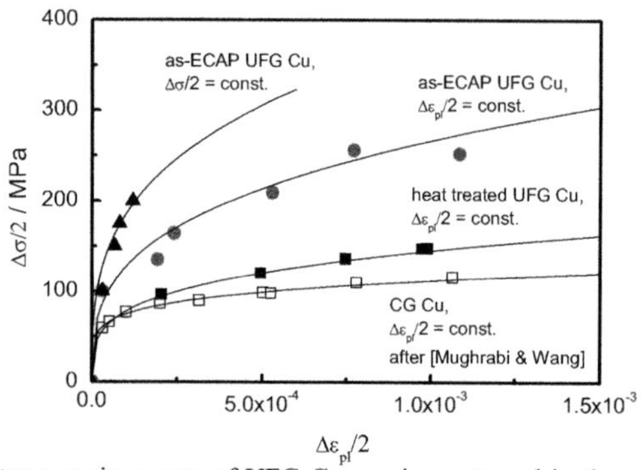

Figure 2. Cyclic stress-strain curves of UFG Cu specimens tested in the as-ECAP-processed state in strain and in stress control and after an annealing treatment in strain control, after [25]. Comparison with cyclic stress-strain curve of CG Cu from [22].

compare [13,26]. In a thorough TEM analysis [26], it could be shown that the dynamic grain coarsening (recrystallization) is a thermally activated process, probably related to the diffusion-controlled migration of GBs. This interpretation has received some support through a very recent molecular dynamics study of cyclic deformation of nanocrystalline copper [27], showing that grain coarsening, attributed to enhanced diffusion in the GBs, occurred in the nanocrystalline material. In commercial less pure materials, GB migration is expected to be impeded by the impurity atoms. Hence, grain coarsening should be less pronounced. This is borne out by the observations on fatigued commercial purity UFG Al and UFG α-brass. In the case of UFG Al, dynamic grain coarsening was not significant, the cyclic softening observed was probably caused mainly by some macroscopic shear banding [16,18,20]. On the other hand, in fatigued UFG α-brass (regarded as a very "impure" material), dynamic grain coarsening and shear banding were absent, and the cyclic hardening curves exhibited true saturation [19].

In view of the pronounced differences between the CSS behaviours of UFG and CG materials, it is of course of interest to gain more insight into the effects of grain size D on the CSSC in the transition from CG to UFG (or nano-) materials. In a study on the cyclic deformation behaviour of Ni specimens of widely differing grain sizes, Klemm [28] has recently obtained some interesting results by analyzing the total saturation stress σ_s for a given plastic strain amplitude $\Delta\varepsilon_{pl}/2$ in terms of the contributions $\sigma_{s,\rho}$ due to dislocations and $\sigma_{s,GB}$ due to grain boundaries:

$$\sigma_s = \sigma_{s,\rho} + \sigma_{s,GB}\,. \qquad (1)$$

Klemm succeeded in determining the first term separately by measuring the dislocation density ρ by X-ray diffraction and could then confirm that the grain size dependence of the second term obeys a Hall-Petch relationship, as had been reported earlier by Thiele et al. [24]. In Fig. 3, such a Hall-Petch plot is shown for a plastic strain amplitude of $2 \cdot 10^{-3}$. Comparing the UFG data with CG data [29], Klemm concluded that, for grain sizes in excess of ca. 4 μm, there is practically no grain size dependence, whereas for smaller grain sizes, a very marked dependence on grain size exists. He gave a convincing reason for this behaviour in terms of the magnitudes of

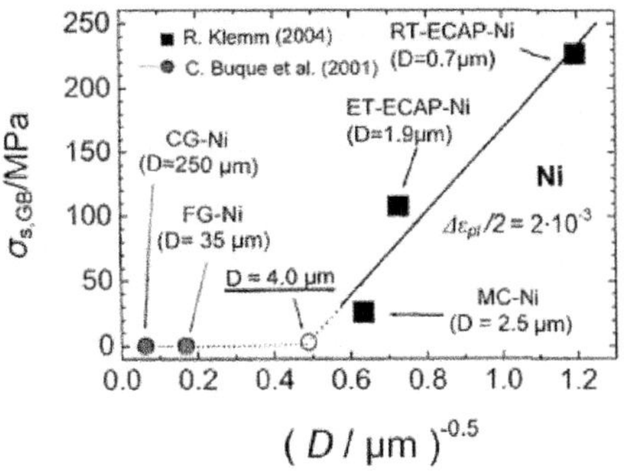

Figure 3. Hall-Petch plot of grain boundary contribution $\sigma_{S,GB}$ to cyclic saturation stresses σ_s of Ni of different grain sizes D (FG: fine-grained, MC: microcrystalline, RT-ECAP and ET-ECAP refer to ECAP at room and at elevated temperature). Adapted from Klemm [28].

the dislocation glide paths L relative to the grain size D. Using a simple dislocation glide model which yielded estimates of L for materials of different grain size D, Klemm could show that grain size effects are observed only, when the dislocations travel over glide paths L that are a substantial fraction (ca. 0.2 to 0.4) of the grain size D. This result implies that when dislocations travel only over small distances compared to the grain size D, they do not "feel" the GBs, and hence the grain size has little or no effect on the CSS response. On the other hand, for grain sizes below some "threshold" value, dislocations will feel the GBs, and a grain-size-dependent CSS behaviour can be expected.

Dislocation patterning and grain-size dependence of CSS behaviour

In CG materials, typical dislocation cell/subgrain structures with characteristic spacings d are observed after cyclic deformation at intermediate or larger amplitudes. Such dislocation "patterns" can only form when the grain size is sufficiently large to allow encounters of dislocations of opposite sign to form wall structures that can be accommodated within the bordering GBs. On the other hand, it is well established that d is inversely proportional to the shear stress $\tau = \sigma/M$, where M is an orientation factor, according to the relation

$$\tau = \frac{\sigma}{M} = \frac{KGb}{d} \tag{2}$$

Here, G is the shear modulus, b the modulus of the Burgers vector and $K \approx 10$, as discussed in ref. [30]. There, it was shown that the minimum hypothetical cell/subgrain size d that could be expected for ECAP-processed UFG Cu with a yield stress of ca. 450 MPa, is about 500 nm. Since this value is larger than the observed UFG grain size of ca. 300 nm, dislocation patterning would be impossible, in agreement with observation [16,26]. Evidently, dislocation patterning requires that at least one cell/subgrain wall must form between the bounding GBs of a grain. In this case, the requirement for dislocation patterning would simply be $D \geq 2$ d. On the other hand, if there existed a limit in which the GBs would take the place of the cell/subgrain walls, one would simply expect $d \approx D$ as a limiting case.

Fatigue lives, fatigue damage and governing microstructural factors

As described in the Introduction, a common observation made on fatigued UFG materials is that, whereas one can always expect a gain in fatigue life in the HCF regime in a Wöhler-type S-N plot, compare [11,13,16], there is usually a reduction of fatigue life in the LCF range, as displayed in a Coffin-Manson plot. Figure 4, taken from [16], summarizes for a number of different UFG materials and their CG counterparts the dependence of fatigue lives N_f on the plastic strain amplitude $\Delta\varepsilon_{pl}/2$ in a Coffin-Manson plot. In the as-ECAP-processed states, all UFG materials investigated by the authors, namely Cu, Al, α-brass and the alloy AA6061 (not shown in the figure) exhibit shorter fatigue lives than their CG counterparts. The same holds for the UFG Al-Ag alloy AA5056 from the work of Patlan et al. [31].

Since it is known that the strength enhancement after ECAP is frequently accompanied by a loss in ductility, mild annealing treatments which restore some of the ductility at the expense of losing some strength have been applied to as-ECAP-processed material to stabilize the UFG microstructure against coarsening and macroscopic shear banding with the goal to improve the LCF performance. Some results of such experiments are also shown in Fig. 4. The success achieved differs from material to material. The best results were obtained in the case of UFG Cu which attained longer fatigue lives than CG Cu after annealing and at significantly higher stress levels [12,13,16,17]. In this case, the annealed UFG microstructure consisted of a "bimodal" grain structure with some larger grains surrounded by the original UFG structure.

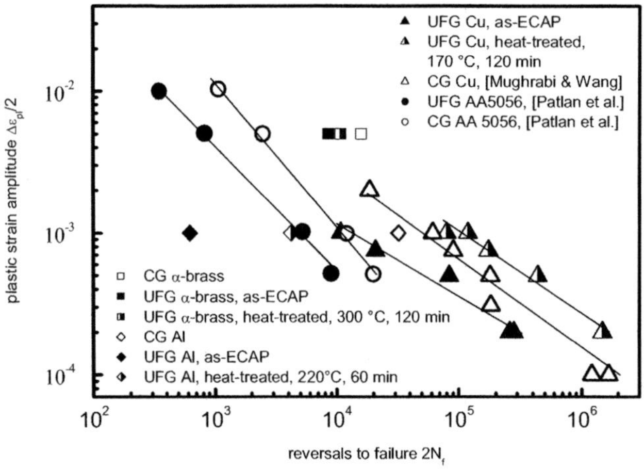

Figure 4. Coffin-Manson plot of fatigue lives of different materials. After [16]. Experimental data from [12,13,18,20,22,26,31].

Bimodal grain structures: Are they readily available? Are they beneficial?

The repeatedly voiced idea that bimodal grain structures could represent an optimum combination of strength (due to the small grains) and ductility (due to the larger grains) in fatigue [12,13,16,17,19] and also in unidirectional deformation [32] has received some support from recent numerical simulations [33]. So far, it has been observed that, in addition to UFG Cu, some other materials such as, e.g., UFG Ni [34] and UFG α-brass [19] also transform into such bimodal structures upon annealing. In the case of UFG Ni, no experimental data exist on the effect of annealing on LCF life. However, as is also shown in Fig. 4, in annealed UFG α-brass, the bimodal grain structure does lead to an improvement of the LCF performance [19], without, however, reaching the fatigue lives of CG material. UFG Al does not develop a bimodal grain structure during annealing but undergoes a slight grain coarsening [18,19], whereupon the LCF life is enhanced (at a lower stress level), compared to that of the as-ECAP-processed state [16,17,18,19,20] but still falls short of the fatigue lives of CG material. Hence, it can be concluded that while annealing seems to always enhance LCF life of UFG materials (irrespective of whether a bimodal grain structure develops), the LCF lives of CG materials frequently still remain superior. The fact that the microstructural changes that occur during an annealing treatment differ from material to material is probably related to the material-specific competition between recovery and recrystallization [16]. Thus, materials with a high stacking fault energy (e.g. aluminium [18,19]) tend to exhibit immediate recovery preceding recrystallization, whereas in materials with lower stacking fault energy such as copper [12,17,19] or nickel [34] recrystallization is unavoidable from the start, compare Stüwe et al. [35]. A more thorough analysis of these competing processes can perhaps provide valuable hints with respect to optimizing the annealing procedures.

Concluding Remarks

The preceding discussion focused on some microstructural features that govern the cyclic deformation and fatigue of UFG materials. Because of lack of space, the effects of grain size on fatigue crack propagation, as studied experimentally, compare [36], or by fracture mechanics [25] were not considered. Considering the overall progress achieved, it can be stated that UFG materials show promise to become future engineering materials. To achieve this goal, however, further systematic microstructure-based experimental work, combined with some modeling, are needed in order to pursue successfully the development and selection of suitable UFG materials possessing microstructural stability and an optimized combination of fatigue strength and ductility, with a good balance between LCF and HCF performance, as the most important features with respect to technological demands.

Acknowledgments

The authors are grateful to Dr. Robert Klemm, for permission to use his unpublished data on UFG Ni (Fig. 3) and to Dr. Ellen Hieckmann (formerly: Thiele) and Professor Carl Holste, all from Technische Universität Dresden, for useful discussions. The financial support of Deutsche Forschungsgemeinschaft is acknowledged gratefully.

References

1. Valiev, R.Z., Islamgaliev, R.K., Alexandrov, I.V., *Prog. Mater. Sci.* 45(2000)103-189.
2. Lowe, T.C., Valiev, R.Z., eds., *Investigations and applications of severe plastic deformation*. Dordrecht: Kluwer Academic Publishers, 2000.
3. Furukawa, M., Horita, Z., Nemoto, M., Langdon, T.G., *J. Mater. Sci.* 36(2001)835.
4. Iwahashi, Y., Wang, Z., Horita, M., Nemoto, M., Langdon, T.G.,. *Scripta Mater.* 35(1996)143-146.
5. Baik, S.C., Estrin, Y., Kim, H.S., Hellmig, R.J., *Mater. Sci. Eng. A* 351(2003)86-97.
6. Alkorta, J., Gil Sevillano, J., *J. Mater. Proc. Tech.* 141(2003)313-318.
7. Buck, O., Essmann, U., *phys. stat. sol.* 4(1964)143-149.
8. Valiev, R.Z., Alexandrov, I.V., Zhu, Y.T., Lowe, T.C., *J. Mater. Res.* 17(2002)5-8.
9. Agnew, S.R., Weertman, J.R., *Mater. Sci. Eng. A* 244(1998)145-153.

10. Hashimoto, S., Kaneko, Y., Kitagawa, K., Vinogradov, A., Valiev, R.Z., *Mater. Sci. Forum*, 28(1999)593-598.
11. Agnew, S.R., Vinogradov, A. Yu., Hashimoto, S., Weertman, J.R., *J. Electron. Mater.* 28(1999)1038-1044.
12. Höppel, H.W., Brunnbauer, M., Mughrabi, H. Valiev, R.Z., Zhilyaev, A.P., *Proc. of Materialsweek 2000.* Available from: http://www.materialsweek.org/proceedings, Munich, 2001.
13. Mughrabi, H., Höppel, H.W., *Structure and Mechanical Properties of Nanophase Materials – Theory and Computer Simulation vs. Experiment*, Mat. Res. Symp. Proc., Vol. 634, eds. D. Farkas, H. Kung, M. Mayo, H. van Swygenhoven, J. Weertman, 2001, B2.1.1-B2.1.12.
14. Lukáš, P., Klesnil, M., *Fatigue of Metallic Materials*, Materials Science Monographs, C. Laird, ed.,Vol. 71, Amsterdam: Elsevier Science Publisher, 1992.
15. Vinogradov, A., Hashimoto, S., *Adv. Eng. Mater.* 5(2003)351-358.
16. Mughrabi, H., Höppel, H.W., Kautz, M., *Scripta Mater.* 51(2004)807-812.
17. Höppel, H. W., Valiev, R.Z., *Z. Metallkd.* 93(2002)641-648.
18. Höppel, H.W., Xu, C., Kautz, M., Barta-Schreiber, N., Langdon, T.G., Mughrabi, H., *Nanomaterials by Severe Plastic Deformation*, Proc. of Int. Conf. NANOSPD2, eds. M.J. Zehetbauer, R.Z. Valiev, WILEY-VCH, 677-683.
19. Mughrabi, H., Höppel, H.W., Kautz, M., Valiev, R.Z., *Z. Metallkd.* 94(2003)1079-1083.
20. Kautz, M., Höppel, H.W., Xu, C. Langdon, T.G., Mughrabi, H., Proc. of *Fifth International Conference on Low Cycle Fatigue*, Berlin, 2003, eds. P.D. Portella, H. Sehitoglu, K. Hatanaka, Deutscher Verband für Materialforschung und –prüfung (DVM), 2004, 113-118.
21. Mughrabi, H., Höppel, H.W., Kautz, M., *Tran. Roy. Soc. South Africa*, in print.
22. Mughrabi, H., Wang, R., *Basic Mechanisms of Fatigue*, Proc. of Int. Colloq., eds. P. Lukáš, J. Polák, Prague: Academia/Elsevier, 1988, 1-13.
23. Haddou, H., Risbet, M., Marichal, G., Feaugas, X., *Mater. Sci. Eng. A* 379(2004)102-111.
24. Thiele, E., Holste, C., Klemm, R., *Z. Metallkd.* 93(2002)730-736.
25. Ding, H.-Z., Mughrabi, H., Höppel, H.W., *Fatigue Fract. Engng. Mater. Struct.* 25(2002)975-984.
26. Höppel, H.W., Zhou. Z.M., Mughrabi, H., Valiev, R.Z., *Phil. Mag. A*, 82(2002)1781-1794.
27. Schiøtz, J., *Mater. Sci. Eng. A* 375-377(2004)975-979.
28. Klemm, R., Doctorate Thesis: *Zyklische Plastizität von mikro- und submikrokristallinem Nickel*, Technische Universität Dresden, 2004.
29. Buque, C., Bretschneider, J., Schwab, A., Holste, C., *Mater. Sci. Eng. A* 319-321(2001)631-636.
30. Mughrabi, H., in ref. [2], 241-253.
31. Patlan, V., Vinogradov, A., Higashi, K., Kitagawa, K., *Mater. Sci. Eng. A* 300(2001)171-182.
32. Wang, Y., Chen, M., Zhou, F., Ma, E., *Nature* 419(2002)912-915.
33. Gil Sevillano, J., Aldazabal, J., *Scripta Mater.* 51(2004)795-800.
34. Thiele, E., Bretschneider, J., Hollang, L., Schell, N., Holste, C., in ref. [2], 173-178.
35 Stüwe, H.P., Padilha, A.F., Siciliano Jr., F., *Mater. Sci. Eng. A* 333(2002)361-367.
36. Hanlon, T., Kwon, Y.-N., Suresh, S., *Scripta Mater.* 49(2003)675-680.

Modelling Mechanical Properties of SPD Materials During and After Severe Plastic Deformation

M.J. Zehetbauer[1*], L.F. Zeipper[1,2] and E. Schafler[1]
[1]Institute of Materials Physics, Vienna University, A-1090 Wien, Austria
[2]Austrian Research Center Seibersdorf GmbH, A-2444 Seibersdorf, Austria

Abstract

As SPD yields ultra fine grained and/or nanograined structures, modelling of properties at first requires a proper reflection of grain refinement process. A model of the author originally developed for conventional large strain hardening has been successfully adapted to conditions of extended hydrostatic pressure which appears as one of the most significant parameters of SPD: It allows for nanofragmentation through additional storage of dislocations. Arguments recently published in literature that the extra hardening arises from geometrically necessary dislocations connected to strong strain gradients can be refuted. The model in question accounts for the fragmentation and hardening characteristics *during* SPD. However, e.g. for application cases, the hardening characteristics *after* SPD ("post-SPD deformation") is important to be simulated, too. In fact, the above mentioned model could be successfully used to quantitatively describe the hardening characteristics in post-SPD deformation, and even the strain dependent evolution of lattice defect densities as well as that of cell size were well simulated. The latter follows a relation being reciprocal to the macroscopic stress, which has been already proven for HPT Cu after deformation. Although no quantitative model for the increased *ductility* of SPD metals at low test temperatures is existing yet, experimental findings of specific decreases of internal strains suggest static recovery processes to occur immediately after SPD, for which the presence of deformation induced vacancies seem to play an important role.

Introduction, and Modelling of SPD during Deformation

Up to date only a few attempts have been made to model the strengthening behaviour and/or the mechanical properties in connection with SPD metals and alloys. Those done by Baik et al. [1] and Enikeev et al. [2] making use of the models of Estrin [3] and Toth [4], and a modified version of them [2] have partially succeeded in describing the evolution of structural parameters for ECAP deformation. However, these simulations do not account for any effects in connection with deformation induced vacancies nor, in particular, for the influence of the hydrostatic pressure to both the strength and the structural parameters which has been made evident by several experiments [5-7]. Fig. 1 shows that the strength *during* SPD must be carefully distinguished from the strength of the same sample *after* SPD, and that the difference between them increases with increasing amount of hydrostatic pressure being present [7]. The model of Zehetbauer which has been originally designed for conventional large strain deformation [8, 9] recently has been adapted for effects of hydrostatic

* Corresponding Author. zehet@ap.univie.ac.at

Y.T. Zhu and V. Varyukhin (eds.), Nanostructured Materials by High-Pressure Severe Plastic Deformation, 217-226.

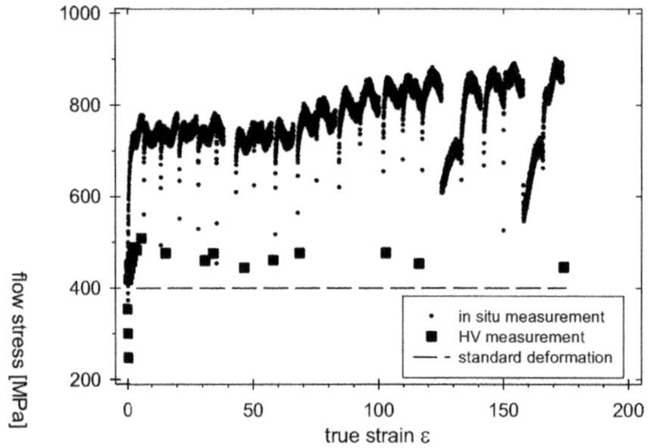

Fig.1. Stress-strain characteristics at HPT deformation of Cu performed at $p = 0.8$ GPa. The highest line represent the flow stress during deformation, the squares represent flow stresses $\sigma = HV/3$ derived from Vickers microhardness HV data (from M. Zehetbauer, TMS 2002 [6]).

pressure during SPD in a successful way [7, 10]. The main issues of this model are the composite Ansatz for individual stresses by Mughrabi [11], the statistical hardening theory by Kocks [12], and the idea that there are essentially two types of dislocations (screws, edges) with individual storage and annihilation behaviour which do not interact with each other. Mobile screws accumulate within cell interiors and store edges within cell walls. The total hardening ($\theta = d\tau/d\gamma$, with τ as the total resolved shear stress and γ as the resolved shear strain) consists of the contributions from cells θ_1 and cell walls θ_2 (Eq. 1 and 2):

$$\theta_1 \equiv \frac{d\tau_1}{d\gamma_1} = C_1 - C_2\tau_1 \quad (1) \qquad \theta_2 \equiv \frac{d\tau_2}{d\gamma_2} = C_3 - C_4[\tau_2 - \tau_2(0)]\,\tau_2^5 \quad (2)$$

The fit parameters C_1 ... C_4 contain physical quantities. C_1 and C_3 describe the storage of dislocations, C_2 and C_4 account for the annihilation of dislocations. $\tau_2(0) = 2\tau_1(0)$ and $\tau_1(0) = \tau_0 / (2-f_1)$ are the corresponding stresses at $\gamma = 0$ due to edges and screws. τ_0 represents the total critical resolved shear stress of the material while f_1 stands for the volume fraction of cell interior i.e. screw dislocation areas. (1) and (2) correspond to a parallel composite with $\gamma = \gamma_1 = \gamma_2$ so that the total resolved shear stress τ is described as:

$$\tau = f_1\,\tau_1 + f_2\,\tau_2 \quad (3) \qquad \tau_i = \alpha_i\, G\, b \sqrt{\rho_i} \quad i = 1, 2 \quad (4)$$

The volume fractions $f_1 = L_1 / (L_1 + L_2)$ represent the cell interiors (areas of screw dislocations with width L_1) and $f_2 = 1 - f_1$ the cell walls (areas of edge dislocations with width L_2). Via the dislocation interaction parameters α_i for screw (i=1)and edge (i=2) dislocations, the shear modulus G, and the Burgers vector b, the stress τ_i is linked to the dislocation density ρ by Eq. 4. The dislocation densities ρ_i for cell interiors and cell walls are weighted by the volume fractions f_i and summed up to the total dislocation density ρ.

Two different fits are carried out. First, by fitting Eq. 3 to the experimental stress-strain data one obtains the fit parameters C_i, τ_0 and f_1. Second, by fitting Eq. 5 to the X-ray dislocation density data, one obtains the interaction parameters α_i.

$$\rho(\gamma) = \frac{1}{(Gb)^2}\left(\frac{f_1\tau_1^2(\gamma)}{\alpha_1^2} + \frac{f_2\tau_2^2(\gamma)}{\alpha_2^2}\right) \quad (5)$$

From the storage parameters C_1 and C_3 , the storage rates β_i are calculated ($\beta_i = (\alpha_i G) / 2C_i$; i = 1,2). The β_i inform on the mean free path P_i of individual dislocations i , in terms of dislocation distance $(1/\sqrt{\rho_i})$, according to $P_i = \beta_i \ (1/\sqrt{\rho_i})$. From C_2 the activation enthalpy for screw annihilation ΔG, is derived (Eq. 6). From C_4, with a given vacancy migration enthalpy ΔH_c^m for core diffusion, the deformation induced vacancy concentration c_v is calculated.

$$\Delta G = -k_B T \ln\left(\frac{C_2\dot{\gamma}}{\nu_D}\right) \quad (6) \qquad c_V = C_4[\tau_2 - \tau_2(0)]\frac{\dot{\gamma}\, k_B T \alpha_2^4 b^2 G^3}{\Omega D_{c,0}}\frac{\pi(1-\nu)}{\sqrt{2}}\exp\left(\frac{\Delta H_c^m}{k_B T}\right)$$

(7)

In Eqs. 6 and 7, k_B denotes the Boltzmann constant, T the temperature, ν_D the Debye frequency, ν Poisson's ratio, Ω the atomic volume and $D_{c,0}$ the diffusibility of core diffusion (Tab. 1). In case of the presence of an elevated hydrostatic pressure p, the expression for the core vacancy migration enthalpy ΔH_c^m changes to

$$(\Delta H_c^m)_{eff} \equiv \Delta H_c^m + p\,\Omega \quad (8)$$

For more details of the model and its fitting see [7] and [10].

The success of the model not only concerns the right description of the stress dependence on the hydrostatic pressure but also the correctly reflected pressure sensitivity of the stored dislocation density, allowing for a continuous lattice fragmentation. At the same time, structural parameters like the dislocation distribution and the volume fraction of cell walls take realistic physical values.

Recently, Gil Sevillano [13] presented an alternative explanation for the pressure-induced increase of dislocation density. In his opinion in case of HPT, extreme strain gradients will be established not at least due to the small thickness of samples used which cause the generation of numerous geometrically necessary dislocations (GND's) in addition to usual deformation induced ones, without any influence of the hydrostatic pressure. However, we stress the fact that with conventional torsion tests of metals under NPT conditions the generated dislocation density was the same as with other deformation modes at the same equivalent strains [14]. Moreover, the effect of hydrostatic pressure on dislocation density has been also observed during HPT of samples with thicknesses up to 0.8mm. Last not least, careful inspection of the HPT experiments by Erbel [15] with gradient free ring samples do exhibit some (though small) raise of the strength level compared with results from conventional torsion tests, in contrast to the related statement of Gil Sevillano.

Zehetbauer's model in the version with pressure dependence has been also applied to Ni deformed by High Pressure torsion at room temperature [10]. Results are somewhat different to those of Cu concerning the level of dislocation density. In HPT Ni, these seem to reach some saturation for still increasing levels of hydrostatic pressure p. In particular, the concentration of vacancies reaches forbidden high values for pressures beyond 4 GPa which itself explains the saturation of dislocation density due to spontaneous annihilation, as a function of p. Inspection of microhardness (Fig. 3) confirms this sight that some dynamic recovery / recrystallization has to be assumed to get prevalent here. It has been also confirmed by investigations of Hebesberger et al. [16] concluding the occurrence of dynamic recrystallization in HPT of Cu from certain extents of p and of shear strains applied.

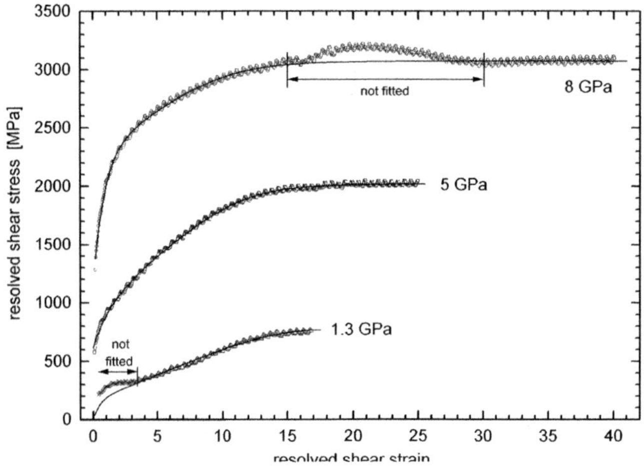

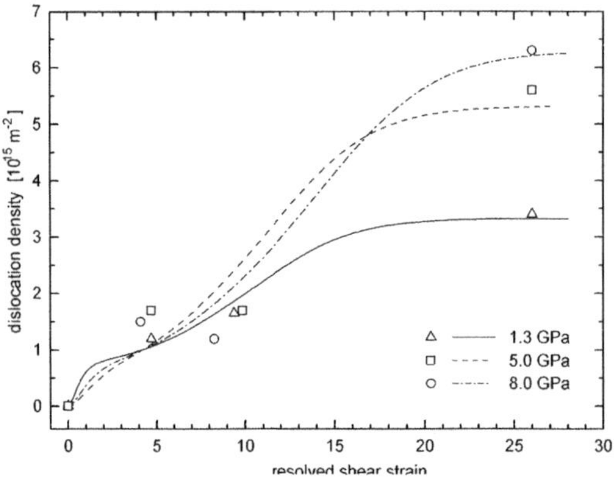

ig. 2.
omparison of model
ts (full lines) with ex-
erimental data points as
function of strain,
alid during HPT of
i at RT. (from [10]).
pper image:
ress values
ower image:
slocation density

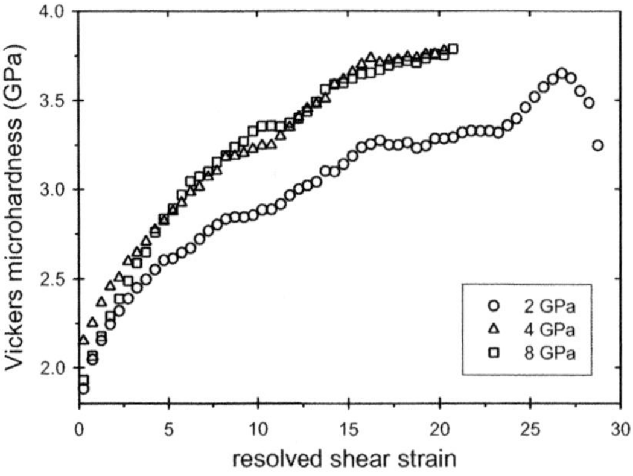

Fig. 3.
Data of microhardness representing the strength values after HPT of Ni. (from [10]).

Table 1: Physical parameters of model fits to hardening of Ni during HPT deformation. Dislocation configuration parameters α_i resulting from fits $\rho(\gamma)$, and annihilation enthalpy of screws δG as well as the pressure dependent concentration of deformation induced vacancies $c_{v\,calc}$ resulting from fits $\tau(\gamma)$, for a volume fraction of cell wall materials of $f_1 = 0.6$, a resolved shear strain rate $d\gamma/dt = 10^{-2}\ s^{-1}$, and a vacancy migration enthalpy $H_m(p = 0) = 0.62$ eV. Note that all values $c_{v\,calc}$ are far higher than that of thermal vacancy concentration under atmospheric pressure, i.e. $c_{th} = 1\cdot10^{-23}$ (for details see [10]).

INPUT		OUTPUT					
p [GPa]	ρ_{sat} [10^{15} m^{-2}]	$H^m{}_{eff}$ [eV]	α_1	α_2	δG [eV]	$c_{v\,calc}$	comment
1.10^{-4}	2.5	0.62	0.22	0.12	0.86	$3.0\cdot10^{-5}$	$c_{v\,calc}$ close to exp. value $c_{v\,exp} = 3\ .\ 10^{-5}$
1.3	3.3	0.71	0.34	0.89	0.68	$1.8\cdot10^{-3}$	$c_{v\,calc}$ close to exp. value $c_{v\,exp} = 3\ .\ 10^{-3}$
5.0	5.3	0.96	1.17	1.37	0.70	$1.0\cdot10^{+2}$	unacceptably high
8.0	6.3	1.16	2.92	1.19	0.69	$1.8\cdot10^{+5}$	unacceptably high

Modelling of SPD Post-Deformation, and Reasons for Enhanced Ductility

The models mentioned above [1-4] have not still applied yet to the post-deformation case i.e. to stress-strain curves connected to any deformation following the treatment by SPD. Recently [17, 18], Zehetbauer's model have been used to simulate the post deformation characteristics of CP-Ti after warm ECAP treatment. In post-compression at cold work conditions, stages II and III reappear [19] which are perfectly reflected by the model simulation. Being a more substantial criterion of the validity of the model, it is seen that both parameters α_i and β_i describing the configuration and storage rates of dislocations I, respectively, reach reasonable values, although the relative magnitudes of the α_i may differ from those usually observed at cold work conditions. This feature seems to arise from the fact that the pre-structure has been formed by medium temperature SPD deformation (see also [18]). The same reason seems to be responsible for the early saturation of edge dislocation density [17, 18] and consequently for that of the deformation induced vacancies (Fig. 5, upper image). A very good agreement of the simulation with the experimentally obtained concentrations of deformation induced vacancies is reached when the self diffusion enthalpy is taken as $\Delta H_b = 1.13$ eV which is a quite realistic value in view of the experimental data given in literature [18]. Following the composite nature of the model, the parameters α_I, β_I should allow to derive the strain dependent evolution of cell size P_I being felt as the mean free path of the screw dislocations after which they store an edge dislocation in the cell wall, by the following equation [9]

$$P_I = \beta_I \quad (1/\sqrt{\rho_I}) = \beta_I \quad (\alpha_I G\, b/\tau_I) \tag{9}$$

In Fig. 5, lower image, it can be seen that within the experimental error the calculations yield the right cell sizes. Eq. (9) also means a reciprocal dependence of the stress on the cell size. As demonstrated by images (a) and (b) of Fig. 6, such a behaviour has been experimentally confirmed by recent XPA investigations of HPT Cu samples which have been deformed under various hydrostatic pressures p [20].

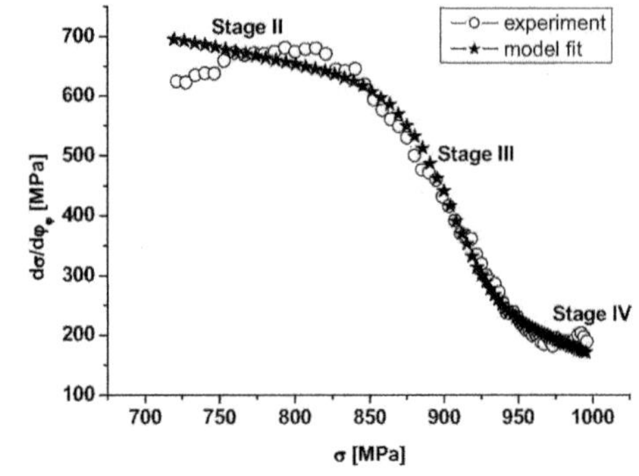

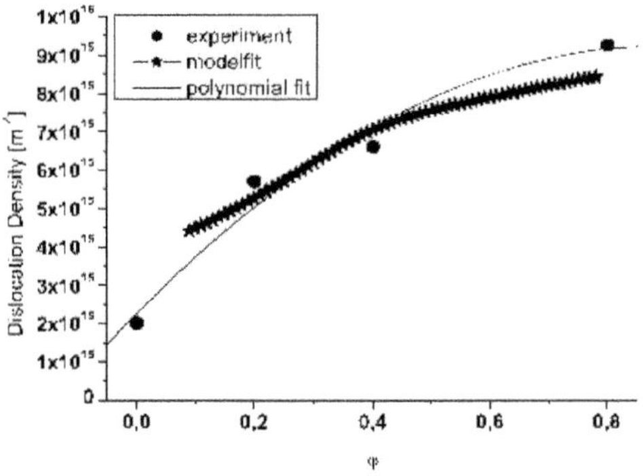

Fig. 4. Fit qualities of simulations of post-ECAP compression done at RT for CP-Ti. **Upper image:** strengthening coefficient, **lower image:** dislocation density, both considered for increasing post-deformation

Table 2: Physical parameters of model fits to post-deformation hardening of ECAP deformed CP-Ti. For the meaning of parameters α_l, δG as well as $c_{v\ calc}$ and $c_{v\ exp}$, see caption of Table 1. From the fits $\tau(\gamma)$, a volume fraction of cell wall materials of $f_1 = 0.97$, a resolved shear strain rate $d\gamma/dt = 10^{-3}\ s^{-1}$, and a vacancy migration enthalpy $H_m = 0.48$ eV have been obtained. Again, the value $c_{v\ calc}$ is far higher than that of thermal vacancy concentration.

INPUT			OUTPUT					
p [GPa]	ρ_{sat} [10^{15} m^{-2}]	H_m^c [eV]	α_1	α_2	β_1	β_2	δG [eV]	$c_{v\ calc}$
1.10^{-4}	9.2	0.48	0.31	0.59	16.3	1.0	0.93	$c_{v\ calc}$ close to exp. value $c_{v\ exp}$=3.5 .10^{-4}

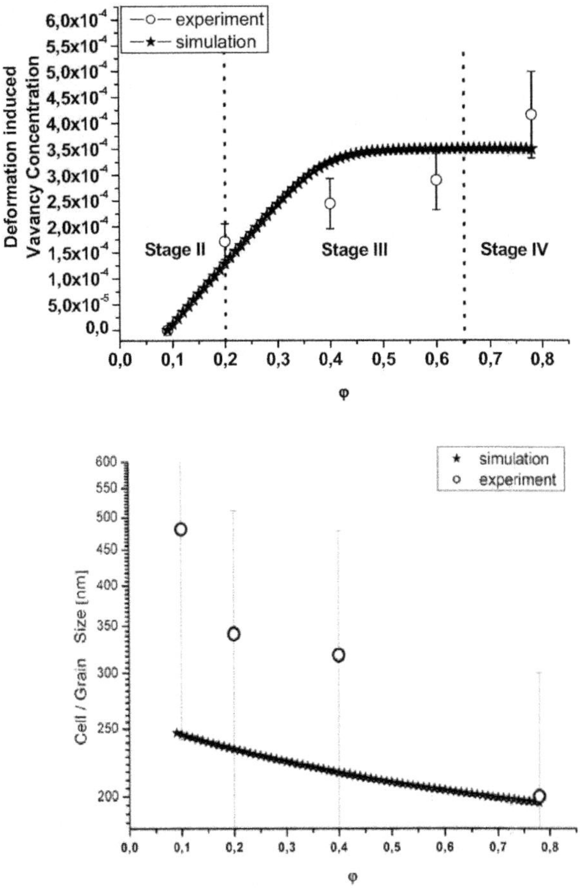

Fig. 5. Comparison of model simulations with experimental data measured as a function of post-deformation after ECAP in CP Ti. **Upper image**: Evolution of vacancy concentration, exdatapoints from residual electrical resistivity. **Lower image**: Evolution of cell size, datapoints from TEM investigations (from [17, 18]).

In contrast to the strengthening characteristics *during* HPT which shows a relative but usually no absolute softening (Fig. 1, [6]), the strength of the material *after* HPT does exhibit such effects (Fig. 6, [20]). Fig. 6 (c) gives the answer why: The internal stresses measured by X-ray diffraction strongly decrease, in parallel to the strength which means that some stress relaxation must have occurred during unloading after HPT. Since the dislocation density does *not* decrease in parallel to the strength decrease [7], this relaxation must arise from pure rearrangements of dislocations rather than from dislocation losses, hinting at the dominance of recovery processes and not of those representing recrystallization. The high concentration of vacancies being present at SPD (see the authors' recent paper [21] and the model predictions achieved in [7, 10]) again lets one believe in dislocation rearrangements

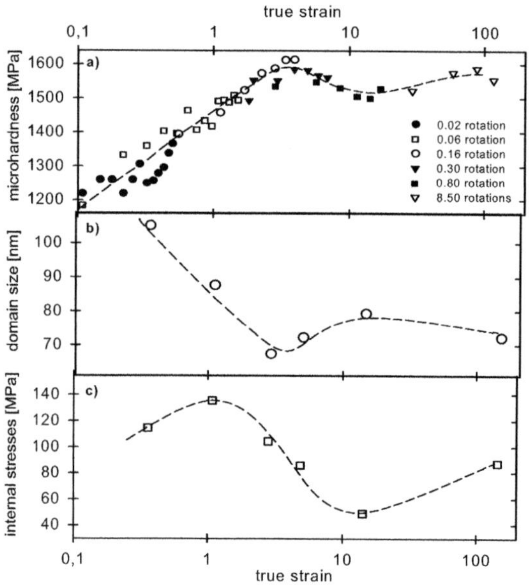

Fig. 6. The microhardness **(a)**, the coherently scattering domain size **(b)**, and the long range internal stresses **(c)** as function of true strain, at the example of HPT deformed Cu performed at a hydrostatic pressure p=4 GPa [20].The dashed lines were attached as a guide for the eye.

especially in case of edge dislocations which need massive concentrations of vacancies when climbing. It seems that many of the grains become "recovered" i.e. they loose internal deformation induced strains, and a certain "recovery strain" adds to the original fraction strain enhancing the ductility. This mechanism fits to the idea of bimodal grain size distribution created by Wang et al. [22] making it responsible for the simultaneous benefits in strength and ductility. The recovery effect mentioned is especially relevant for post-deformation after ECAP: Although the hydrostatic pressure will not exceed 1 GPa with this method, there are many unloading events occuring between single ECAP passes. Each of it allows recovery processes to operate causing a certain recovery strain which gets larger with increasing number of passes and thus enhances ductility. This feature also explains the occurrence of absolute softening mentioned above which happens when the recovery strain (i.e. the strain necessary to reach the original stress-strain curve) exceeds that of a single ECAP pass (for details see also [23]).

Alternatively, there has been often discussed the possibility that deformation induced vacancies may be primarily stored in the grain boundaries, and may be capable of triggering boundary edge dislocations' climb, i.e. Coble creep thus leading to an enhancement of ductility. Thanks to a recent work published by the group of Blum [24], however, this explanation can be excluded at least for deformation temperatures $T \leq 0.2\ T_m$ (T_m being the melting temperature in K).

Conclusions

(1) The model of Zehetbauer does work well for simulation of strengthening during SPD deformation. It accounts for the effect of increased hydrostatic pressure during SPD which suppresses annihilation of edge dislocations and allows for increased formation of new grain boundaries i.e. fragmentation to ultrafine grained and/or nanograined microstructures. Arguments favouring geometrically necessary dislocations to be responsible for the fragmentation can be disproved.

(2) The model of Zehetbauer also applies to post-deformation characteristics after SPD even exhibiting a good agreement with the magnitude and evolution of structural parameters such as dislocation density, vacancy concentration and the cell size. The model also gives a physical background for the reciprocal relation between total external stress and the cell size.

(3) Compared to SPD strength models, those for ductility are developed only poorly yet. Nevertheless, there is clear experimental evidence that, not at least due to the presence of high concentrations of deformation induced vacancies, massive dislocation rearrangements take place immediately after SPD deformation which lead to recovery in a number of grains. Therefore a certain recovery strain adds to the total fracture strain giving an extended ductility of SPD metals. Recent literature showed that Coble creep being an alternative mechanism for increased ductility can be excluded.

Acknowledgements

This work has been supported by the Austrian Science Funds FWF under project P17095N02.

References

1. Baik, S.C., Estrin, Y., Hellmig, R.J., Kim, H.S., "Modeling of Deformation Behavior and Texture Development in Aluminium under Equal Channel Angular Pressing", *Nanomaterials by Severe Plastic Deformation*, ed. M. Zehetbauer and R.Z. Valiev. Wiley-VCH, Weinheim, Germany; 2004, 233-238
2. Enikeev, N.A., Alexandrov, I.V., Kim, H.S., "Deformation Behaviour of ECAP Cu as described by a Dislocation-Based Model", *Nanomaterials by Severe Plastic Deformation*, ed. M. Zehetbauer and R.Z. Valiev. Wiley-VCH, Weinheim, Germany; 2004, 245-250
3. Estrin, Y., Toth, L.S., Molinari, A., Bréchet, Y., *Acta Mater.* 46 (1998), 5509-5522
4. Toth, L.S., Molinari, A., Estrin, Y., *J. Eng. Mater. Tech.* 124 (2002) 71-77.
5. Hebesberger, T., Pippan, R., Stuewe. H.P., "Effect of Pressure on the Final Grain Size after High Pressure Torsion", *Ultrafine Grained Materials II*, ed. Y.T. Zhu et al., 2002 TMS Annual Meeting, Seattle, Washington, USA; 2002, 133-140
6. Zehetbauer, M., "Features of Severe Plastic Deformation as Compared to Conventional Deformation Modes",*Ultrafine Grained Materials II*, ed. Y.T. Zhu et al., 2002 TMS Annual Meeting, Seattle, Washington, USA; 2002, 669-678
7. Zehetbauer, M.J., Stüwe, H.P., Vorhauer, A., Schafler, E., Kohout, J., *Adv.Eng.Mater.* 5 (2003), 330-337
8. Zehetbauer, M., *Acta metall.mater.* 41 (1993), 589-599
9. Les, P., Zehetbauer, M., *Key Eng. Mater.* 97-98 (1994), 335-340
10. Zehetbauer, M.J., Kohout, Schafler, E., Sachslehner, F., Dubravina, A., *J.Alloys Comp.* 378 (2004), 329-334

11. Mughrabi, H., Acta Metall. 31 (1983) 1367-1379
12. Kocks, U.F., *Trans.ASME "H" (J.Eng. Mater. Technol.)* 98 (1976), 76-83
13. J. Gil Sevillano, "Strengthening by Plastic Work: From LPS to SPD. A 25 Years Perspective", *Evolution of Deformation Microstructures in 3D*, ed. C.Gundlach et al. Riso National Laboratory, Roskilde, Denmark; 2004, 1-11
14. Zehetbauer, M., Seumer, V., *Acta metall.mater.* 41 (1993), 577-588
15. Erbel, S., *Met.Technol.* 6 (1979), 482-486
16. Hebesberger, T., Stüwe, H.P., Vorhauer, A., Wetscher, F., Pippan, R., *Acta mater.* 53 (2005), 393-402
17. Zeipper, L.F., Ph.D. Thesis, University of Vienna, Austria (2004)
18. Zeipper, L.F., Zehetbauer, M., Holzleithner, Ch., *Mater.Sci.Eng.A*, The Langdon Symposium - 2005 TMS Annual Meeting, submitted
19. Zeipper, L.F., Gemeinboeck, G., Zehetbauer, M., Korb, G.., "Compression Tests to approximate the ECAP Deformation and Post-Processing of Coarse-Grained and NanoSPD Grade2-Titanium", *Ultrafine Grained Materials III*, ed. Y.T. Zhu et al., 2004 TMS Annual Meeting, Charlotte, North Carolina, USA; 2004, 541-546
20. Dubravina, A., Zehetbauer, M., Schafler, E., Alexandrov, I., *Mater.Sci.Eng. A* 387-389, (2004) 817-821
21. Schafler, E., Steiner, G.., Korznikova, E., Kerber, M., Zehetbauer, M., *Mater.Sci.Eng.A,* The Langdon Symposium, 2005 TMS Annual Meeting, submitted
22. Z Wang, Y., Chen, M., Zhou, F., Ma, E., *Nature* 419 (2002), 912-915
23. Zehetbauer, M., Trattner, D., *Mater.Sci.Eng.* 89 (1987), 93-101
24. Zheng, X.M., Li, Y.J., Blum, W., *phys.stat.sol. (a)* 201 (2004), R114-R117

Unique Deformation Behaviors of the Ultrafine Grained Aluminum Alloys Fabricated by Accumulative Roll Bonding

N.Tsuji*

Department of Adaptive Machine Systems, Osaka University, Suita, Osaka 565-0871, Japan.

Abstract

Aluminum alloys sheets having various ultrafine grain sizes were fabricated by the accumulative roll bonding (ARB) process followed by annealing. The materials were provided for mechanical testing to clarify the deformation behaviors of the ultrafine grained materials systematically. Unique mechanical behaviors of the ultrafine grained materials were found through the investigations. Unexpected similarity was found in the flow behaviors between pure aluminum and iron (ultralow carbon IF steel). Ultrafine grained pure aluminum showed yield-drop phenomenon and Lüders deformation. The ultrafine grained materials performed very high strength. Hall-Petch relationship standed even down to the grain size of 100 nm, but it was far above the extrapolated curve of the conventionally grain sized materials. Although the elongation generally decreased with increasing the SPD strain, it was found in certain kinds of materials that the total elongation significantly increased with increasing the ARB strain.

Introduction

Severe plastic deformation (SPD) of metallic materials has been one of the hottest subjects in the field of materials science and engineering in the last decade. Various kinds of unique SPD processes have been developed and the ultrafine grained (UFG) bulk materials whose mean grain sizes were several tens nm to 100 nm have been realized by them [1]. Generally the UFG materials perform very high strength at low temperature and excellent superplasticity at elevated temperature. However, the mechanical properties of the UFG materials have not yet been systematically clarified, which has been a barriar for practical application of the UFG metals as structural materials. One of the reasons for this lack of the systematic data is the difficulty to fabricate enough amount of UFG specimens with bulky dimensions for systematic mechanical tests. The accumulative roll bonding (ARB) [1], which was originally invented by the present author and his colleagues [2], is a SPD process applicable to continuous production of bulky materials. It is also possible to produce enough amount of specimens having various UFG sizes by means of the ARB process followed by annealing [3,4]. Using this advantage, the author has been studying the mechanical properties of the UFG materials systematically as a function of the grain size [5-7]. In the present paper, typical and unique deformation behaviors of the UFG materials clarified through that research are introduced.

* Corresponding Author. tsuji@ams.eng.osaka-u.ac.jp

Y.T. Zhu and V. Varyukhin (eds.), Nanostructured Materials by High-Pressure Severe Plastic Deformation, 227-234.

Experimental Procedures

Figure 1 illustrates the principle of the ARB process [1,2]. In the ARB process, 50% rolled material is cut into two, stacked to be the initial dimensions and then rolled again. In order to obtain one-body solid materials, the rolling in the ARB is not only a deformation process but also a bonding process (roll-bonding). To achieve good bonding, the surface of the materials were degresed by acetone and wire-brushed before stacking. The stacked sheets were sometimes heated at elevated temperature below recrystallization temperature of the material and then immediately roll-bonded, in order to realize sound bonding and to reduce the rolling force. The ARB process has been applied most of the major metals and alloys and the UFG bulky sheets have been successfully fabricated by the ARB [1]. The details of the ARB processing have been described in the previous papers [2,5,7].

In the present study, the results of 1100-Al, 8011-Al, and ultralow carbon interstitial free (IF) steel for comparison are shown. The aluminum alloys were roll-bonded at either room temperature or 200°C, and the IF steels at 500°C. Typical dimensions of the ARB sheets were 1^t x 40^w x 200^l (mm^3). In every case, pancake shaped UFGs elongated along RD were obtained. The mean grain thicknesses were about 200 nm [5]. The ARB processed sheets were then annealed at various temperatures to change the grain sizes. The specimens having various uniform grain sizes ranging from 200 nm to 20 μm were fabricated through the ARB and annealing process. The specimens were provided for tensile test at RT. The details of the investigations were shown in elsewhere [5-7].

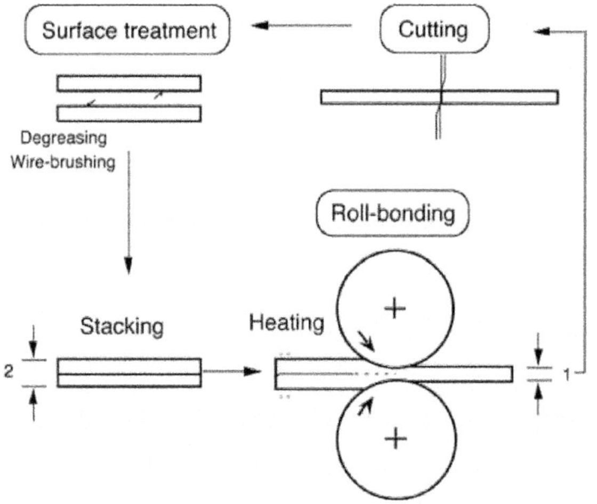

Figure 1 Schematic illustration showing the principle of the ARB process.

Results and Discussion

Microstructure of the ARB processed and Annealed Materials

It has been clarified [4] that the mechanism of the UFG formation during the ARB is explained in terms of *grain subdivision* [8]. As a result, the elongated ultrafine structures, which are quite similar to so-called lamellar boundary structure [8] in

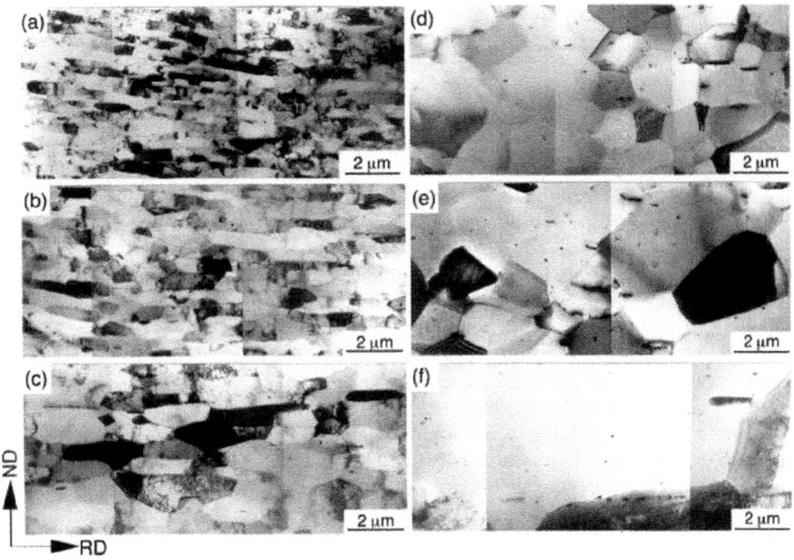

Figure 2 TEM microstructures of the 1100 Al ARB processed by 6 cycles at 200°C and then annealed for 1.8 ks at (a) 100°C, (b) 150°C, (c) 200°C, (d) 225°C, (e) 250°C and (f) 300°C. Observed from TD

heavily cold-rolled FCC metals, generally form in the ARB processed sheets. After 5 or 6 cycles ARB, the sheets are uniformly filled with the elongated ultrafine structures [9]. Most of the elongated or pancake-shaped regions (grains) are surrounded by high-angle grain boundaries whose misorientations are much larger than 15°, and the fraction of high-angle boundaries in the microstructure is above 70-80% [9-11]. In case of pure Al and IF steel, subsequent annealing can continuously changes the grain size of the ARB processed specimens [3,4]. Figure 2 shows the TEM microstructures of the 1100 Al ARB processed by 6 cycles at 200°C and then annealed at various temperatures for 1.8 ks [5]. The 100°C annealed specimen (Fig.2(a)) showed the elongated or pancake shaped ultrafine lamellar structure which is nearly the same as that of the as-deformed specimen. The elongated morphology was maintained up to 200°C (Fig.2(c)), but recovery at grain interior proceeded to decrease the dislocation density within the grains and the grain boundary migration occurred to increase the grain size continuously. The 225°C annealed specimen (Fig.2(d)) showed nearly equiaxed grain structure free from dislocations inside. It is quite difficult to distinguish this microstructure from those obtained by conventional deformation and recrystallization, though the grain size in Fig.2(d), about 1.5 μm, is very fine as this material. Above 250°C, a kind of normal grain growth occurred to change the grain size continuously. Such annealing behaviors can be understood as *continuous recrystallization* [12] which is greatly different from conventional discontinuous recrystallization characterized by nucleation and growth [4]. It should be emphasized, anyhow, that bulky sheet materials having various mean grain sizes ranging from 200 nm to 20 μm could be obtained by means of the ARB followed by annealing. Nearly the same microstructural changes happened in the IF steel as well [4,5]. By using these samples, systematic investigations on the mechanical properties of the UFG materials were realized.

Unexpected Similarity between Al and Fe

Stress-strain curves of the 1100-Al and IF steel with various grain sizes are shown in Fig.3 [5]. The specimens were fabricated by the ARB process followed by annealing. The curves show true stress and true strain assuming uniform deformation, and the curves after necking are also indicated purposely. Firstly it is very interesting that both materials showed quite similar change in the stress-strain behaviors with changing the grain sizes, although they are completely different materials having different crystal structures. The submicrometer grain-sized specimens performed very high strength which is two to three times higher than that of the starting materials having conventional grain sizes. However, the specimens showed their maximum strength at early stage of the tensile test, followed by necking. As a result, the uniform elongation was limited within a few percents, and the total elongation was below 10%. The flow stress decreased with increasing the mean grain size, but the uniform elongation recovered finally after the grain size became larger than 1 μm (for both materials, surprisingly). It should be also noted that the difference in the value of strength between (a) and (b) nearly coincides with the difference in elastic modulus between two materials (E^{Fe}:E^{Al} = 200 GPa:70 GPa = 2.86:1). This unexpected similarity in the flow behaviors between aluminum and iron suggests that the relatively macroscopic deformation behaviors in the ultrafine grained materials are primarily determined by the microstructures itself.

The limited elongation in the ultrafine grained pure aluminum and iron was explained by plastic instability expressed as [5],

$$\sigma > d\sigma/d\varepsilon$$

where σ and ε are true stress and strain, respectively. The ultra-grain refinement greatly increases the strength (σ), especially the yield strength, of the materials, as was shown in Fig.3. On the other hand, in Al (FCC metal having high stacking fault energy) and Fe (BCC metal), dynamic recovery due to cross-slip of screw dislocations is easy to occur, so that the strain hardening rate (dσ/dε) of these materials are originally low. Moreover, the strain hardening rate tends to decrease with decreasing the grain size, as was found in Fig.3, though the exact mechanism

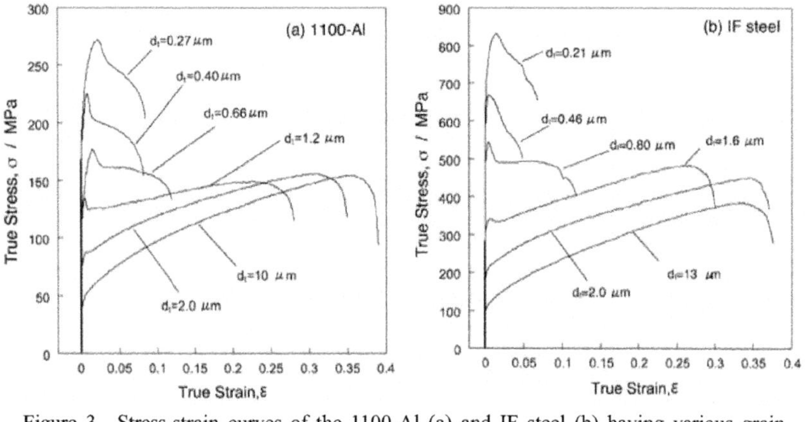

Figure 3 Stress-strain curves of the 1100-Al (a) and IF steel (b) having various grain sizes. The specimens were fabricated by the ARB process followed by annealing.

and reason for it have not yet been clarified. Consequently, the plastic instability condition expressed in the above formula is achieved at early stage of tensile test in the UFG materials. It should be noted, here, again that the present two materials (Al and Fe) have relatively low strain hardening ability originally. Actually, FCC metals and alloys having relatively low stacking fault energy, such as Cu and austenitic steel, showed larger elongation even after the ARB process [13]. Furthermore, adequate uniform elongation can be maintained even in the UFG microstructures keeping high strength, when, for example, fine second particles distribute in the UFG matrix to increase the strain hardening rate [14].

Another curious phenomenon can be seen in Fig.3(a). Yield-drop is obviously recognized even in pure aluminum, when the grain size is smaller than 2 μm. The yield-drop phenomenon in the IF steel (Fig.2(b)) should be also emphasized, because discontinuous yielding does not happen usually in this material where solute carbon and nitrogen are fixed as carbides and nitrides (interstitial free). Therefore, this unexpected discontinuous yielding is considered as a characteristic feature of the UFG materials, though the detailed mechanisms for it have not yet been clarified. It has been also clarified that Lüders band deformation accompanies with the discontinuous yielding even in Al.

Increasing Ductility with SPD Strain

In most of the materials, strength significantly increases with increasing SPD strain, while elongation largely decreases [13]. This is not only the case in the ARB but also in other SPD processes, such as ECAE [15]. However, in a certain kind of Al alloy, it was found that total elongation significantly increased with increasing the ARB strain. Figure 4 shows the stress-strain curves of the (a) 1100-Al (99% commercial purity aluminum) and (b) 8011 (Al-0.63wt%Si-0.72%Fe) ARB processed by various cycles at RT. The 1100-Al (a) performed typical deformation behaviors of the ARB processed materials. The strength increased with increasing the number of the cycles, i.e., strain, while the elongation greatly dropped by one ARB cycles and kept nearly constant value. The tensile strength of the 1100 Al reached up to 340 MPa, which was a surprising value for this material. In case of the 8011-Al, on the other hand, the strength did not significantly increase after 3 cycles. The elongation greatly decreased by 1 cycle of the ARB as well as the 1100, however it significantly increased with increasing further strain and finally recovered up to 20% after 9 cycles. This is an unusual but curious behavior which has not been found in the SPD/UFG materials previously. The detailed mechanism of this

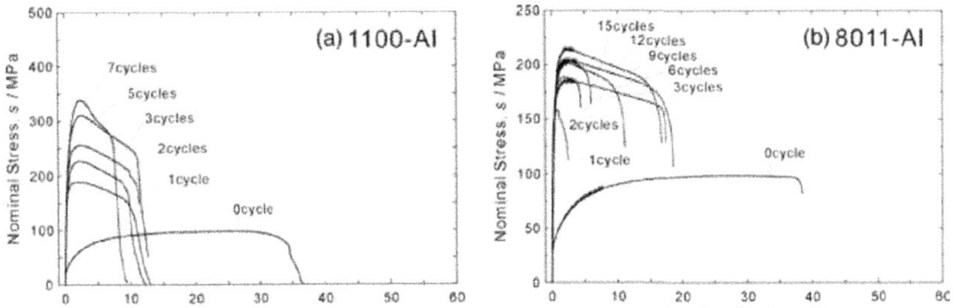

Figure 4 Nominal stress-strain curves of the 1100-Al (a) and 8011-Al (b) ARB processed by various cycles at RT.

increasing ductility is still unclear. However, the author and colleagues have found that the strain rate sensitivity of the 8011 significantly increased with increasing strain, which should be attributed to the increase in the post-uniform elongation [16].

Hall-Petch Relationship

It is well-known that the (yield) strength of the metallic materials proportionally increases to the minus square root of the grain size ($d^{-1/2}$), which is called Hall-Petch relationship. The 0.2% proof stresses of the 1100-Al having various ultrafine grain sizes are plotted as a function of $d^{-1/2}$ in Fig.5. Though there are certain degrees of deviations, the strength shows linear relationship with $d^{-1/2}$. That is, Hall-Petch relationship stands even in the submicrometer grain size ranges. Although so-called inverse Hall-Petch relationship has been sometimes reported in nanocrystalline materials [17], the strength in the present materials constantly increased with decreasing the grain size down to around 200 nm. However, it also showed a significant difference from the conventionally grain sized materials. In Fig.5, the data for the conventionally coarse-grained 5N (99.999%) Al reported by Hansen [18] are also plotted. The specimens were prepared by conventional deformation and recrystallization. It is obviously recognized that the strength of the UFG materials fabricated by the ARB and annealing deviate far above the Hall-Petch slope of the coarse-grained Al. This suggests that the UFG materials fabricated by SPD have special strengthening factor(s) in their structures different from conventionally recrystallized materials. One of the possibilities is a large portion of low-angle grain boundaries in the ARB processed and annealed specimens [11], which may act as the source of dislocation hardening [19]. From this viewpoint, the strength of the ARB processed and annealed Al has been re-evaluated from the quantitative and detailed microstructural parameters [20]. Such evaluation has explained the gap between the UFG materials and the conventionally grain-sized materials in Fig.5 to some extent, but it has not been perfectly dissolved [20]. That

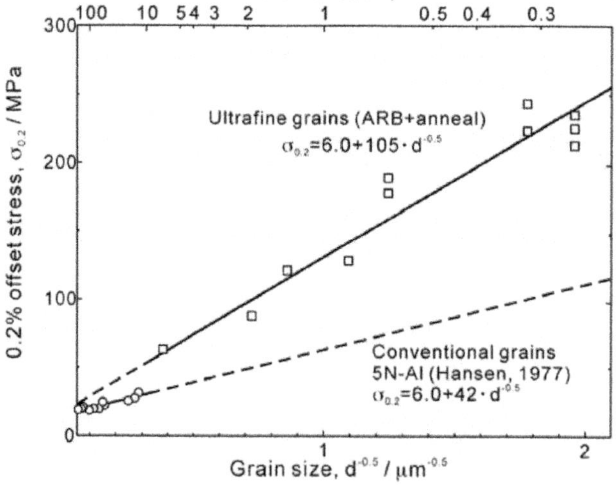

Figure 5 Hall-Petch plot of the 0.2% proof stress of the UFG 1100-Al fabricated by the ARB and annealing process. The data for conventionally grain-sized 5N-Al reported by Hansen [18] are also plotted.

is, there are still unknown strengthening factors in the SPD/UFG materials besides grain boundary strengthening and dislocation hardening. Possible reasons for this extra hardening may be due to dissolution of second phase and their re-precipitation during SPD, and large internal stress.

Conclusion

It has been shown that the ARB process is an advantageous SPD process not only for continuous production of bulky materials but also for fundamental studies to clarify the unknown properties of the UFG materials systematically. The ARB followed by annealing could fabricate the sheets having various ultrafine grain sizes ranging from 200 nm to 20 μm. By using these specimens, mechanical properties of the UFG materials have been studied systematically as a function of the grain size. Various kinds of unique and sometimes surprising deformation behaviors were shown in the UFG materials fabricated by the ARB process. Fundamental research works on such curious behaviors are necessary in the future.

Acknowledgements

The author thanks his colleagues, Emeritus Prof. Y.Saito, Prof. Y.Minamino of Osaka University, Dr. R.Ueji of Kagawa University, Dr. X.Huang, Dr. N.Hansen of Risø National Laboratory, Denmark, and Dr. H.W.Kim of KIMM, Korea, for their fruitful discussions on the subjects described in this paper. The present work was financially supported by the 21st Century COE Program, "the Center of Excellence for Advanced Structural and Functional Materials Design" in Osaka University, through MEXT Japan, and the support is gratefully appreciated.

References

1. Tsuji N., Saito Y., Lee S.H. and Minamino Y., *Adv. Eng. Mater.* 5 (2003), 338-344.
2. Saito Y., Tsuji N., Utsunomiya H., Sakai T. and Hong R.G., *Scripta Mater.* 39 (1998), 1221-1227.
3. Tsuji N., Ito Y., Nakashima H., Yoshida F. and Minamino Y., *Mater. Sci. Forum*, 396-402 (2002), 423-428.
4. Tsuji, N., Kamikawa, N. and Minamino Y., *Mater. Sci. Forum*, 467-470 (2004), 341-346.
5. Tsuji N., Ito Y., Saito Y. and Minamino Y., *Scripta Mater.* 47 (2002), 893-899.
6. Tsuji N., Okuno S., Matsuura T., Koizumi Y. and Minamino Y., *Mater. Sci. Forum*, 426-432 (2003), 2667-2672.
7. Tsuji N., Okuno S., Koizumi Y. and Minamino Y., *Mater. Trans.* 45 (2004), 2272-2281.
8. Hansen N., *Metall Mater. Trans. A*, 32A (2001), 2917-2935.
9. Kamikawa N., Tsuji N. and Minamino Y., *Sci. Tech. Adv. Mater.* 5 (2004), 163-172.
10. Tsuji N., Ueji R. and Minamino Y., *Scripta Mater.*, 47 (2002), 69-76.
11. Huang X., Tsuji N., Hansen N. and Minamino Y., *Mater. Sci. Eng. A*, A340 (2003), 265-271.
12. Humphreys F.J., Prangnell P.B. and Priestner R., *Current Opinion in Solid State and Mater. Sci.*, 5 (2001), 15-21.
13. Tsuji N., Kamikawa N., Kim H.W. and Minamino Y., *Ultrafine Grained Materials III*, TMS, Ohio (2004), 219-226.

14. Tsuji N., Ueji R., Minamino Y. and Saito Y., *Scripta Mater.*, 46 (2002), 305-310.
15. Horita Z., Furukawa M., Oh-ishi K., Nemoto M. and Langdon T.G., *Proc. of the 4th Int. Conf. On Recrystallization and Related Phenomena*, JIM, Tsukuba (1999), 301-308.
16. Kim H.W., Kang S.B., Tsuji N. and Minamino Y., submitted to *Acta Mater.*
17. Suryanarayana C., *Int. Mater. Reviews*, 40 (1995), 41-64.
18. Hansen N., *Acta Metall.*, 25 (1977), 863-869.
19. Hansen N., Huang X., Ueji R. and Tsuji N., *Mater, Sci. Eng. A*, (2004), in press.
20. Ueji R., Huang X., Hansen N., Tsuji N. and Minamino Y., *Mater. Sci. Forum*, 426-432 (2003), 405-410.

Microstructure and Mechanical Properties of Long, Ultrafine-Grained Ti Rods

I.P. Semenova[1*], V.V. Latysh[2], G.H. Sadikova[1], Y.T. Zhu[3], T.C. Lowe[3], R.Z. Valiev[1]
[1] Institute of Physics of Advanced Materials, Ufa State Aviation Technical University, Ufa, Russia
[2] Scientific Design Technical Office, «Iskra», Ufa, Russia
[3]Los Alamos National Laboratory, Los Alamos, NM 87545, USA

Abstract

In this paper, we report our recent studies on the effect of processing routes on the microstructure and mechanical properties of long ultrafine-grained Ti rods. CP Ti (Grade 2) rods were processed into long, ultrafine-grained (UFG) rods with a dimension of 6.5 mm in diameter and more than 1000 mm long), via equal-channel angular pressing followed by thermo-mechanical treatment. This paper presents the UFG microstructures in the processed rods and evolution of mechanical properties at various stages of the processing. It was established that the formation of homogeneous UFG structure with a grain size about 70 nm enhanced the ultimate strength by 2.5 times in comparison with the initial annealed state, while maintaining a good ductility of 12%.

Introduction

Commercially pure (CP) Ti has better corrosion resistance and biocompatibility than Ti alloys. Because of this advantage, CP Ti is widely used to fabricate various medical implants and devices. However, its strength is much lower than that of Ti alloys. Investigators have shown that the grain refinement in metal and alloys leads to significant enhancement in mechanical properties of metallic materials, such as strength, fatigue resistance, and superplasticity [1-5]. Our earlier studies [6-11] revealed the possibility to form UFG structure in bulk billets out of CP Ti by equal-channel angular pressing (ECAP), and further strengthening by cold rolling. However, in those studies we used a laboratory ECAP die set and the size of samples (especially the length) was limited by the die-set geometry. The production of commercial high-strength UFG rods of the specified dimensions represents a great challenge. Another issue is the homogeneity of the UFG structure and mechanical properties along the billet length. The technology to produce long, UFG Ti rods has been recently developed as the next step of the new processing route [12]. It consists of ECAP processing and further thermo-mechanical treatment by special rolling.

This paper presents the microstructural features and mechanical properties of UFG Ti rods with a diameter of 6.5 mm and up to 1 m long. The Ti rods were first processed by ECAP, and then by several stages of thermo-mechanical treatment.

* Corresponding Author. Semenova-ip@mail.ru

Y.T. Zhu and V. Varyukhin (eds.), Nanostructured Materials by High-Pressure Severe Plastic Deformation, 235-240.

Experimental Procedures

The investigation was carried out using hot-rolled Grade-2 CP-Ti rods (purchased from a US supplier) with a dimension of 40 mm in diameter. The composition of the CP-Ti is (wt%): Ti – base, C – 0.07%, O_2 – 0.25%, Fe – 0.3%, N – 0.05%. The as-purchased Ti rods have an equiaxed grain structure with grain size in the range of 25-30 μm in both cross- and longitudinal sections (Fig. 1). Small quantities of particles are located both along the grain boundaries and in the grain interior.

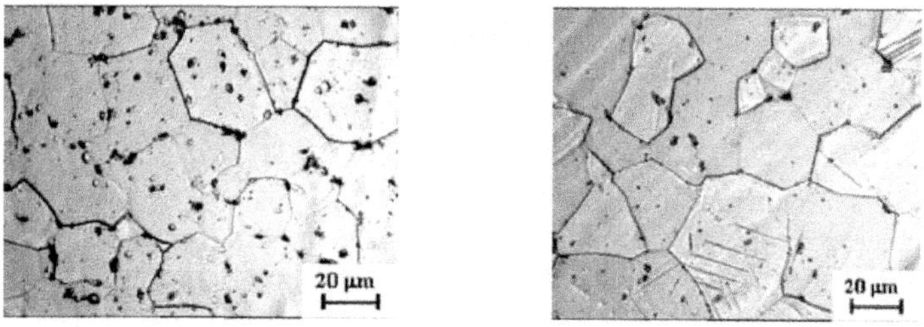

Figure. 1. Optical micrographs of Grade 2 CP Ti in initial state: a) cross section, and b) longitudinal section.

The billets were subjected to ECAP by the following processing parameter: T = 450°C, 90° die angle, route Bc, 4 passes [5]. This was followed by further - mechanical treatment, which resulted in a total accumulated-mechanical strain of 80%.

The microstructures of the billets at various stages of process were investigated using optical metallopgraphy and transmission electron microscopy (TEM).

Mechanical tensile tests were carried out at room temperature on an Instron machine at a loading speed of 1mm/min on samples with a gage diameter of 3mm and a gauge length of 15 mm.

Fig. 2 shows the outer appearance of rods processed by ECAP and by rolling for various strains.

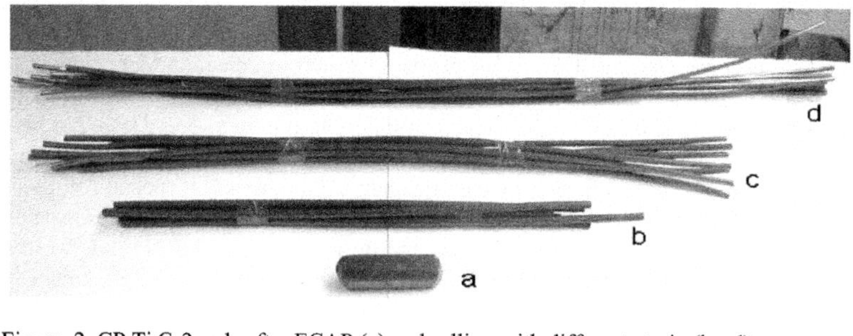

Figure. 2. CP Ti Gr2 rods after ECAP (a) and rolling with different strain (b - d)

Results and Discussion

The experimental results by optical metallography and TEM study are shown in Fig. 3-5. It was found that the structure after ECAP was considerably refined. However, a non-uniform etching testified to non-homogeneous strain and microstructure in the billets (Fig. 3 a, b) in the micrometer scale. Meanwhile, TEM study of the microstructure processed by 4 ECAP passes showed the formation of equiaxed UFG structure in both cross and longitudinal section with grain sizes in the range of 0.3-0.7 μm (Fig. 3 c, d). Lattice dislocations in some small grains are absent.

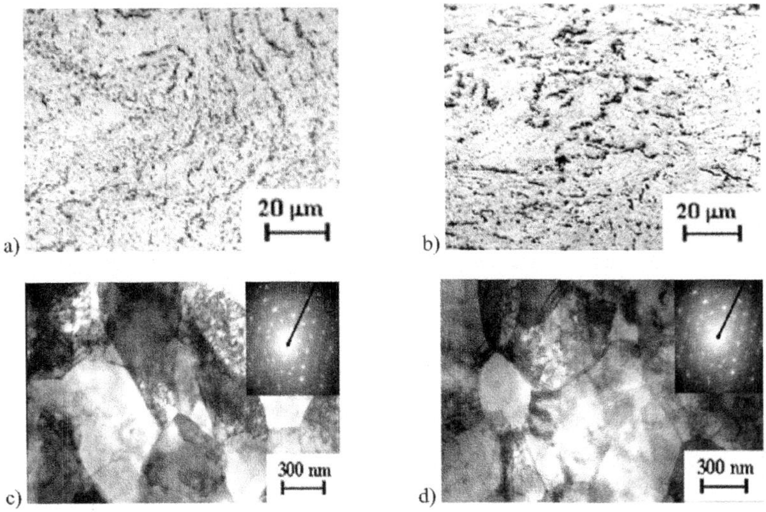

Figure. 3. Microstructure of Grade 2 CP Ti after ECAP for 4 passes: a) cross section (optical); b) longitudinal section (optical); c) cross section, TEM; d) longitudinal section, TEM.

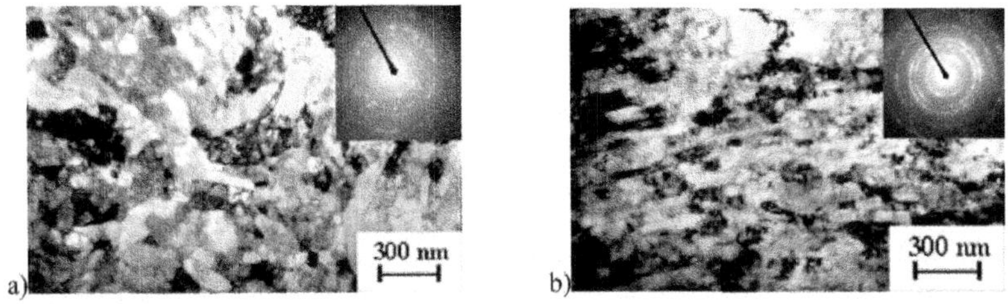

Figure. 4. TEM micrographs of Grade 2 CP Ti after ECAP, 4 passes + thermo-mechanical treatment (60% strain): a) cross section; b) longitudinal section.

Thermo-mechanical treatment after ECAP made the microstructure in the cross section more homogeneous (Fig. 4 a), while elongating the grains in the longitudinal section (Fig. 4 b). At the first stage of the thermo-mechanical treatment processing, additional grain refinement took place and grain-subgrain structure was formed. Size of some grains are about 150 nm. The analysis of SAED patterns revealed that both low-angle and high-angle boundaries were present in the structure considered; the azimuthal spreading of the diffraction spots testified to the increase in internal stress.

After the next stage of thermo-mechanical treatment the microstructure became more homogeneous and equiaxed in the cross section, while remained elongated along the strain direction in the longitudinal section (Fig. 5 a, b). Further refinement of grain-subgrain structure also took place. In longitudinal section grains in the billets were strongly elongated along strain direction, while in the cross-section it had equiaxed form with a mean grain size of about 70 nm. The grains are mostly separated by high-angle boundaries. Dislocation density is very high and it seems impossible to measure its value from TEM micrographs.

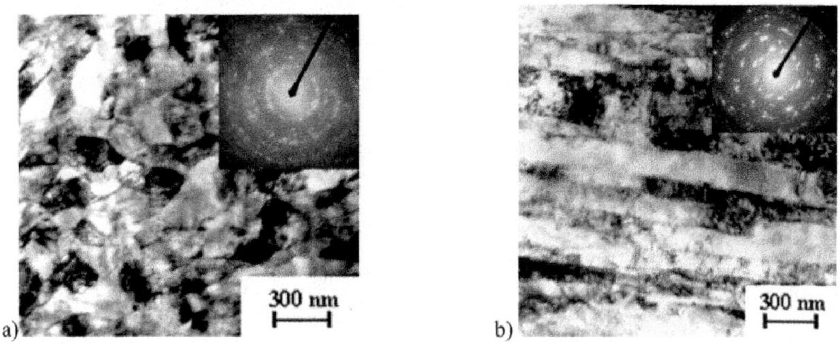

Figure. 5. TEM micrograph of Grade 2 CP Ti after ECAP for 4 passes + thermo-mechanical treatment for 80% strain: a) cross section; b) longitudinal section.

Table 1 Mechanical properties of Ti billets

State	σ_B, MPa	YS, MPa	δ, %	ψ, %
Initial	440	370	38	60
ECAP 4 passes	630	545	22	51
ECAP 4 passes + TMT* ε=60%	1030	845	12	51
ECAP 4 passes + TMT ε=80%	1150	1100	11	56

*TMT = thermo-mechanical treatment

Table 1 presents the mechanical properties of Ti billets at various processing stages. It reveals that the formation of equiaxed structure with grain size decreasing from 25 to 0.7 μm by ECAP resulted in considerable strengthening. The further thermo-mechanical treatment with a total accumulated strain of 60% further enhanced strength up to 1030 MPa, which is 2.5 times higher than the strength of the as-purchased CP Ti, due to additional microstructural refinement (down to grain size of 150 nm) and accumulation of crystalline defects. At the same time, the ductility of the CP Ti decreased to 12 %, which is still fairly good. The reduction of cross-sectional area at fracture ψ is slightly lower, which is caused by a higher deformation localization in UFG samples under tension at room temperature.

To examine the homogeneity of mechanical properties of the processed Ti rod, we studied the mechanical properties on samples cut along the length of a semi-finished rod. The yield and ultimate tensile strengths scatter over a range of 40 MPa, and the elongation in tension scattered over a range of 2%. These results demonstrates a good homogeneity in the mechanical properties along the length of the semi-finished Ti rod.

It has been reported that the yield strength of materials can be represented in the form of superposition of a few independent strength mechanisms [13]:

$$\sigma_{0.2} = \sigma_{s.s} + \Delta\sigma_b + \Delta\sigma_{def} + \Delta\sigma_{s.ph.}$$

In this equation $\sigma_{s.s}$ is the solid-solution strengthening; σ_b is the polycrystalline (grain-boundary) strengthening, σ_{def} is the dislocation strengthening, $\sigma_{s.ph}$ is the multiphase strengthening. The contribution of the solid-solution strengthening ($\sigma_{s.s}$), caused by impurities dissolved in Ti, can be regarded as a constant value for the CP Ti. Structural refinement by ECAP, rendered by the formation of barriers to dislocation slip in the form of new grain and subgrain boundaries during ECAP, in accordance with Hall-Petch equation, contributes considerably to strengthening [12]. Apparently, thermo-mechanical treatment contributes to the formation of a greater amount of crystalline lattice defects, which in turn contributes to additional strengthening of the material.

Conclusion

We have studied the microstructural evolution and mechanical properties of Ti rods at various processing stages. The study shows that ECAP processing contributed significantly to the strength enhancement of Grade 2 CP-Ti, which was rendered by the grain refinement. The preservation of considerable ductility in UFG Ti after ECAP allowed us to conduct further thermo-mechanical processing, which contributed to additional grain refinement and higher dislocation density. As a result, homogeneous UFG structure with grain size about 70 nm was formed in a semi-finished CP Ti rod, resulting in the enhancement of strength by 2.5 times in comparison with its annealed state as well as a good ductility of 11 %.

Acknowledgements

The present work was carried out within the frames of ISTC project 2398p, which was funded by the US DOE IPP program office.

References

1. Valiev R.Z., Islamgaliev R.K., Alexandrov I.V. *Progr. Mater. Sci.* 45(2) 2000, 103-189.
2. Valiev R.Z. *Proc NATO ARW*; 2000.
3. Iwahashi Y., Horita Z., Nemoto M., Langdon T.G. *Acta Mater.* 46 (1998), 3317.
4. Iwahashi Y., Horita Z., Nemoto M., Langdon T.G. *Metall. Mater. Trans.* 29A (1998), 2503.
5. Horita Z., Furukawa M., Nemoto M., Barnes A.J., Langdon T.G. *Acta Mater.* 48 (2000), 3633.
6. Valiev R.Z., Stolyarov V.V., Latish V.V., Raab G.I., Lowe T. C., Zhu Y.T. *9th Int. Conf. Titanium-99*, V3(2000), 1569-1572.
7. Stolyarov V.V., Latish V.V., Valiev R.Z., Zhu Y.T., Lowe T. C. *Investigations and Aplications of Severe Plastic Deformation*, 2000, 367-372.

8. Stolyarov V.V., Zhu Y.T., Alexandrov I.V., Lowe T. C. and Valiev R.Z. *Mater. Sci. Eng.* A 299, 2001, 59.
9. Stolyarov V.V., Zhu Y.T., Alexandrov I.V., Lowe T. C. and Valiev R.Z. *Mater. Sci. Eng.* A 303, 2001, 82.
10. Stolyarov V.V., Zhu Y.T., Alexandrov I.V., Lowe T. C. and Valiev R.Z. *MRS.* A343 (2003), 43-50.
11. Zhu Y.T., Huang J.Y., Gubicza J., Ungar T., Wang Y.M., Ma E.and Valiev R.Z. *MRS.* Vol.18 №8, 1908-1916.
12. Patent of USA № 6,399,215. MKI[7] C22C14/00; C22F001/18. Ultrafine-grained titanium for medical implants. - Published 4.07.2002.
13. Goldshtein M.I., Litvinov V.S., Bronfin M.F. *Metallophisics of high- strength alloys.* Moscow: Metallurgia, 1986, p. 312.

Improving the Mechanical Properties of Ti-6Al-4V Alloy by Equal Channel Angular Pressing

L.R. Saitova[1], I.P. Semenova[*1], G.I. Raab[1], R.Z. Valiev[1],
T.C. Lowe[2], Y.T. Zhu[2]
[1]Ufa State Aviation Technical University, Ufa, Russia
[2]Los Alamos National Laboratory, Los Alamos, NM 87545, USA

Abstract

In this study, we attempted to enhance the strength of the two-phase Ti-6Al-4V alloy using severe plastic deformation (SPD) and thermo-mechanical treatment. It was established that structure refinement by ECAP significantly improved the strength of this alloy. The variation of the initial structure before ECAP was found to significantly affect the microstructures and mechanical properties after ECAP. Lower ECAP temperature resulted in finer microstructure, higher density of crystalline lattice defects, and consequently higher strength. Up to 40% of improvement in strength has been obtained in ECAP-processed Ti-6Al-4V alloy, compared with its annealed state.

Introduction

The Ti-6Al-4V alloy is widely used in various industries as a structural material due to its high specific strength and excellent corrosion resistance. Its strength can be enhanced further by conventional thermal and thermo-mechanical treatments. However, although capable of improving the strength of Ti-6Al-4V alloy, these treatments also usually sacrifice the ductility, which is one of the most important properties for structural applications.

It is well-known that ultrafine-grained (UFG) structures with grain sizes less than 1 μm could significantly improve mechanical properties [1,2]. One of the most promising methods to produce UFG structure in metals and alloys is equal-channel angular pressing (ECAP), which allows to preserve the same geometrical dimensions of billets [1-3]. Processing two-phase Ti-6Al-4V by ECAP is a complicated task. First, this alloy has very low deformability at room temperature. To overcome this difficulty, we have to do the ECAP processing at elevated temperatures and to use large intersection angles of die channels [4-6]. Moreover, the phase components and the morphology of metastable phases precipitating during the ECAP have great influence on the workability of the alloy.

This paper presents the results of the study of ECAP and thermo-mechanical treatment effect on the UFG microstructure formation in the Ti-6Al-4V alloy and its mechanical properties.

Experimental Procedures

The investigations were carried out on medical grade Ti-6Al-4V ELI alloy billets with a diameter of 40 mm (Intrinsic Devices Company). Its composition is (in wt%):

[*] Corresponding Author. semenova-ip@mail.ru

Y.T. Zhu and V. Varyukhin (eds.), Nanostructured Materials by High-Pressure Severe Plastic Deformation, 241-246.

Ti – base, Al – 6.0%; V – 4.2%; Fe – 0.2%; C – 0.001%; O_2 – 0.11%; N_2 – 0.0025%; H_2 – 0.002%. The temperature of its phase transformation is 960°C. The billets were processed by ECAP route Bc in the temperature range of 600-700°C using a die set with a channel intersection angle of 120° [4,5]. After ECAP, multi-pass extrusion was carried out at 300°C with drawing ratio equal to λ=4.9 (relation between initial area of billet cross section to the final one) (Treatment 1). One of the billets was subjected to thermal treatment before ECAP and extrusion: the billet was heated in β-phase region (1000°C) for 40 minutes with water quenching and further aging at 675 °C for 4 hours (Treatment 2).

The as-processed microstructure was studied by optical metallography and transmission electron microscopy (TEM). Mechanical tensile tests were carried out at room temperature (20°C) on an Instron machine at a displacement rate of 1 mm/min. The test samples have a diameter of 3 mm and a length of 15 mm in the gage section. They were cut out from the central part of a rod along its longitudinal axis. Three samples were tested for each state.

Results and discussion

Treatment 1

The structure of billets in as-purchasedf state was of α+β type, consisting of globular grains of α-phase with the sizes of 5-10 μm in the cross section and 20-25 μm in the longitudinal section. Interphase areas are dispersed with transformed β-phase structure (Fig. 1).

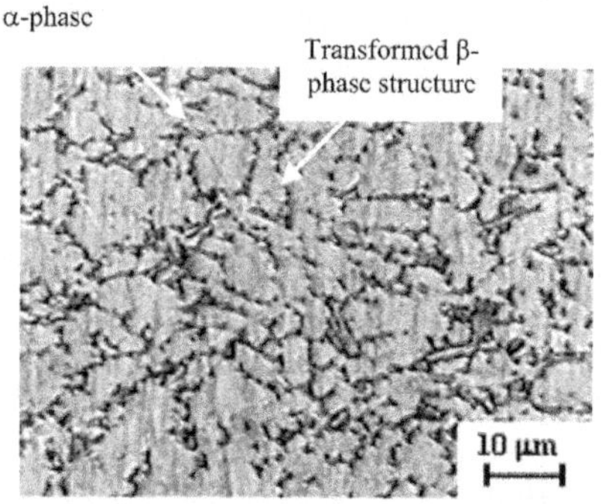

Fig. 1 Optical microscopy showing microstructures of the Ti-6Al-4V ELI alloy in initial state. Cross section.

Fig. 2 presents typical TEM images of the Ti-6Al-4V ELI alloy microstructure after ECAP (4 passes, e=2.8). The ECAP resulted in the fragmentation of globular grains of α-phase, which was caused by dislocation slip and twinning processes.

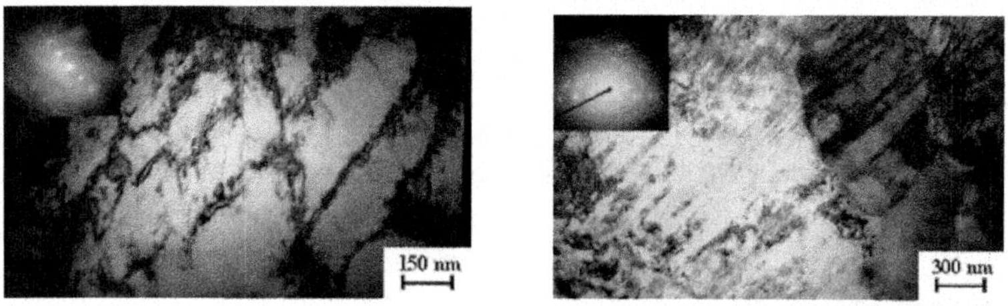

a) b)

Fig. 2 TEM micrographs of the Ti-6Al-4V ELI alloy with initial equiaxed structure subjected to ECAP, cross section: a) formation of α-phase subgrains; b) grains with microtwins.

Fig. 2a presents the formation of subgrains with low-angle boundary misorientations of about 200 nm in size with lattice dislocations nearly absent in them. Single grains contain fine microtwins of about 50 nm thick (Fig. 2 b). According to SAED patterns low-angle boundaries of grains/subgrains prevail in the investigated structure.

Further ECAP at a lower temperature of 300°C led to further structural refinement via intense twinning. Fig. 3 presents a typical image of microstructure after ECAP and warm extrusion. Grain/subgrain size of α-phase varies in the range of 200 to 500 nm. SAED patterns reveal the increase in the amount of diffraction spots, uniformly distributed along circles, which attests to the formation of high-angle grain boundaries. Strong azimuth and tangential scattering of spots is also observed, testifying to the presence of high internal macro- and microstress. However, the ECAP-processed grain/subgrain structure is not sufficiently refined, containing grains with thin twins, almost completely dislocation-free areas together with areas of high-dislocation density (up to $10^{13}cm^{-1}$) (Fig. 3). A considerable elongation of structural elements along extrusion direction is observed in the cross section of billets.

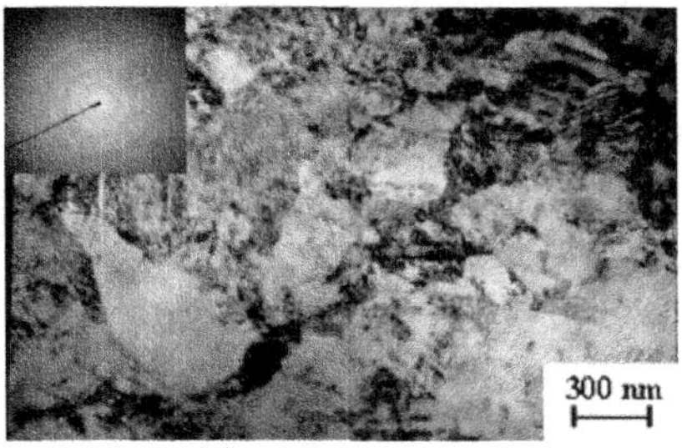

Fig. 3 Microstructure of the Ti-6Al-4V ELI alloy subjected to ECAP and extrusion. Cross section, bright field, TEM.

Treatment 2

Billet structure, after quenching and further aging, is observed in the form of β-transformed lamellar structure, consisting of α-phase lamella packages (Fig. 4). Billets with lamellar structure remained integral during 2 ECAP passes with total accumulated strain equal to e=1.4, while as-delivered billets endured 4 passes (e=2.8) due to their higher density, connected with globular component of α-phase.

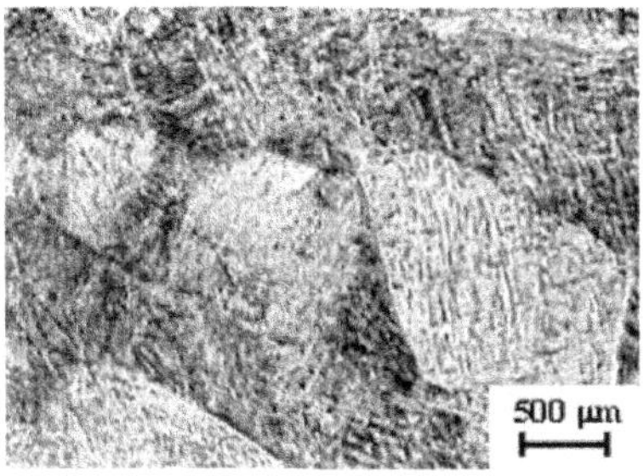

Fig. 4 Optical microscopy showing microstructures of the Ti-6Al-4V ELI alloy after water quenching from 1000°C and overaging at 675°C for 4 hours.

Fig. 5 presents a typical TEM image of billet microstructure with initial lamellar structure after ECAP and further warm extrusion. The structure reveals the fragmentation and curving of α-phase lamellas as a result of ECAP. Thickness of α-phase lamellas is nearly the same and mean size of its fragments is about 300 nm. Internal structure of deformed lamellas is characterized by high dislocation density and in certain cases by the presence of thin twins. SAED patterns show clear spots distributed along several circles. This can be due to the existence of both low- and high-angle grain boundaries. Weaker azimuth spreading of spots in comparison with the previous sample (see Fig. 3), is possibly caused by a relatively lower level of micro- and macro-stress in the structure formed.

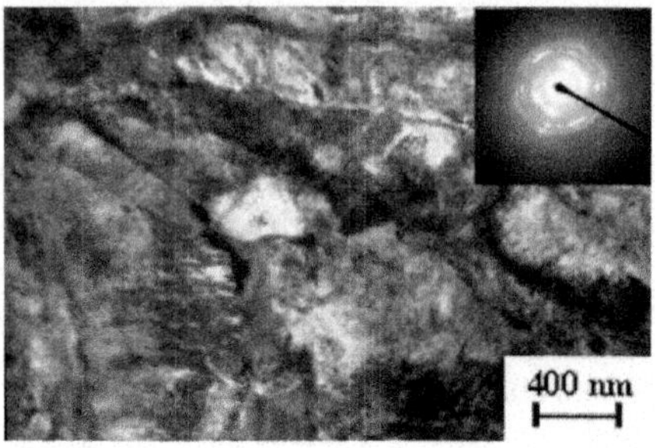

Fig. 5 Microstructure of the Ti-6Al-4V ELI alloy with initial fine lamellar structure subjected to ECAP and extrusion. Cross section, bright field, TEM.

The strength of the alloy increased by nearly 20% due to strong microstructural refinement (Table 1). Warm extrusion of billets led to additional microstructural refinement and possibly increased the total amount of large-angle grain boundaries due to slip processes and twinning. Finally, by treatment 2 and 1 increased the strength of the billets to 1360 and 1300 MPa, respectively. The ductility after treatment 2 slightly increased to δ=12% (δ=8% after treatment 1). Obviously, the better ductility in the alloy with initial lamellar structure (treatment 2) at room temperature are connected with the more uniform strain during ECAP and extrusion. This is proved by the structural investigations, which revealed relatively low level of micro- and macro-stress in billets after treatment 2.

The yield strength of metals and alloys can be described in the form of superposition of several strengthening mechanisms: $\sigma_{0.2} = \sigma_{ss} + \Delta\sigma_b + \Delta\sigma_{def} + \Delta\sigma_{s.ph.}$, where σ_{ss} is the contribution from solid-solution strengthening, σ_b is polycrystalline (grain-boundary) strengthening, σ_{def} is dislocation strengthening, $\sigma_{s.ph}$ is multiphase strengthening [8].

The estimate of their quantitative relationship with some simplification [8,9] enables one to find the contribution from each active strengthening mechanisms. The starting data for these calculations are the alloy chemical composition, quantitative microstructure parameters, phase composition (see Tab. 2).

Table 1 The mechanical properties of Ti-6Al-4V ELI alloy in different processing states.

State	UTS, MPa	0.2YS, MPa	δ, %	ψ, %
Initial	940	840	16	45
ECAP, 600°C, 4 passes	1090	1030	9	40
ECAP, 600°C, 4 passes + extrusion 300°C, λ=4.9	1300	1230	8	35
Preliminary thermal treatment + ECAP, 600°C, 2 passes + extrusion 300°C, λ=4.9	1360	1250	12	39

Table 2 Analysis of strengthening components in Ti-6Al-4V ELI alloy.

State	The calculated values of yield stress $\sigma_{0.2}$, MPa					The experimental value of 0.2 YS, MPa
	The strengthening components				Sum	
	$\sigma_{s.s}$	$\Delta\sigma_b$	$\Delta\sigma_{def}$	$\Delta\sigma_{s.ph}$	$\Sigma\sigma_{0.2}$	
Initial	550	150	---	130	830	840
ECAP	550	500	20	130	1200	1030
ECAP + extrusion	550	520	100	130	1300	1230

It is seen that the contribution of solid-solution strengthening (σ_{ss}), which is caused by dissolving all alloying elements and, can be regarded as constant for all the states of the alloy. The contribution of grain-boundary strengthening constitutes up to 40% of total strength, which is caused by the formation of barrier for dislocation motion in the form of new grain and subgrain boundaries during ECAP [7,8]. The further extrusion gives additional contribution to the increase of yield stress of the alloy by the dislocation strengthening mechanism.

Conclusion

This study has allowed us to establish that ECAP in combination with thermal and thermomechanical treatment contributes to the formation of UFG structure in the Ti-6Al-4V alloy. Strength was increased up to 1360 MPa while retaining a ductility of about 12%.

The microstructural refinement produced by ECAP is the main cause for the improved strength. The subsequent thermomechanical treatment gives additional contribution to the alloy strength due to dislocation density increase and further microstructural refinement.

It was shown that one more way to enhance strength and ductility of the deformed billets was the prior thermal treatment to form fine-lamellar microstructure before ECAP.

Acknowledgements

This work was carried out within the frame of ISTC project 2398p. The project was supported by US DOE Initiative on Proliferation Prevention (IPP) program office.

References

[1] Valiev R.Z., Islamgaliev R.K., Alexandrov I.V. *Progr. Mater. Sci.* 45(2) 2000, p. 103-189.
[2] *Ultrafine Grained Materials III, TMS* (The Minerals, Metals and Materials Society). 2004.
[3] Segal V.M. The development of essentials of material processing by severe plastic deformation. Metally, (1), 2004, p. 5-14.
[4] Ko Y.G., Jung W.S., Shin D.H., Lee C.S. *Scripta Mater.* 2003. V. 48 P. 197-202.

[5] Semenova I.P., Raab G.I., Saitova L.R. *Book of abstracts 13 International conference on the strength of materials*. 2003, p. 378.
[6] Semenova I.P., Zhu Y.T., Raab G.I., Lowe T.C., Valiev R.Z. *Ultrafine Grained Materials III, TMS* (The Minerals, Metals and Materials Society). 2004, p. 463-466.
[7] Materials Properties Handbook: Titanium Alloys / editors, R. Boyer, G. Welsch, E.W. Collings, 1998.
[8] Goldshtein M.I., Litvinov V.S., Bronfin M.F. *Metallophisics of high- strength alloys*. Moscow: Metallurgia, 1986, p. 312.
[9] Popov A.A. *Proceedings of the conference "Nanomaterials by severe plastic deformation – NANOSPD2"*. 2002, p. 817.

Structural Features, Strength, and Mechanisms of Deformation of Nanocrystalline Materials

N. I. Noskova*
Institute of Metal Physics of UrD Russian Academy of Science,Ekaterinburg, 620219, Russia

Abstract

Results of recent studies of structure and properties of nanocrystalline metals and alloys produced by severe plastic deformation and by nanocrystallization of amorphous alloys are considered. High resolution transmission electron microscopy, scanning electron microscopy, and *in situ* deformation in the column of an electron microscope were used to analyze the structures and the mechanisms of plastic deformation of nanocrystalline materials.

Introduction

Nanostructured materials can be produced by different processing methods, namely, evaporation-condensation of nanosized particles with further compaction, mechanical activation, super fast quenching from a melt with further crystallization, and severe plastic deformation (SPD) [1]. It is commonly accepted that a nanocrystalline grain in a nanostructural multiphase material does not exceed a nanophase size ($\leq$ 100 nm). Due to their unique strength and functional properties nanocrystalline materials undoubtedly raise much interest among material researchers. The present paper is aimed to analyze structural features, strength characteristics and plastic deformation mechanisms in a number of nanocrystalline materials.

Experimental Procedures

Nanocrystalline metals (Ni, Cu, Fe, Ti) and alloys (Al+0,1%Zr, Al+1%Hf, Al+0.5%Ce, Al+0.2%Nb, Al+0,5%Re, Al+0.2%Sn, Al+10%Mg, Al+1%Hf+0.2%Nb+0.2%Sn, Al+0.5%Ce+0.1%Zr+0.5%Re) produced by severe plastic deformation (SPD) and alloys ($Fe_{73.5}Cu_1Nb_3Si_{13.5}B_9$, $Fe_{60}Cr_{18}Ni_7Si_3B_{12}$ and $Fe_{60}Cr_{18}Ni_7Si_5B_{10}$) produced by nanocrystallization of amorphous alloys were studied. The structure of nanocrystalline metals and alloys are analyzed by transmission electron microscopy JEM –200KX. Mechanical properties and the hardness of nanocrystalline materials will be studied. *In situ* deformation in the column of an electron microscope was used to analyze the mechanisms of plastic deformation in nanocrystalline metals and alloys.

* Corresponding Author: noskova@imp.uran.ru

Y.T. Zhu and V. Varyukhin (eds.), Nanostructured Materials by High-Pressure Severe Plastic Deformation, 247-252.

Results and Discussion

Structural features of nanocrystalline metals and alloys

Pure Metals: The investigation by transmission electron microscopy and by high resolution transmission electron microscopy have shown that FCC nanocrystalline metals (Ni, Cu, Fe, Ti) processed by SPD, are characterized by a high level of internal distortions within a nanosized grain body, which increase with approaching the grain boundary, and a non-equilibrium state of grain boundaries attributed to high dislocation density [1-5]. Nanocrystalline Mo and W processed by SPD possess "wider" grain boundary interlayers with high dislocation density than FCC-metals [5], which is most likely due to low dislocation mobility in BCC metals. Finally, nanocrystalline SPD processed Ti [6]and Co have two types of boundaries in their structure, interphase and grain ones, that, in turn, results in different dislocation densities in them. Unlike the spectrum of disorientation of a conventional crystal the spectrum of disorientation of nanosized grains is shifted to a low angle range.

Alloys. The structure of nanosized grains and grain boundaries of nanocrystalline multiphase SPD processed alloys has been investigated in a number of publications [1-4, 7,8]. Different processing schemes (modes) for formation of nanosized grains and phase generation in such alloys been proposed in [1-4,7,8]. It is shown [9-10] that plastic deformation within the range of high strains results in formation (transformation) of internal structure of different alloys by at least three ways (mechanisms), namely, continuous refining of structure, non-monotonous refining with a "threshold" of internal elastic distortions overcoming of which results in structure refining and combination of phase formation and structure refining. According to these publications, the crystal lattice of nanosized grains of alloys with internal elastic distortions similar to that of pure metals is distorted most of all at nanosized grain boundaries. Evidently, only nanosized phases of the smallest size (4-6 nm) may be free of defects. As to grain boundaries of nanosized grains and phases, they can be quasi-amorphous, heavily-distorted crystalline, of a twinning type or ones with incompatibility dislocations similar to a transition layer with another chemical composition. It is assumed that a wide range of specific features is apt to the structure of nano-phases and boundaries in nano-phase alloys.

Mechanical properties (strength and plastic), the hardness of nanocrystalline materials

Mechanical properties of nanocrystalline materials are determined mainly by the nanograin size *d* and the state of nanograin boundaries. Thus, with decreasing *d* from 1500 to 5 nm the microhardness *H* of pure metals may increase by a factor of 2 - 6 (Table 1) [1,4]. The yield strength $\sigma_{0.2}$ and ultimate tensile strength σ_B of nanocrystalline metals also grow compared to those of the coarse grain state, while the low temperature plasticity δ decreases (Table 1). Tensile diagrams of SPD processed pure nanocrystalline metals in "stress-strain" coordinates differ from that of conventional polycrystals. Fig.1 demonstrates typical tensile diagrams for coarse grain and nanocrystalline copper, armco-iron and titanium with FCC, BCC and HCP crystal structure, respectively.

Table 1. Microhardness (H), yield strength ($\sigma_{0.2}$) and ultimate tensile strength (σ_B) of SPD obtained submicro- and nanocrystalline pure metals with different grain size (d) [1]

Metal	d change, nm (from – to)	H change, GPa (from – to)	$\sigma_{0.2}$ change, MPa (from – to)	σ_B change, MPa (from – to)	δ change, % (from – to)
Ti	50000 - 40	0.8 - 6.7	275 - 980	420 - 1310	29 - 5
Fe	10000 - 80	1.8 - 4.5	375 - 1010	480 - 1100	25 - 15
Cr	1500 - 70	-	313 - 780	485 - 960	-
Al	1500 - 200	0.3 - 1.2	13 -26	42 - 176	20 - 2
Mo	500 - 50	1.0 - 6.0	-	-	-
W	200 - 40	3.0 - 6.2	-	-	-
Cu	200 - 70	0.9 - 1.4	60 -365	120 - 650	60 - 30
Cu	70 -20	1.4 – 1.8	-	-	-

For nanocrystalline metals extension and slope angle of the linear elastic portion of the diagram increase drastically (Fig. 1a,b,c). For nanocrystalline Fe with a grain size of 40 nm the "yield tooth" and the "yield flat" disappear and the extension of the linear elastic portion of the tensile diagram increases (Fig. 1b). For nanocrystalline titanium with a grain size of 40 nm the extension and slope angle of the linear elastic portion of the tensile diagram increase as well (Fig. 1c). In the majority of cases the plasticity of pure nanocrystalline metals at low testing temperatures is lower than that of conventional polycrystals (Table 1).

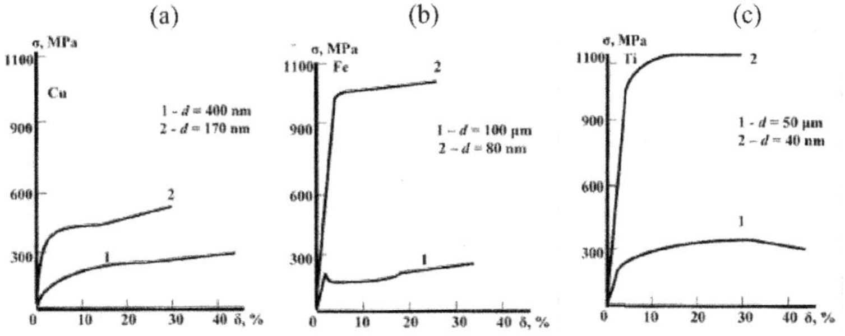

Fig.1.Uniaxial tensile diagrams for coarse-crystalline and nanocrystalline (SPD processed) metals: a – Cu, b – Fe, c – Ti.

Mechanical properties of some nanocrystalline multiphase alloys are presented in Table 2 and Fig.2. It is seen (Table 2) that in alloys, similar to pure metals, the transition to a structure with sub-micro- or nanosized grains is accompanied by the increase in microhardness and strength. Besides, there is observed an increase in plasticity in the alloys in the state of superplasticity at high temperatures [1].

Table 2. Nanograin size (*d*), microhardness (*H*), yield strength ($\sigma_{0.2}$), ultimate tensile strength (σ_B) and relative elongation (δ) of submicro- and nanocrystalline SPD obtained alloys [1]

Alloy	***d*** change, nm from–to	***H*** change GPa from–to	$\sigma_{0.2}$ change, MPa from – to	σ_B change, MPa from–to	δ change, % from–to
FeCuNbSiB	200-6	6,0- 15,0	140-2180	140-2280	0-0.8
CoFeSiB	100-25	10,5-13,0	945-1880	950-2100	0-2.6
-«-	25-8	13,0- 9,0	-	-	-
VT-6	12000-40	-	950-1080	1050-1350	9 - 7
Fe-12%Cr-18%Ni-10%Ti	3000-150	-	500-1600	-	-
Fe-12%Cr-25%Ti	2000-50	1.8-6.5	-	1500-2500	-
AlNiCeFe	2000-80	-	-	120-1560	-
Al-10%Mg	2500-900	0,3-0,5	120-170	180-275	0,5-1,1
Al-1%Re	2000-440	0,3-0,6	30-150	35-180	
Al-1%Hf	2500-760	0,3-0,4	30-160	40-180	
Al-1%Hf-0,2%Nb-0,2%Sn	2000–110 110-60	0,4-0.8 0,8-1,8	40–140 -	60–200 -	0,3-4,8 -
Al-0,5%Ce-0.5%Re-0,1%Zr	2000–150 120-40	0,4-0,65 1,0-1,9	40–180 -	80–220 -	0,1-5,2 -

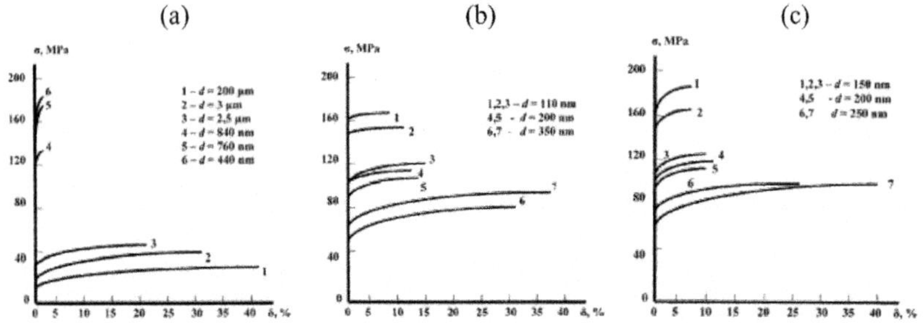

Fig. 2. Uniaxial tensile diagrams for coarse-crystalline and nanocrystalline (SPD processed) alloys:a- 1-Al ; 2-Al+0.5%Re; 3-Al+1%Hf; 4- Al+0,2%Sn; 5-Al+0.5%Ce; 6-Al+0.1%Zr; b–Al+1%Hf+0.2%Nb+0.2%Sn, c– Al+0.5%Ce+0.12%Zr+0.5%Re

Similar to pure nanocrystalline metals, the Hall-Petch dependence for nanophase alloys is valid not within the whole nanophase size range [1,2,7]. In most cases the Hall-Petch law is not valid for alloys with a grain size of 30 nm and less, the coefficient k having a negative value. The possible causes for violation of the Hall-Petch law for nanocrystalline alloys are the same as for pure metals, namely, modification of the plastic deformation mechanism and the weakening of nanograin boundaries.

Plastic deformation mechanism of nanocrystalline materials

The results of "*in situ*" deformation investigations [1,5,6,9-12] show that tensile deformation of nanocrystalline nickel with the grain size of 40-70 nm is realized both through dislocation sliding and nanograin turn. There arise dislocation pile-ups near grain boundaries in coarser grains. In fine nanograins dislocations are immobile and nanograins undergo the turn which, as a rule, is accompanied by the growth of density of grain boundary dislocations and the formation of a deformation band along grain boundaries of a number of nanograins.

Figures 3a-c and Figures 4 a-c demonstrate transmission electron microscope images of one and the same site of microstructure of Al + 1% Hf (Fig.3) and Al-0,5%Ce-0.5%Re-0,1%Zr (Fig.4) [11,12] nanocrystalline alloys subjected to tensile deformation at different strain values increased successively ("*in situ*" method). As seen from Fig. 3 and Fig.4 deformation is realized by the dislocation-rotation mechanism (with increasing strain value there occurs a turn of nanograins and the density of garin boundary dislocations increases). Particles of a phase , no more than 10 nm in size, are observed at nanograin boundaries. With strain, phase particles, on the one hand, evidently, retard the movement of dislocations, and, on the other hand,

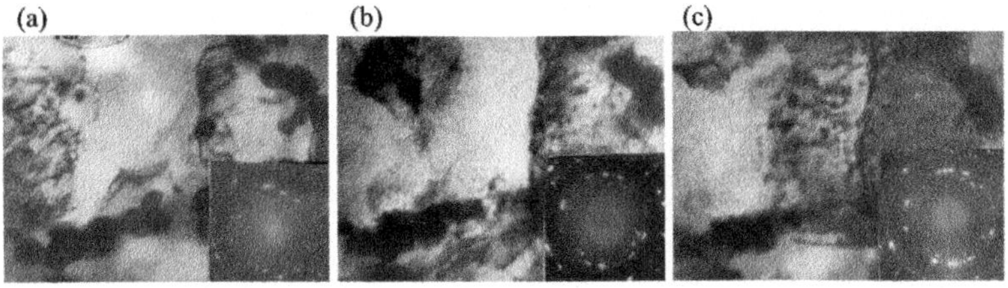

Fig.3 TEM successive images of one and the same area of nanocrystalline Al + 1% Hf alloy (d = 60 nm) deformed in the electron-microscope column: a– δ=9.0; b– δ=14.5; c– δ=24.5%.

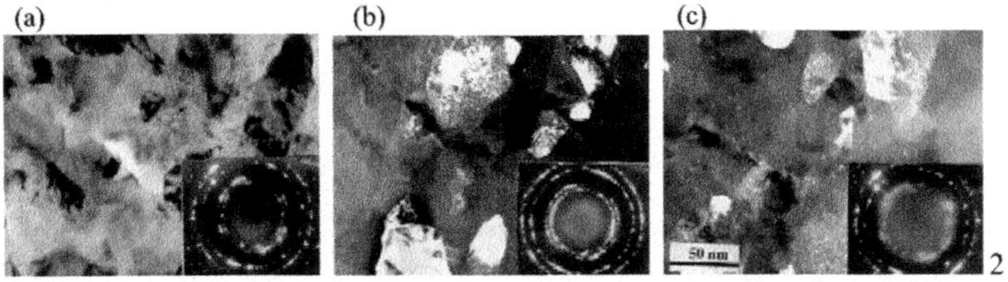

Fig.4.TEM successive images of one and the same area of nanocrystalline Al-0,5%Ce-0.5%Re-0,1%Zr alloy (d= 40 nm) deformed in the electron-microscope column: a – δ = 9.0; b – δ = 24.5; c– δ = 67.5%.

interacting with grain boundary dislocations these particles undergo "dissolution". Concurrently there occurs formation of disperse phases, no more than 1 nm in size, within the nanograin interior that surely results in strengthening of nanograins.

The view of deformation process in a nanocrystalline material through nanograins turn and nanograins displacement along boundaries is apparently the closest one to the real mechanism of deformation. One should take into account dependence of strength and plasticity on concentration of the second nanophase in a nanocrystalline alloy forming in the process of its deformation, because elastic modulus changes appreciably with growth of a quantitative ratio of matrix and the second phase concentration.

Conclusion

The grain size refinement to several tenths of nanometers in nanocrystalline metals and alloys is accompanied by the considerable growth of their strength and hardness. The *"in situ"* study of deformation of FCC, BCC and HCP nanocrystalline materials made it possible to conclude that during tensile deformation of a nanocrystaline material resulting in a decrease of a nanograin size there occur rotation modes which due to their cooperative character cause the development of mesoscopic deformation shifts at grain boundaries. Development of the rotation deformation modes in nanocrystalline materials leads to initiation of rotation plasticity in the form of grain boundary mesoscopic shifts.

Acknowledgment

This work was supported by the Russian Foundation for Basic Research, grants, No 02-03-96413, No 04-02-17674

References

1. Noskova N.I., and Mulyukov R.R. *Submicrocrystalline and Nanocrystalline Metals and Alloys.* ***UD RAS,*** 2003, p. 269. (in Ressian).
2. Morokhov I.D., Trusov V.I., Lapovok L.I. *Physical phenomena in ultra-fine environments.* Moscow, Nauka,. 1984. (in Russian).
3. Akhmadeev N.A., Valiev R.Z., Kopylov V.I. *Izvestiya. Akademii Nauk SSSR, Metallyr.,* 1992 ,Vol.5, p.96 (in Russian).
4. Noskova N., Korshunov L., Potapov A., Tchernenko N. *Proc. "Investigation and Applications of Severe Plastic Deformation",* Kluwer Academic Publishers. 2000, p.115.
5. Noskova N.I., *The Physics of Metals and Metallography.,* 2002 Vol.94.Suppl. 1, p.S119.
6. Noskova. N.I.,Pereturina I.A, Stolyarov V.V.,.Elkina O.A. *The Physics of Metals and Metallography,* 2004, Vol.97, No 5, p.531.
7. Noskova N.I., Korznikov A.V. *The Physics of Metals and Metallograph,* 2002. V.94. Suppl. 1, p.S24-S29
8. .Noskova N.I., Ponomareva E.G., and Myshlyaev M.M.,. *The Physics of Metals and Metallography,* 1997, Vol. 83 (5), p.511
9. Noskova N.I., Volkova E.G. *The Physics of Metals and Metallography,* 2001, Vol.91(6), p.629.
10. Noskova N.I.,.Volkova E.G. *The Physics of Metals and Metallography,* 2001, Vol.92 (4).
11. Noskova N.I.,.Vildanova,.Churbaev R.V, *The Physics of Metals and Metallography* 2004, in presse
12. Noskova N.I., Vildanova N.F.,.Pereturina I.A., Filippov Yr.I. *.The Physics of Metals and Metallography,* 2004, in presse

Response of Mechanical Properties across the Sample to ECAP

A.I.Korshunov[1*], I.I.Vedernikova[1], L.V.Polyakov[1], T.N.Kravchenko[1], A.A.Smolyakov[1], P.N.Nizovtsev[1]
[1]Russian Federal Nuclear Center – VNIIEF
37 Mira Ave. Sarov, Nizhniy Novgorod region, 607190 Russia

Abstract

Heterogeneity in mechanical properties across a sample depending on the number of ECAP passes was studied using annealed M1 copper (min. Cu content 99.9 %) as a test material. The samples under examination had a square cross-section with a side length of 8 mm and were cut from an 8-mm thick sheet along the rolling direction.

Microhardness distribution across the sample was determined. Mechanical properties in tension (tensile strength, conventional offset yield strength, and elongation to rupture) were determined at 9 points across the sample using small-size specimens, 1.5 mm in diameter, cut out along the pressing direction.

The investigations were carried out after one and two passes for routes B_C and C.

Introduction

Creating materials with qualitatively new properties attracts a significant scientific and practical attention. Severe plastic deformation, and in particular, equal channel angular pressing (ECAP), is one of the methods for fabricating ultrafine materials [1].

In creating ultrafine materials it is important not only to produce unique macroproperties, but also to ensure uniformity of physical and mechanical properties throughout the sample, which is required for parts manufactured later to be able to perform effectively. Results of ECAP macromodeling [2,3] suggest that there is considerable heterogeneity in accumulated strain across the sample. This means that the same heterogeneity should be observed in mechanical properties.

Experimental Procedures

Annealed 8-mm thick sheet steel was chosen as a test material.

ECAP samples had a size of 8×8×60 mm^3 and were cut so that the 60-mm side was oriented along the rolling direction. At ECAP, the sheet's upside was facing the outlet die channel. ECAP was performed in the rolling direction. When pressed using route B_C, the sample was turned clockwise if viewed from the side of load application.

ECAP of copper was performed at room temperature with a die moving at a speed of 50 mm/min; lead was used to extrude the material from the horizontal channel. Pressing was performed using a lubricant based on natural oil with molybdenum disulfide as additive.

Standard tension test specimens having a nominal diameter of 5 mm and a gage

* Corresponding Author. korshun1@sar.ru

Y.T. Zhu and V. Varyukhin (eds.), Nanostructured Materials by High-Pressure Severe Plastic Deformation, 253-258.

length of 25 mm were cut out from the ECAP samples. The specimens were tested at room temperature at the strain rate $\dot{\varepsilon} \approx 10^{-3} s^{-1}$. The following mechanical properties were determined as a result of the tests: tensile strength $\sigma_{в}$, offset yield strength $\sigma_{0,2}$, and elongation to rupture δ_5 [4].

In order to identify points and sections for studying mechanical properties distribution in the ECAP sample, a coordinate system shown in Fig. 1a was used. The origin was situated in the geometric center of the sample. Nine small-size tensile specimens having a nominal diameter of 1.5 mm and a gage length of 7.5 mm were cut out from each ECAP sample along the *X* axis so that the specimens' axes matched with the points indicated in Fig. 1b. The *X* coordinate of each specimen's geometric center was the same, *viz.* 10 mm. The same mechanical properties as in standard test specimens were measured.

The Vickers hardness was measured by means of a TPP-2 device in the *Y0Z* plane at *X*=0 (Fig. 1a).

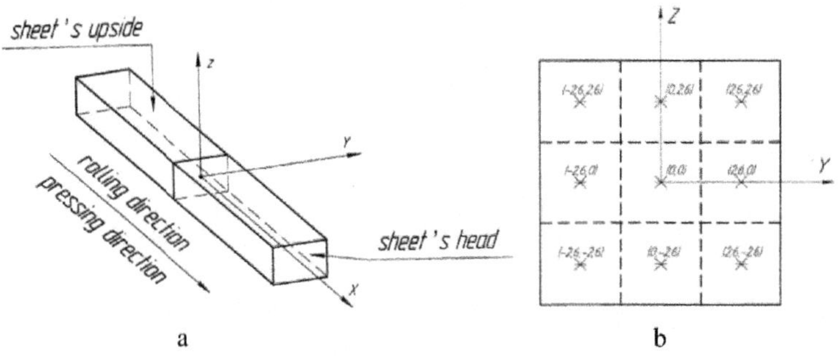

Fig. 1. Coordinate system used (a) and coordinates of small-size specimens' axes (b).

Results and Discussion

Figure 2 graphically represents the dependencies of mechanical properties of the test material on the number of ECAP passes, as determined in standard test specimens.

Distributions of mechanical properties across the sample after ECAP are shown in Figs. 3–5. The same figures also show initial distributions of mechanical properties in copper.

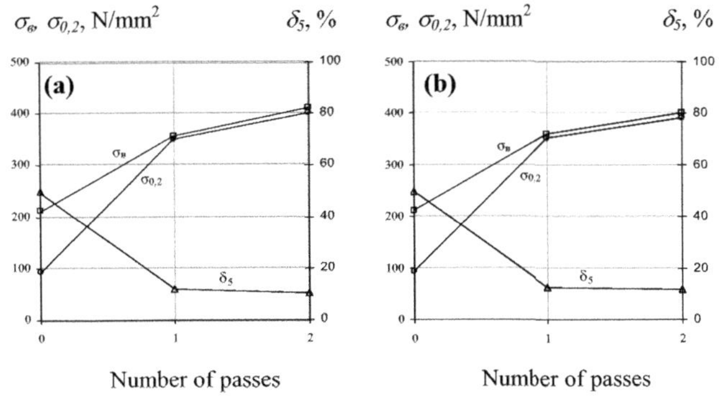

Fig. 2. Dependencies of mechanical properties of M1 copper on the number of ECAP passes for routes B_C (a) and C (b)

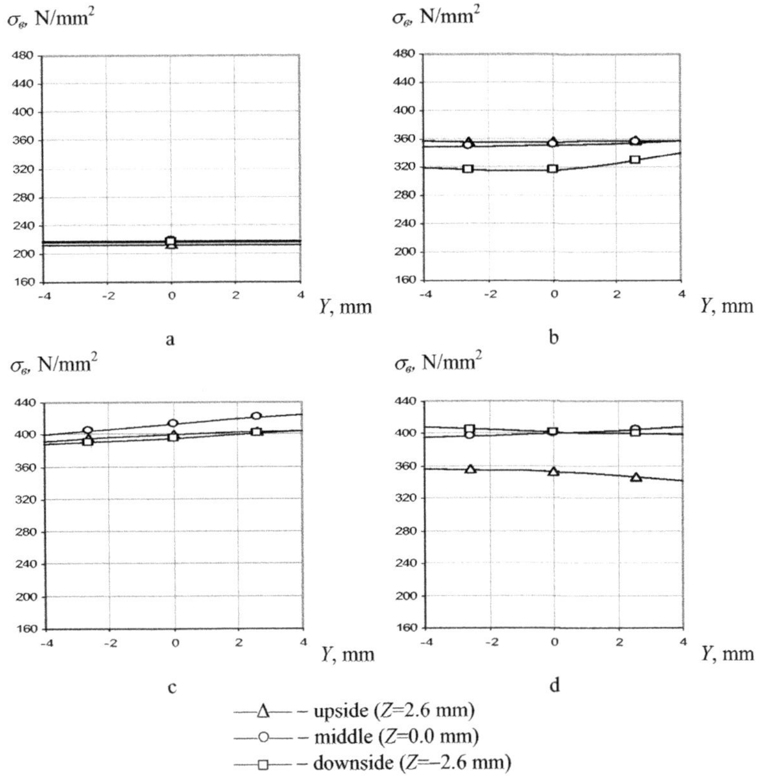

Fig. 3. Tensile strength (σ_B) distribution across the sample: initial (a), after one pass (b), after two ECAP passes by routes B_C (c) and C (d)

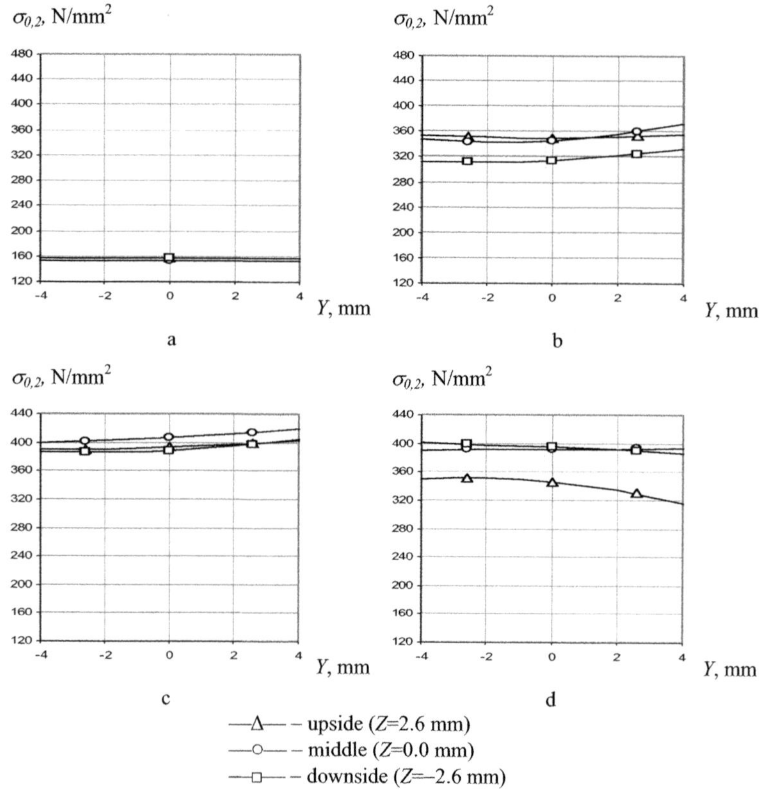

Fig. 4. Offset yield strength ($\sigma_{0.2}$) distribution across the sample: initial (a), after one pass (b), after two ECAP passes by routes B_C (c) and C (d)

When determined in standard test specimens, post-ECAP strength properties were found to increase, and ductility to decrease, all three mechanical characteristics of interest being practically identical for both routes B_C and C.

At the same time, one can observe a considerable difference in the distribution of mechanical properties across the sample, as determined using small-size specimens. As follows from Figs. 3 and 4, strength properties ($\sigma_в$, $\sigma_{0,2}$) across the sample can be observed to be distributed heterogeneously after the first pass and the second pass by route C (~12÷15%). After the second pass by route B_C, there is almost no heterogeneity in these properties (~5%), just as in the initial state.

Data on the distribution of elongation to rupture (δ_5) across the sample (Fig.5) point at some heterogeneity after the first pass, which almost completely disappears after two passes by both of the routes, B_C and C.

Vickers hardness distribution across the sample after the second pass by both loading routes is given in Fig. 6. In all cases one can observe a middle zone of about 5 mm with uniform hardness.

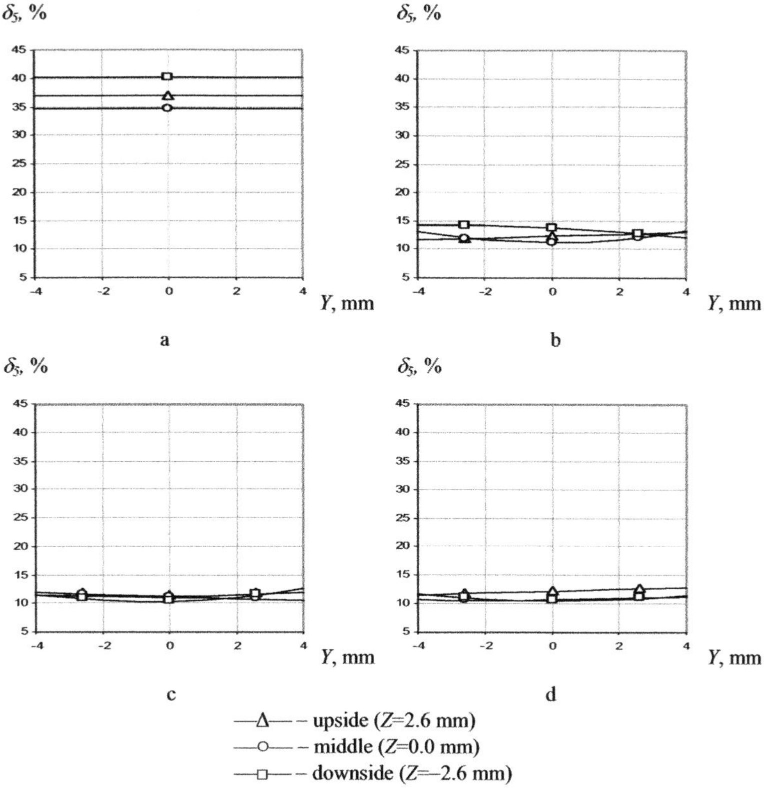

Fig. 5. Elongation to rupture (δ_5) distribution across the sample: initial (a), after one pass (b), after two ECAP passes by routes B_C (c) and C (d)

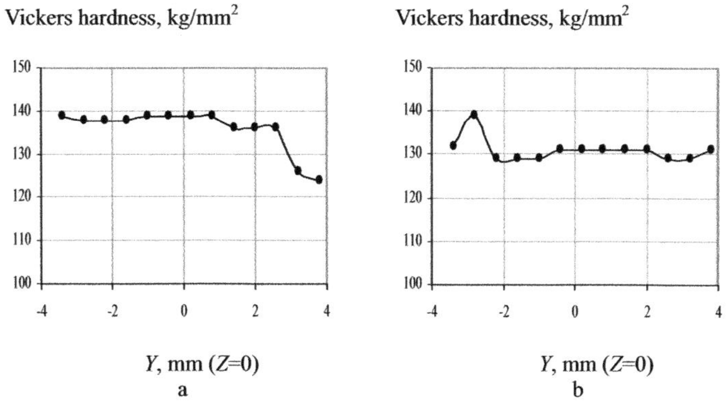

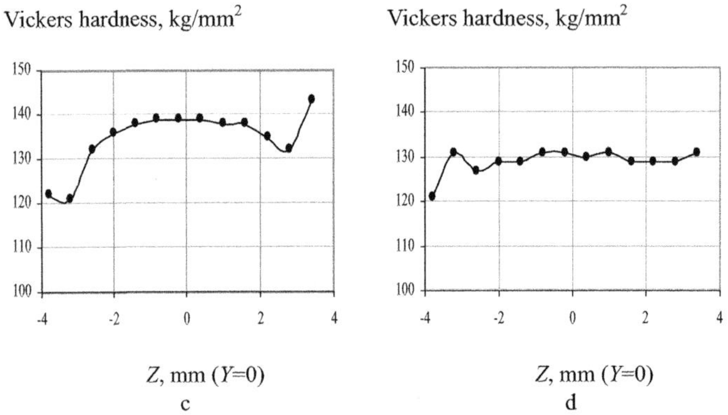

Fig. 6. Hardness distribution across the sample after two ECAP passes by routes B_C (a,c) and C (b,d).

Conclusions

1. Strength properties ($\sigma_{в}$, $\sigma_{0.2}$) and elongation (δ_5) in initial-state M1 copper are distributed across the sample rather homogeneously.
2. One can observe considerable heterogeneity in all mechanical properties after the first ECAP pass.
3. There is almost no heterogeneity in the distribution of all strength ($\sigma_{в}$, $\sigma_{0.2}$) and plastic (δ_5) properties across the sample after the second pass by route B_C.
4. Heterogeneity in all properties after the second ECAP pass by route C was found to be more pronounced than that after pressing by route B_C and quite similar to the heterogeneity after the first pass.
5. A middle zone with constant hardness can be observed in all cases of pressing M1 copper by ECAP. Minor hardness variation can be observed at the ends of a workpiece.

Acknowledgements

This research has been funded by LANL-IPP project #T2-0197.

References

1. Valiev R. *Nature Materials*, vol. 3, august (2004), 511-516.
2. Roven H.J., Dumoulin S. and Werenskiold J.C., "Measuring 3D Strains in Equal Channel Angular Pressing", *Ultrafine Grained Materials III*, ed. Y.T. Zhu, T.G. Langdon, R.Z. Valiev, S.L. Semiatin, D.H. Shin, T.C. Lowe, TMC, 2004, 117-124.
3. Beyerlein I.J., Li S., Alexander D.J., Necker C.T., Tome C.N. and Bourke M.A., "Heterogeneity in Texture Development in Single Pass Equal Channel Extrusion", *Ultrafine Grained Materials III*, ed. Y.T. Zhu, T.G. Langdon, R.Z. Valiev, S.L. Semiatin, D.H. Shin, T.C. Lowe, TMC, 2004, 185-192.
4. Korshunov A.I., Kravchenko T.N., Saveleva O.M., *Strength Mater.* (USA), vol 14, 9 (1982), 1268-1271.

Deformation and Failure of Ti-6Al-4V Nanostructured Alloy at 300-4.2 K

E. D. Tabachnikova[1*] V. Bengus[1], V. Natsik[1], K. Csach[2], J.Miskuf[2], V. Stolyarov[3], R. Valiev[3]
[1]B. Verkin Institute for Low Temperature Physics & Engineering
Kharkov, Ukraine
[2]Institute of Experimental Physics, Kosice, Slovakia
[3]Ufa State Aviation Technical University, Ufa, Russia

Abstract

This work is devoted to comparative experimental study of the low temperature mechanical behavior of the nanostructured (NS) Ti-6Al-4V alloy processed by severe plastic deformation equal channel angular pressing (ECAP) with an average grain size d = 0.5 μm and polycrystalline Ti-6Al-4V alloy (d = 6 μm). It is discovered that decreasing of grain size by ECAP processing of Ti-6Al-4V alloy from 6 μm to 0.5 μm have resulted in 25% increasing of the yield stress under compression at 300-4.2 K. Obtained high strength state of ECAP processed Ti-6Al-4V alloy (with the 1.28 GPa yield stress at 300 K and 2.47 GPa at 4.2 K) is combined with a good plasticity down to 4.2 K. Deformation was of a serrated character at 4.2 K. At 77 and 4.2 K the failure took place by catastrophic plastic shear sliding-off of one part of the sample relative to another at the compressive failure stresses up to 2.8 GPa. But as a distinction from the nanostructured CP Ti, shear failure surfaces of the nanostructured Ti-6Al-4V alloy were almost free of "veins" – ductile dimples (only 20-30% of the shear surface contains islands of dimples). Established low temperature features are analyzed with taking into consideration of non-uniform spatial structure of the Ti-6Al-4V two-phase alloy.

Introduction

At room temperature the yield stress of ECAP-produced nanostructured (NS) commercially pure titanium is more than two times higher than that of coarse-grained (CG) titanium [1, 2] and approaches the theoretical shear strength of single crystal titanium. Previously we had observed that the yield stress of NS Ti [3] is comparable to the yield stress in the strongly alloyed VT-6 (Russian name of the Ti-6Al-4V alloy) coarse-grained alloy, and concluded that the effectiveness of ECAP technology is in certain respects comparable to that of alloying. On the other hand, it has long been known that CG titanium and titanium alloys have high plasticity at room temperature, which persists or even increases on cooling to 77 K [3, 4]. In this context it is interesting to investigate the mechanical properties of NS Ti-6Al-4V alloy at low temperatures and estimate its potentialities for cryogenic applications.

* Corresponding Author. tabachnikova@ilt.kharkov.ua

Y.T. Zhu and V. Varyukhin (eds.), Nanostructured Materials by High-Pressure Severe Plastic Deformation, 259-263.

Experimental Procedures

Two structural states were investigated: initial–polycrystalline Ti-6Al-4V alloy with the average grain size $d = 6$ μm, hot pressing, and ECAP alloy with the average grain size $d = 0.5$ μm, produced by 12 ECAP passes at 700°C of the initial alloy.

The mechanical characteristics were studied under uniaxial compression in a deformation machine with the 10 kN/mm rigidity, at the relative strain-rate about $4 \cdot 10^{-4} s^{-1}$ (quasi-static deformation), at 300 K, 77 K (liquid nitrogen) and 4.2 K (liquid He^4). The compression samples were 2×2×7 mm right-angle prisms. The accuracy of stress and strain measurement was ±3 MPa and ±0.1%, respectively.

The samples from polycrystalline alloy were cut from bars both parallel (||) and perpendicular (⊥) to the bar axis. The samples from ECAP alloy were cut from bar both parallel (||) and perpendicular (⊥) to the ECAP axis. The following characteristics of the samples were measured:

- the strain diagrams *P(t)* (*P* is a deforming load, *t* is the time after the onset of deformation) were recorded;
- work-hardening curves $\sigma(\varepsilon)$ (σ is the deforming stress, ε is the plastic deformation) were obtained from the stress-strain diagrams in the approximation of uniform deformation, and an ultimate plastic deformation ε_f before failure was registered;
- conventional microscopic yield stress σ_0 was found by the method of intersections of the tangents at the initial and steady-state region of the stress-strain diagram *P(t)*;
- the type of fracture was specified, and the fracture surface morphology was studied with the TESLA BS-300 scanning electron microscope.

Quantitative characteristics were obtained by averaging over the strain and failure data for five and more samples.

Results and Discussion

Deformation curves of NS Ti-6Al-4V alloy under compression at 300-4.2 K had a single stage of linear work hardening. The expression $\sigma(\varepsilon) = \sigma_0 + \theta\varepsilon$, were θ is the work hardening coefficient, fit the dependence $\sigma(\varepsilon)$ quite well. At room temperature there was no hardening: $\theta \sim 0$. Coefficient θ increases with decreasing temperature: θ (77K) = 6 GPa and θ (4.2) = 9 GPa. This phenomenon is similar to the one previously observed in the ECAP – processed NS Ti [3]. It is discovered that decreasing of grain size by ECAP processing of Ti-6Al-4V alloy from 6 μm to 0.5 μm have resulted in 25% increasing of the yield stress at 300 - 4.2 K. The attained high strength state of the ECAP-processed Ti-6Al-4V alloy (with the 1.28 GPa yield stress at 300 K, 1.81 GPa at 77 K and 2.17 GPa at 4.2 K) is accompanied by a rather high ductility of the material. The yield stress σ_0 and failure stress σ_f of the ECAP processed and the initial Ti-6Al4V alloy at 300, 77 and 4.2 K are indicated in the Table 1. At 4.2 K deformation was of a serrated character starting at the yield stress. The characteristic amplitudes and frequency of jumps depend appreciably on the grain size and compression axis orientation (similar of NS Ti [3]).

Table 1. The yield stress σ_0 and failure stress σ_f of the ECAP processed and the initial Ti-6Al-4V alloy at 300, 77 and 4.2 K.

State	T = 300 K		T = 77 K		T = 4.2 K	
	σ_0, GPa	σ_f, GPa	σ_0, GPa	σ_f, GPa	σ_0, GPa	σ_f, GPa
Initial ‖	1.02	1.34	1.60	2.04	1.96	2.50
ECAP ‖	1.29	1.47	1.82	2.31	2.02	2.31
Initial ⊥	1.03	1.24	1.63	2.25	2.03	2.58
ECAP ⊥	1.37	1.45	1.95	2.37	2.36	2.47

Quantitative estimation of ductility was impossible at 300 K because, being rather ductile, specimens usually flattened under compression. But at 77 K and 4.2 K the "sliding-off" the failure mode was observed, similar to available in NS Ti [1, 3]. The failure took place by the catastrophic plastic shear sliding-off of one part of the sample relative to another at the rather large compressive failure stresses (see Table 1). Shear failure surfaces were oriented nearly at 45 degree to the compression axis (Fig. 1).

Figure 1. General view of the nanostructured Ti-6Al-4V alloy sample after the "sliding-off" failure at 4.2 K.

At these conditions it became possible to measure ε_f. The ε_f values at 77 and 4.2 K are indicated in the Table 2. It is seen that ε_f decreases almost twice under cooling from 77 to 4.2K.

Table 2. An ultimate plastic deformation ε_f before a sliding-off failure of initial and ECAP processed Ti-6Al-4V alloy at 77 and 4.2 K.

State	ε_f, %	
	T = 77 K	T = 4.2 K
Initial ‖	23	15
ECAP ‖	19	11
Initial ⊥	24	13
ECAP ⊥	13	7

SEM fractography revealed that failure surfaces are almost free of "veins" - ductile dimples - (only 20-30% of the shear surface contains "islands" of dimples) (Fig. 2).

This phenomenon is unlike nanostructured CP Ti, where 80-90% of the shear failure surface contains "veins" - ductile dimples. Such the non-uniform space distribution of "veins" and its dependence on the structural state of samples is considered as a consequence of a non-uniform space structure of the Ti-6Al-4V two-phase alloy. Probably "islands" of the dimples may be connected with the trails of colonies of the Widmanstätten structures [5] left after ECAP of the Ti-6Al-4V alloy. For better ECAP these colonies must be removed from the initial Ti-6Al-4V alloy by the proper thermal treatment before ECAP.

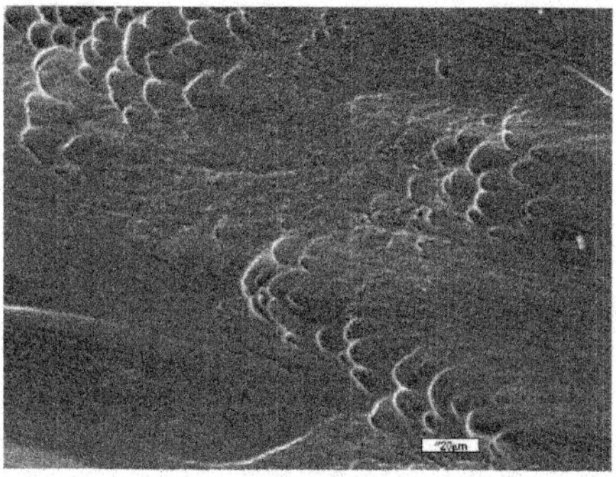

Figure 2. Surphace morphology of the shear failure surface of nanostructured Ti6Al4V alloy, failed under the uniaxial compression at 4.2 K. "Islands" of the dimples are observed, surrounded by surface of a free sliding-off.

The observed phenomenon of the catastrophic plastic shear failure can be explained in terms of adiabatic catastrophic plastic shear along grain boundaries [1]. Observed non-uniform space distribution of islands of dimples reflects non-uniform adiabatic heating along the surface of shear and failure in studied specimens.

Main regularities of low temperature mechanical behavior observed in this study for the nanostructured Ti-6Al-4V alloy are common with those of nanostructured Ti, which had been studied earlier in our investigations [1-4].

Conclusions

1) Mechanical properties of the nanostructured Ti-6Al-4V alloy with the average grain size 0.5 μ, produced by severe plastic deformation, using equal-channel angular pressing (ECAP), have been experimentally studied under uniaxial compression at 300, 77, and 4.2 K. The results have been compared with properties of the polycrystalline Ti-6Al-4V alloy (average grain size 6 μ). It has been established that ECAP leads to the yield stress increasing to 25 % in the 300-4.2 K temperature interval, with preserving of the sufficient plasticity level.
2) At 77 and 4.2 K the failure under uniaxial compression occurs by catastrophic plastic sliding-off of one part of the sample relative to another under compressions up to 2.8 GPa.

Acknowledgement

This work have being carried out with a partial financial support by the project № 3 –026/2004 of the NASU complex program "Nanosystems, nanomaterials and nanotechnologies" and by the INTAS-01-0320 Project.

References

1. Bengus V.Z., Tabachnikova E.D., Csach K., Miškuf J., Stolyarov V.V., Valiev R.Z., Ocelík V., De Hosson J.Th.M., "Observation of the shear failure mode in nanostructured titanium under quasistatic uniaxial compression at 300 – 4.2 K", *Science of Metastable and Nanocrystalline Alloys. Structure, properties and Modelling*, ed. A.R. Dinesen, M. Eldrup et al., Riso National Laboratory, Roskilde, Denmark, 2001, 217-222.
2. Tabachnikova E.D., Bengus V.Z., Natsik V.D., Valiev R.Z., Stolyarov V.V., Alexandrov I.V., "Low temperature plasticity of bulk nanostructured Cu, Ni and Ti processed by severe plastic deformation", *Science of Metastable and Nanocrystalline Alloys. Structure, properties and Modelling,*, ed. A.R. Dinesen, M. Eldrup et al., RISO National Laboratory, Roskilde, Denmark, 2001, 389 394.
3. Bengus V.Z., Tabachnikova E.D., Natsik V.D., Mishkuf J., Csach K., Stolyarov V.V., Valiev R.Z. , *Low Temperature Physics* 28 (2002), 864 – 874.
4. Tabachnikova E.D., Bengus V. Z., Natsik V.D., Podolskii A.V., Smirnov S. N., Valiev R.Z., Stolyarov V.V., Alexandrov I.V., *Nanomaterials by Severe Plastic Deformation, Proceedings of the Conference "Nanomaterials by Severe Plastic Deformation - NANOSPD2" (December 9 – 13, 2002, Vienna, Austria)*, ed. Michael Zehetbauer and Ruslan Z. Valiev, J. Wiley VCH, Weinheim, 2004, 207-213.
5. Kim S.M., Kim J., Shin D.H., Ko Y.G., Lee C.S. Semiatin S.L., *Scripta Materialia* 50 (2004), 927-930.

Improvement of Mechanical Properties in a Nanostructured Fe-Ni-Mn Steel Prepared by Heavy Plastic Deformation

S. Hossein Nedjad[1], M. Nili Ahmadabadi[1*], R. Mahmudi[1]

T. Furuhara[2] and T. Maki[2]

[1] Department of Metallurgy and Materials Engineering, Faculty of Engineering, University of Tehran, P. O. Box: 14395-731, Tehran, Iran.
[2] Department of Materials Science and Engineering, Kyoto University,Yoshida-Honmachi, Sakyo-ku, Kyoto 606-8501, Japan

Abstract

Fe-Ni-Mn martensitic steels show excellent age hardenability but suffer from severe embrittlement after aging. Discontinuous coarsening of grain boundary precipitates resulting in the formation of precipitate free zone (PFZ) along prior austenite grain boundaries (PAGBs) was found as the main source of embrittlement. Heavy cold rolling for 85% was carried out on the solution annealed material with subsequent aging at 753 K. Hardness measurement, tensile test and transmission electron microscopy were used to study the effect of heavy cold rolling and aging on the microstructure, aging behavior and mechanical properties of the steel. It was found that grain refinement to a nanometer scale takes place at the initial stages of aging and the discontinuous coarsening of grain boundary precipitates is suppressed. The ultrafine grained steel showed tensile strength of about 1940 MPa while conventional steel shows premature fracture strength of 824 MPa.

Introduction

Fe-Ni-Mn martensitic steels show excellent age hardenability by the formation of nanometer sized intermetallic precipitates such as f.c.t. θ-NiMn during aging. Fe-Ni-Mn steels are sufficiently ductile in the solution annealed condition but suffer from severe embrittlement along prior austenite grain boundaries (PAGBs) after aging. Discontinuous coarsening of grain boundary precipitates resulting in the formation of precipitate free zone (PFZ) along PAGBs was found in Fe-10Ni-7Mn steel [1].

Grain refinement has been known as an effective way to improve strength and toughness of metallic materials. Various techniques have been introduced to produce ultrafine grained materials which mainly consist combination of heavy plastic deformation and subsequent annealing [See Ref. 2 for a review]. Severe plastic deformation (SPD) of metallic alloys has been found to realize submicrocrystalline or nanocrystalline materials and various SPD techniques have been developed, e.g. equal channel angular pressing (ECAP) [3], high pressure torsion (HPT) [4], accumulative roll bonding (ARB) [5]. Ueji et al. [6] have reported ultrafine grain formation in plain carbon steel by cold rolling and annealing of lath martensite. The mechanism of ultrafine grain formation in SPD materials is taught to take place by subdivision of initial grains by intense deformation and subsequent recovery which has been called continuous or in-situ recrystallization [7, 8].

Heavy cold rolling was performed to investigate effect of heavy plastic

* Corresponding author; nili@ut.ac.ir

Y.T. Zhu and V. Varyukhin (eds.), Nanostructured Materials by High-Pressure Severe Plastic Deformation, 265-269.

deformation and aging treatment on the discontinuous coarsening of grain boundary precipitates and PFZ formation. The aim of this paper is to report effect of heavy cold rolling and subsequent aging treatment on the aging behavior, tensile strength and microstructure of Fe-10Ni-7Mn steel.

Experimental Procedure

A 6 kg ingot was melted in a vacuum induction melting furnace under 10^{-2} Torr and cast into an iron mold. A 200 g bar cut from the cast material was remelted under an inert argon atmosphere in a vacuum arc remelting furnace with a water cooled copper mold. Chemical composition of the steel is given in Table 1. The bar was encapsulated in a quartz tube and argon gas was purged after evacuation to 10^{-5} Torr. Homogenizing treatment was performed at 1473 K for 172.8 ks followed by water quenching and subzero treating at 77 K for 3.6 ks. Cold rolling for 85% was carried out at room temperature followed by direct aging, hereafter referred to cold rolled steel (CRS) or solution annealing at 1223 K for 3.6 ks, hereafter referred to conventionally treated steel (CTS). Schematic illustration of processing routes are shown in Fig. 1. Aging treatment was performed at 753 K in a neutralized salt bath. Hardness measurement was made by Vickers method with a load of 9.8 N. Tensile specimens of 0.6 × 2.5 × 8.5 mm dimension cut in longitudinal direction were tested by Shimadzu tensile test machine at room temperature under an initial strain rate of 2×10^{-3} s^{-1}. Transmission electron microscopy observations were made on RD section by PHILIPS CM200-FEG microscope operating at 200 kV.

Table 1. Chemical composition of the steel studied (wt. %)

Fe	Ni	Mn	C	S	P	N	Al
Base	10.38	6.88	0.006	0.007	0.005	0.005	0.003

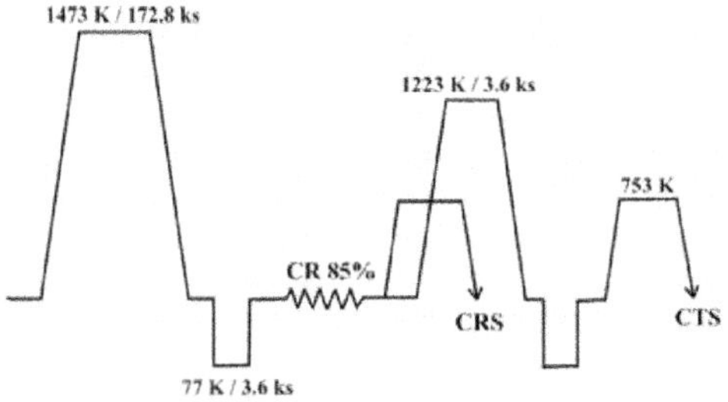

Figure 1. Schematic illustration of processing routes.

Results

Figure 2 shows changes in the hardness of steels during isothermal aging at 753 K. CTS has a hardness of 260 HV in the solution annealed condition which increases to 310 HV after cold rolling for 85%. Table 2 shows longitudinal tensile properties corresponding to aging times of 0.36, 3.6 and 86.4 ks. CTS shows premature failure at 0.36 and 3.6 ks aging times at strengths of 824 and 1104 MPa, respectively, with zero tensile ductility. However, at longer aging time (86.4 ks) strength increases and tensile ductility is restored to about 3% in total elongation. CRS shows UTS of about 1940 MPa at 0.36 and 3.6 ks with minor tensile ductilities while at longer aging time (86.4 ks), UTS decreases and ductility increases. Figure 3 shows TEM bright field (BF) image of CTS aged for 0.36 and 3.6 ks indicating discontinuous coarsening of grain boundary precipitates and formation of PFZ along PAGBs. The mean size of prior austenite grains in CTS was measured around 20 μm. TEM BF image and corresponding selected area diffraction pattern (SADP) of CRS aged for 0.36 is shown in Fig. 4a indicating formation of ultrafine grains of about 100 nm wherein discontinuous coarsening and PFZ formation have been prevented. Figure 4b shows a TEM BF image of CRS aged for 3.6 ks showing a lamellar structure of two different precipitate sizes.

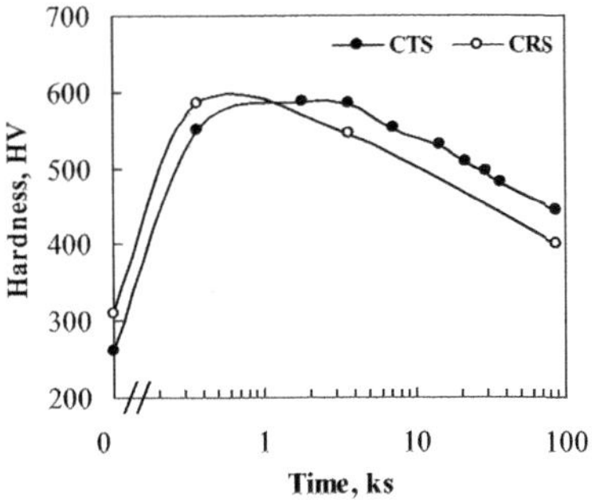

Figure 2. Changes in hardness of steels during isothermal aging at 753 K.

Discussion

Hardness increasing during isothermal aging was found similar in CTS and CRS indicating that cold rolling did not result in considerable hardness increment. However, accelerated aging of CRS was identified showing the effect of cold rolling on the kinetics of aging. It has been found that cold deformation prior to aging does not affect hardness increment in 18Ni maraging steels remarkably [9].

Fe-Ni-Mn maraging steels suffer from severe embrittlement after aging. The source of embrittlement in Fe-10Ni-7Mn steel was found to be discontinuous coarsening of grain boundary precipitates and formation of PFZ along PAGBs as shown in Fig. 3. However, CRS showed UTS of about 1940 MPa indicating considerable improvement of UTS in comparison to CTS which shows premature failure at strength of about 820 ~ 1100 Mpa. Figure 4a shows that CRS has an ultrafine grained structure of about 100 nm grain size at the early stages of aging, furthermore, the discontinuous coarsening of precipitates and resulting PFZ has been prevented. Figure 4b shows CRS aged for 3.6 ks indicating a lamellar structure of two different precipitate size. It seems that at longer aging times, discontinuous coarsening takes place on the recrystallizing boundaries while the recrystallized areas are not continuous.

Table 2. Tensile properties of CTS and CRS at different aging times at 753 K

Steel	CTS			CRS		
Aging time, ks	0.36	3.6	86.4	0.36	3.6	86.4
YS, MPa	---	---	1301	1864	1766	1275
UTS, MPa	[824]	[1104]	1337	1942	1939	1401

[ ] indicates premature failure.

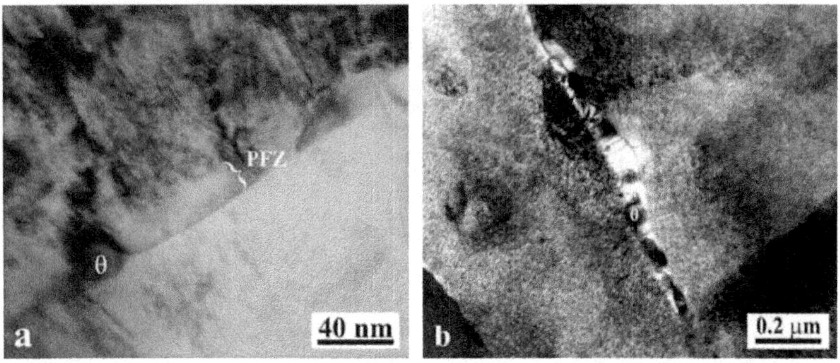

Figure 3. TEM BF images of CTS showing coarsening of grain boundary precipitates and formation of PFZ along PAGBs. (a) aged for 0.36 ks, (b) aged for 3.6 ks.

Figure 4. TEM BF images and corresponding SADPs of CRS. (a) aged for 0.36 ks showing ultrafine grain formation, (b) aged for 3.6 ks.

The prevention of precipitate coarsening and PFZ formation along PAGBs at early stages of aging and discontinuity of recrystallized areas at longer aging times are taught to be the main reasons for increasing of UTS. The mechanism of prevention has not been well understood yet and more research is to be carried out in order to clarify the structural evolutions at the early stages of aging. However, It is suggested that the pronounced precipitation and coarsening of grain boundary precipitates has been suppressed in cold deformed structure due to enormous deformation-induced nucleation sites and concurrent recovery process. Increasing of grain boundary area in ultrafine grained steel may also affect grain boundary properties, e. g. segregation level, diffusivity, precipitation kinetics, etc.

The mechanism of ultrafine grain formation in the early stages of aging in CRS is taught to be the recovery of deformation-induced cell structure, however, SADP shows a widely distributed spots to be resulted from high misorientation across boundaries.

Conclusion

Cold rolling and aging of Fe-10Ni-7Mn steel resulted in the formation of nanostructured material of about 100 nm grain size. Considerable improvement in tensile properties was achieved which is taught to be due to prevention of discontinuous coarsening of grain boundary precipitates and PFZ formation along PAGBs at the early stages of aging.

References

1. Hossein Nedjad S, Nili Ahmadabadi M, Mahmudi R, Furuhara T, Farhangi H and Maki T, Submitted for publication.
2. Maki T, Proc. of 2nd International Symposium on Ultrafine Grained Structures, Australia, 2003.
3. Zhu Y. T. and Lowe T.C., Mater. Sci. Eng. A291 (2000), 46-53.
4. Valiev R.Z. and Alexandrov I.V., Nanostructured Materials, 12 (1999), 35-40.
5. Tsuji N, Saito Y, Utsunomiya H and Tanigawa S, Scripta Mater. (40) 1999, 795-800.
6. Ueji R, Tsuji N, Minamino N and Koizumi Y, Acta Mater. 50 (2002), 4177-4189.
7. Belaykov A, Sakai T, Minura H and Tsuzaki K, Phil. Mag. A 81 (2001), 2629-2643.
8. Tsuji N, Ueji R, Ito Y and Saito Y, Proc. of 21st Riso Int. Symp. on Materials Science - Recrystallization, 2000, 607-616.
9. Decker R.F., ed., *"Source book on maraging steels"*, ASM, OH, 1979, p.

Surface nanostructure and tribological properties of metallic materials

N.B. Efros[1*], L.G. Korshunov[2], B.M. Efros[1] and N.L. Chernenko[2]

[1]Donetsk Physics & Technology Institute of NASU, Donetsk, 83114, Ukraine

[2]Institute of Metal Physics of the RAN, Ekaterinburg, 620219, Russia

Abstract

Duirng sliding friction of metals, in the surface layer (≈10 μm) of metals subjected to sliding friction, ultra disperse structures are formed, which to a considerable extent determine the tribological properties of the surfaces. The structures can be considered to be nanocrystalline (NC): micro-fragments with large-angle boundaries are 0.05 to 0.2 μm in size.

The objects of this investigation were the nitrogen-containing Fe-Mn-(Cr) alloys (wt.%): Fe-19% Cr-10% Mn-2% Si-0.7% N-0.07% C and Fe-18% Cr-20% Mn-0.8% N-0.08% C. To compare the obtained results, we also made use of the tribological characteristics of different alloys: Fe-10% Mn-0.05% C, Fe-20% Mn-0.05% C, Fe-11% Cr-13% Mn-0.03% C, Fe-40%Mn-0.05% C and Fe-18% Cr-12% Mn-0.5% N-0.07% C, differing in phase composition, structure and the stacking fault energy (SFE). Tribological tests were done at laboratory-scale plants in conditions of sliding friction for the steel-alloy and abrasive-alloy pairs.

The NC HCP ε-alloys and NC nitrogen-containing FCC γ-alloys possess low friction adhesion coefficients K and high resistance of abrasive wear ε. In these alloys, the developed basal or planar slip characterized by a smaller quantity of the slip systems condition a decrease in the resistance of NC fragments as a result of rotation and, thus, the improvement of tribological properties.

Thus, the essential lowering of friction adhesion coefficient K and intensity of adhesion wear Ih, as well as the increase in the abrasive wear ε of NC HCP ε-Fe-Mn alloy and nitrogen-containing NC FCC γ-Fe-Mn-Cr alloys are mainly related to decreasing the quantity of slip systems in nanocrystals of these alloys ($n \to \leq 3$). As a result, the dislocations travel to the boundaries of nanocrystals more freely.

Introduction

In the surface layer (≈ 10 μm) of different metals, the sliding friction induces ultra-disperse structure, which to a considerable extent the tribological properties of the surfaces [1,2]. The structures can be considered to be nanocrystalline (≈ 50 - 200 nm) micro-fragments with large-angle boundaries. The experiments show that in different materials the initiated frictional NC structure with relatively close levels of dispersion, disorientation of fragments and average density of dislocations ($\rho \approx 5 \cdot 10^{12}$ cm^{-2}), differ very much in their strength and tribological properties. It is obvious that further investigations are needed to reveal structural factors that highly influence the tribological properties of frictional NC structure.

---

* Corresponding Author. efrosn@kinetic.ac.donetsk.ua

Y.T. Zhu and V. Varyukhin (eds.), Nanostructured Materials by High-Pressure Severe Plastic Deformation, 271-274.

Experimental Procedures

The objects of investigation were the nitrogen-containing γ-alloys $Fe_{68.23}Cr_{19}Mn_{10}Si_2N_{0.7}C_{0.07}$, $Fe_{63.43}Cr_{18}Mn_{18}N_{0.5}C_{0.07}$ and $Fe_{61.12}Cr_{18}Mn_{20}N_{0.8}C_{0.08}$. To compare the obtained results, the use was also made of the tribological characteristics of alloys $Fe_{89.95}Mn_{10}C_{0.05}$, $Fe_{79.95}Mn_{20}C_{0.05}$, $Fe_{75.97}Mn_{13}Cr_{11}C_{0.03}$ and $Fe_{59.95}Mn_{40}C_{0.05}$ differing in phase composition, structure and value of the SFE [1,2].

Tribological studies were done at laboratory-scale plants in conditions of sliding friction for the steel-alloy and abrasive-alloy pairs. The steel-alloy pairs were tested according to the finger (alloy)-plate (steel $Fe_{99.55}C_{0.45}$, HRC=50) scheme, in air at the velocity of sliding V = 0.07 m/s, under load F=294 N. Increase of the average (bulk) temperature due to the frictional heating in the zone of specimen friction did not exceed 20°C. The NC structure_was formed in a less than 10 μm thick surface layer in conditions of preferentially adhesive interaction of metallic surfaces. The abrasive-wear investigation were done for Armco-Fe upon specimen (alloy) sliding by the fixed abrasive, namely the abrasive paper.

Procedures of testing and determination of the strength and tribological properties are described in articles [1,2]. Structure of layers of the investigated alloy specimens was studied by the X-ray, electron-microscopic and metallographic methods.

Results and Discussion

The Table lists the values of microhardness H_μ, shear strength's τ, friction coefficient K, intensity of adhesion wear Ih, abrasive wear resistance ε for NC structures formed on friction surfaces of the investigated alloys having the BCC, FCC and HCP structures. It is seen that the HCP ε-alloys $Fe_{79.95}Mn_{20}C_{0.05}$ and $Fe_{75.97}Mn_{13}Cr_{11}C_{0.03}$ as well as the nitrogen-containing FCC γ-alloys $Fe_{63.43}Cr_{18}Mn_{18}N_{0.5}C_{0.07}$, $Fe_{68.23}Cr_{19}Mn_{10}N_{0.7}C_{0.07}$, and $Fe_{61.12}Cr_{18}Mn_{20}N_{0.8}C_{0.08}$ are characterized by the minimal values of K (0.25-0.28), Ih ($3\cdot10^{-7}$-$3.5\cdot10^{-7}$) and by the increased values of ε (1.5-2.2).

Table Strengthening and tribological properties of NC structures_of investigated alloys under adhesion and abrasive wear

steel	phase composition	H_μ, MPa	τ, MPa	K	Ih× 10^7	ε
$Fe_{89.95}Mn_{10}C_{0.05}$	α (BCC)	6300	3450	0.55	20	0.5
$Fe_{79.95}Mn_{20}C_{0.05}$	ε (HCP)	5600	1570	0.28	3	1.5
$Fe_{75.97}Mn_{13}Cr_{11}C_{0.03}$	ε (HCP)	6250	1560	0.25	3.5	1.9
$Fe_{59.95}Mn_{40}C_{0.05}$	γ (FCC)	6650	3000	0.45	6	0.8
$Fe_{63.43}Cr_{18}Mn_{18}N_{0.5}C_{0.07}$	γ (FCC)	8100	4650	0.25	2.1	2.2
$Fe_{68.23}Cr_{19}Mn_{10}Si_2N_{0.7}C_{0.07}$	γ (FCC)	6700	3100	0.27	5.4	1.9
$Fe_{61.12}Cr_{18}Mn_{20}N_{0.8}C_{0.08}$	γ (FCC)	7250	2400	0.25	4.7	1.8

The X-ray phase analysis has shown that on the friction surface of alloys $Fe_{79.95}Mn_{20}C_{0.05}$ and $Fe_{75.97}Mn_{13}Cr_{11}C_{0.03}$ the structure of HCP ε-martensite is formed as a result of (γ+ε)→ε - deformation martensite transformation (DMP) (≈ 95 %

ε-phase in alloy $Fe_{79.95}Mn_{20}C_{0.05}$ and ≈ 90% ε-phase in alloy $Fe_{75.97}Mn_{13}Cr_{11}C_{0.03}$), whereas on the friction surface of alloys $Fe_{63.43}Cr_{18}Mn_{18}N_{0.5}C_{0.07}$, $Fe_{68.23}Cr_{19}Mn_{10}N_{0.7}C_{0.07}$, and $Fe_{61.12}Cr_{18}Mn_{20}N_{0.8}C_{0.08}$ the structure of FCC γ-phase is formed (concentration α-phase at DMP < 5 %). They are responsible for tribological properties of the above alloys. It should be noted that the value of strain hardening (microhardness H_μ) for ε-phase occurring on the friction surface of alloy $Fe_{75.97}Mn_{13}Cr_{11}C_{0.03}$ is noticeably higher than that for ε-phase formed in alloy $Fe_{79.95}Mn_{20}C_{0.05}$ under similar conditions of straining. This fact can be explained by a positive influence of chromium on capacity of ε-martensite for strain hardening [2]. A decrease of quantity H_μ in the NC structure of the nitrogen-containing γ-alloys with the growth of nitrogen concentration is evidently due to the suppression of $\gamma \rightarrow \alpha$ - DMP (concentration C_α at DMP decreases from ≈ 5 to 1.5 %).

NC BCC α-alloy $Fe_{89.95}Mn_{10}C_{0.05}$ and fcc γ-alloy $Fe_{59.95}Mn_{40}C_{0.05}$ possess higher values of K (0.45-0.55) and Ih ($6\cdot10^{-7}$-$2\cdot10^{-6}$) and lower values of ε (0.5-0.8) in spite of their improved strength characteristics (H_μ, τ) with respect to NC studied of HCP ε-alloys.

The NC HCP ε-phase possesses approximately the same tribological properties as the hcp α-cobalt under analogous friction conditions (K≈0.25, Ih≈10^{-7}). The low friction coefficient and the increased resistance to adhesive wear of cobalt are explained by the development of easy basal slip and formation of the friction texture in the surface layer of cobalt. It is characterized by the $(0001)_\alpha$-orientation of basal planes parallel to friction surface [3].

The electron-microscopic investigations have shown that structure of the as-quenched alloy $Fe_{79.95}Mn_{20}C_{0.05}$ consists of the intersecting plates of cooling HCP ε-phase with the $\{111\}_\gamma$ plane (C_ε≈55 %) and of the residual FCC γ-phase (C_γ≈45 %). In austenite there are stacking faults, which is due to a low enough value SFE of the γ-phase [4]. Sliding friction results in the formation of a NC structure at the surface consisting mainly of HCP ε-phase crystals, 0.01-0.1 μm in size. Apart from the ε-phase, in the NC structure there is also a small quantity of austenite. With the increase in distance from the friction surface to ≥10 μm, the size of NC fragments essentially grows due to the decreasing intensity of plastic deformation, the depth of the affected surface layer being increased. At a distance of ≈20 μm from friction surface, the ε-phase crystals become larger, they are in the form of usual deformed plates, the same as in the case of traditional (less severe) methods of deformation [4].

According to the obtained results (see the Table), it is seen that the NC structure of the ε-phase, having no noticeable texture in the investigated alloys, possesses approximately the same tribological properties as the textured surface of cobalt. This means that the basal slip in nanocrystals of the ε-phase isomorphism to basal slip α-cobalt reduces the shear resistance of NCS in direction of action of the external tangential force (τ≈1600 MPa). Based on the results of articles [4,5], it can be assumed that the basal slip characterized by a small (~3) number of sliding systems makes the passage of dislocations through the nanocrystalline body to the boundaries easier, thus reducing stresses necessary for the rotation of structure fragments under the influence of external contact forces.

In contrast to NC γ-alloy $Fe_{59.95}Mn_{40}C_{0.05}$, the NC nitrogen-containing γ-alloys $Fe_{63.43}Cr_{18}Mn_{18}N_{0.5}C_{0.07}$, $Fe_{68.23}Cr_{19}Mn_{10}N_{0.7}C_{0.07}$, and $Fe_{61.12}Cr_{18}Mn_{20}N_{0.8}C_{0.08}$ have a friction coefficient K (0.25-0.27) as low as that of the NC ε-alloys $Fe_{79.95}Mn_{20}C_{0.05}$ and $Fe_{75.97}Mn_{13}Cr_{11}C_{0.03}$ (see the Table).

The electron-microscopic investigations have shown that in the structure of quenched nitrogen-containing FCC γ-alloys $Fe_{63.43}Cr_{18}Mn_{18}N_{0.5}C_{0.07}$, $Fe_{68.23}Cr_{19}Mn_{10}N_{0.7}C_{0.07}$, and $Fe_{61.12}Cr_{18}Mn_{20}N_{0.8}C_{0.08}$ there are annealing twins, some dislocations, dislocation dipoles and multipoles. The presence of dislocation multipoles evidences that the investigated alloy with nitrogen-containing γ-phase tends to planar slip [6,7]. Sliding friction results in the formation of a NC structure in the surface layer (the micro fragments are 0.01-0.1 μm in size). The structure is typical of the nitrogen-containing FCC γ-alloys deformed by shear under pressure [8]. When distance from the friction surface increases to ≥10 μm, the NC fragments in the nitrogen-containing γ-alloys becomes much larger.

The obtained results show that one of the main reasons for the improvement of tribological properties of the NC nitrogen-containing FCC γ-alloys is, evidently, the suppression of cross slip and, thus, the planar slip activation in nanocrystals of γ-phase. As a result, the quantity of the acting slip systems in austenite crystals is essentially limited. Under the action of rotational mechanisms of plasticity, near the friction surface, the decrease in the quantity of slip systems in austenite makes the passage of dislocations through the body of fragments to their boundaries easier, thus reducing the resistance to relative micro fragments-boundary slip of the nitrogen-containing NC alloy.

Thus, the essential lowering of friction coefficient K and intensity of adhesion wear Ih, as well as the increase in the resistance abrasive wear ε of NC HCP ε-alloy and NC nitrogen-containing FCC γ-alloy are mainly related to decreasing the quantity of slip systems in nanocrystals of these alloys ($n \rightarrow \leq 3$). As a result, the dislocations travel to the boundaries of nanocrystals more freely. The Table demonstrates that the NC HCP ε-phase shows increased in intensity of adhesion wear which is 3-3.5 times as high as that of the textured surface of α-cobalt in NC material. It is evident that the extremely developed system of boundaries possessing excessive energy makes the adhesion interaction of metallic surfaces more intensive.

Conclusion

The NC HCP ε-alloys and NC nitrogen-containing FCC γ-alloys_possess low friction coefficients K and high resistances abrasive wears ε. In the alloys, the developed basal or planar slip is characterized by a smaller quantity of slip systems, which leads to a lowering of the resistance of the NC fragments against rotation and, as a result, to the improvement of tribological properties.

References

1. Korshunov L.G., Fiz.met. Metalloved., 1992, no. 8:3-21.
2. Efros B.M., Studenok E.S., Loladze L.V., Friction and wear. 14 (1992), 730-735.
3. Backley D., Surface phenomena upon adhesion and frictional action. Moscow:Machine Building, 1986, p. 360.
4. Efros B.M., HPPE. 8 (1987), 82-96.
5. Korshunov L.G. and Chernenko N.L., Fiz.met. Metalloved. Volume 63 (1987) no. 2, p. 319-328.
6. Afanas'ev N.D., Gavrilyk V.G., Duz V.A., Nadutov V.M., Fiz.met. Metalloved. (1990) no. 8, p. 121-127.
7. Efros B.M., Metals, 1999, no. 3:95-102.
8. Teplov V.A., Pilyugin V.P., Talutz G.G., Metals, 1992, no. 2:109-115.

V. PROPERTIES AT ELEVATED TEMPERATURES

Elevated Temperature Deformation Characteristics of Nanocrystalline Materials

A.V.Sergueeva[1], N.A.Mara[1], R.Z.Valiev[2], A.K.Mukherjee[1*]
[1]Chemical Engineering & Materials Science Department, University of California, One Shields Avenue, Davis, CA 95616
[2]Ufa State Aviation Technical University, 12 K. Marks str, Ufa, 450000, Russia

Abstract

This paper reviews the current understanding of the effects of microstructural characteristics on mechanical behavior of materials produced by severe plastic deformation. The elevated temperature deformation behavior of metals and intermetallics subjected to severe plastic deformation was analyzed in order to establish possible deformation mechanisms and microstructural effects.

Introduction

The technique of severe plastic deformation (SePD) under pressure has been demonstrated to be an effective technique for the production of truly bulk, fully dense and contamination-free metals with ultrafine (ufg) and nanocrystalline (nc) structures [1-4]. High pressure torsion (HPT) [1,2] is the method that provides the conditions under which the formation of truly nc structures with grain sizes below 100 nm can be attained.

Despite the intensive investigations of deformation behavior of nc materials at room temperature, very little is known about creep behavior and the actual deformation mechanisms in such materials at elevated temperatures. The technologically significant prediction for high temperature nc material performance is their enhanced superplasticity. Superplasticity in microcrystalline (mc) materials is well established as a grain size dependent phenomenon that can be described by the constitutive equation for conventional superplasticity [5]:

$$\dot{\varepsilon}=A\frac{DGb}{kT}\left(\frac{b}{d}\right)^{p}\left(\frac{\sigma}{E}\right)^{2} \tag{1}$$

where $\dot{\varepsilon}$ is the strain rate, D is the appropriate diffusivity (lattice or grain boundary), G is the shear modulus, b is the Burger's vector, k is the Boltzmann's constant, T is the test temperature, d is the grain size, p is the grain size exponent (usually 2 for lattice diffusion controlled flow and 3 for grain boundary diffusion controlled flow) and σ is the applied stress. Behavior described by this equation has been observed for metals, intermetallics and ceramic materials. According to this relationship between grain size, superplastic strain rate, and superplastic temperature, a shifting of the optimum strain rate to higher values and/or the optimum temperatures to lower values with microstructural refinement has been predicted.

Mechanical response of ultrafine grained structures at elevated temperatures

A number of features observed during elevated temperature deformation behavior in

* Corresponding Author. akmukherjee@ucdavis.edu

Y.T. Zhu and V. Varyukhin (eds.), Nanostructured Materials by High-Pressure Severe Plastic Deformation, 277-284.

nc materials are different from the superplastic behavior of mc materials, which are highlighted with examples below.

Excessive strain hardening: Figure 1 shows flow curve of SePD 1420-Al and Ni_3Al alloys [6-8]. Excessive strain hardening at all strain rates is quite clear. It is important to note that the strain hardening in 1420-Al alloy increases with increasing strain rate. We have observed a similar trend in other SePD processed materials as well, e.g., Zn-Al, 2124-Al alloy [9,10]. Notable grain growth was observed in these nc materials during superplastic deformation. Because of the grain size dependence in the constitutive relationship for superplastic flow, strain hardening during superplasticity has been conventionally explained in terms of grain growth. However, grain growth alone cannot account for the entire strain hardening in all these materials. At constant temperature, the stress dependence n, and grain size dependence p, have been experimentally determined for nc Ni_3Al during the present investigation. Both of these values are equal to 2. An examination of Eqn.(1) will reveal that at constant T and constant $\dot{\varepsilon} : \sigma \propto (d)^{p/n}$; with n = 2 and p = 2, $\sigma \propto d$, i.e., the increase in stress should be directly proportional to increase in grain size. However, the results on Ni_3Al alloy at 650°C and 1×10^{-3} s^{-1} strain rate show a flow stress increase by a factor of ~5 where the grain size increased only by a factor of ~2.

High flow stresses: Figure 2a shows a comparison of the flow curves during superplasticity in Ni_3Al alloy. This alloy was tested both in microcrystalline and

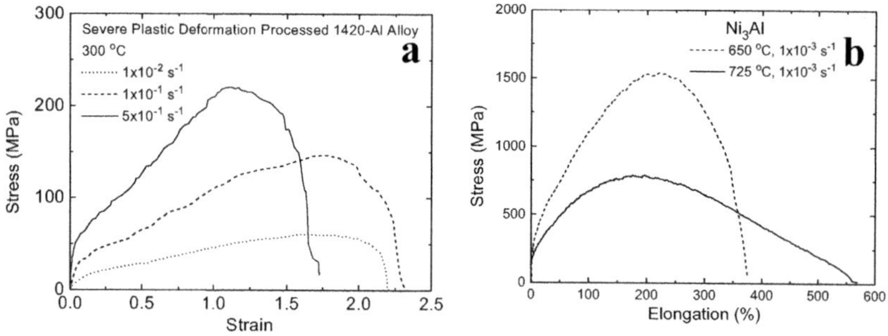

Figure 1. Flow curves for (a) nanocrystalline 1420-Al alloy at different strain rates and (b) nanocrystalline Ni_3Al alloy at different temperatures show excessive strain hardening. Note the increase in strain hardening with increasing strain rate in 1420-Al alloy.

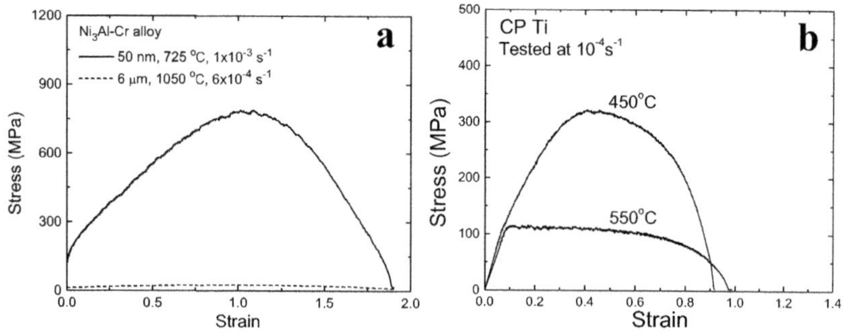

Figure 2. (a) A comparison of flow curves for Ni_3Al alloy in microcrystalline and nanocrystalline state; (b) A comparison of flow curves for CP Ti tested at different temperatures. Note the higher flow stress in nanocrystalline state, although specimens showed similar ductility.

nanocrystalline state. The high flow stresses during superplasticity of $ncNi_3Al$ alloy are apparent. Although the test temperature and grain size is different for the two microstructural conditions, the comparison is interesting because the overall ductility in both cases is similar. One of the features of conventional superplasticity in mc material is low flow stresses. The observation of superplasticity with high flow stresses in nanocrystalline material needs a new approach to explain the origin of such high flow stresses. First, the starting flow stress is around 170 MPa. Second, after significant strain hardening the flow stress reaches to the level of 780 MPa at 725 °C. At 650 °C, the flow stress values are even higher (Fig. 1b).

Despite titanium's reputation as a relatively low strain-hardening material [11], high flow stresses and extensive strain hardening were observed in this material at temperatures where nc microstructure is stable. At higher temperatures, when extensive grain growth was observed, a transition to steady-state flow has been revealed. Figure 2b presents flow curves of commercially pure (CP) Ti after HPT tested at 450° and 550°C and at a strain rate of $1x10^{-4}s^{-1}$. Even though similar elongations were reached in both cases, the shape of the flow curve has changed dramatically with temperature. The existence of relatively stable nc structure during testing results in high flow stress.

Figure 3 shows the variation of normalized stress with temperature and grain size compensated strain rate for Ti-6Al-3.2Mo alloy. The raw stress-strain rate data were taken from Salischev et al. [3,4]. Also, the data for a mc Ti-6Al-4V alloy [12] and the expected behavior from equation (2a) is included. Titanium alloys exhibit slip accommodated, lattice diffusion controlled grain boundary sliding. Two things are apparent from Fig.3: (a) the data for microcrystalline alloy agree well with the expected trend for slip accommodated, lattice diffusion controlled grain boundary sliding, and (b) the data for the submicrocrystalline alloy and nanocrystalline material shows higher flow stress or slower kinetics (in this paper we would refer to this as higher flow stress).

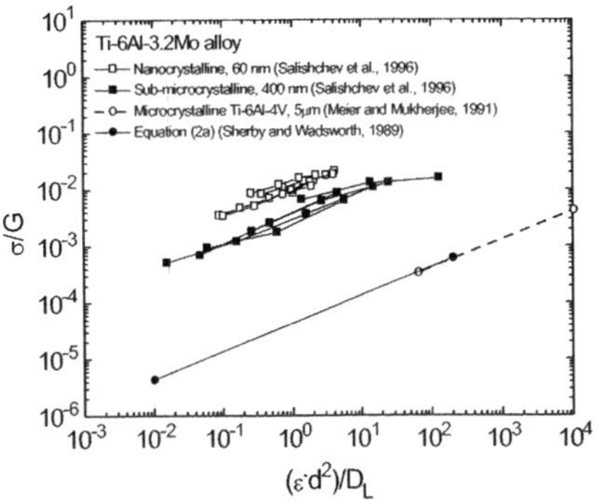

Figure 3. The variation of normalized flow stress with temperature and grain size compensated strain rate for Ti alloys.

The observed higher flow stress for submicrocrystalline and nanocrystalline material suggests a transition in micromechanism or additional difficulty in grain boundary sliding. This is also valid for superplasticity in nanocrystalline Zn-Al eutectoid [9], where the superplastic flow is grain boundary diffusion controlled. So, regardless of grain boundary or lattice diffusion controlled grain boundary sliding, the flow stresses for superplasticity in nc materials are higher. This is opposite to some of the earlier expectations of enhanced superplasticity in nc materials.

Parametric dependencies: Constitutive equations are very helpful in understanding the deformation mechanism. For superplasticity, it is customary to evaluate the stress exponent, activation energy and grain size exponent. Based on the values of these parametric dependencies, one can decide on the dominant deformation micro mechanism. Stress exponent and activation energy data obtained for superplastic flow in some nc materials are listed in Table I. We note that during some of the tests used to obtain the values mentioned below, significant grain growth occurred. The extent to which the results summarized below are influenced by concurrent grain growth is not clear.

Table 1. Parametric dependencies for superplastic flow in nanocrystalline SePD materials

Material and Reference	Grain Size* (nm)	Strain rate (s^{-1})	T (°C)	Flow stress (MPa)	Elong. (%)	Deformation parameters	
						Stress exponent	Activation energy (kJ/mol)
Ni_3Al alloy [6]	50	$1x10^{-3}$	650 725	1530 790	380 560	2	ND
1420-Al alloy [6]	100	$1x10^{-1}$	250 300	154 146	330 850	2.5	90
Ti-6Al-4V [13]	70	$1x10^{-4}$	650 725	65 130	530 400	2.7	270
CP Ti [14]	120	$1x10^{-4}$ $1x10^{-2}$	450 550 650	350 110 75	145 170 215	5.8	270
NiTi [14]	30	$1x10^{-4}$ $1x10^{-3}$	400 500	1500 900	65 230	3	210

* Denotes the starting average grain size.

Note: All materials showed significant coarsening during heating, so the microstructure during testing was not nanocrystalline.

From the evaluation of the behavior of nc materials, it is clear that a nc grain size does not appear to be a sufficient condition for superplasticity. For instance, nc Ni_3Al, 1420-Al and Ti-6Al-4V alloys have exhibited superplasticity at elevated temperatures. At particular temperatures and strain rates very high elongation (more than 500%) was reached (Table 1). The stress exponent values are close to the theoretically expected value of 2 for grain boundary sliding mechanisms. The activation energy in some cases is significantly higher than the activation energy for grain boundary diffusion. This is intriguing, because with the abundance of grain boundaries in nc materials, an activation energy corresponding to grain boundary related deformation mechanism is expected at elevated temperatures. In all cases, extensive grain growth occurred during superplastic flow.

At the same time, other nc materials subjected to HPT did not reveal superplastic flow at elevated temperatures [14] and was deformed either by dislocation creep (nc Ti, m=0.2) or viscous dislocation glide (nc NiTi, m=0.3) as seen in Table 1.

Correlation between microstructural instability and enhanced plasticity: The driving force for grain growth is quite high in nc materials because of the high interfacial area per unit volume. A nice correlation between the temperature for microstructural instability and onset of enhanced plasticity in nc materials was observed. In some ways this is to be expected because both grain boundary migration and grain boundary sliding involve diffusion. An important implication of this is that one needs to establish the narrow temperature range in which there is enough thermal activation

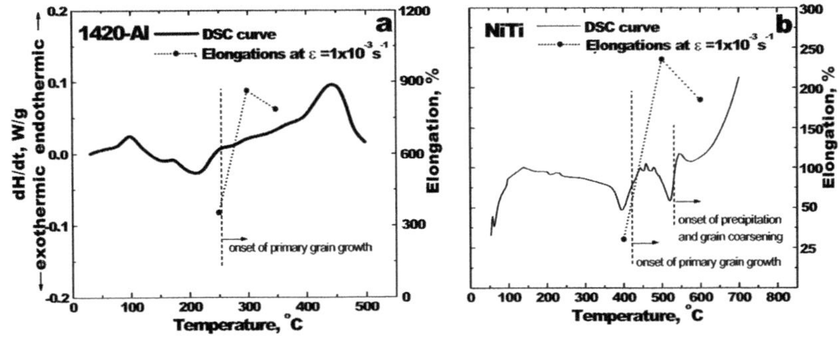

Figure 4. Combined plots showing variation of DSC signal and elongation with temperature for SePD (a) 1420-Al alloy and (b) NiTi alloy.

for grain boundary sliding but the grain growth is minimal. This is important for observing the grain boundary sliding regime without significant influence of grain growth, particularly from the viewpoint of obtaining parametric dependencies for constitutive modeling. The correlation between Differential Scanning Calorimeter (DSC) signal for thermal stability and increase in ductility is shown for SePD processed 1420-Al and NiTi alloys in Fig.4.

Effect of short annealing: A unique experimental observation during the present program has been the effect of a short anneal on the flow behavior. An example of this is shown in Fig. 5. The annealing of SePD processed nc Ni_3Al does not change the grain size significantly, but it has a dramatic effect on the flow behavior. The initial flow stress increases and the shape of flow curve changes. The overall ductility reduces from 350 % in the as received condition to ~100 % after the short anneal.

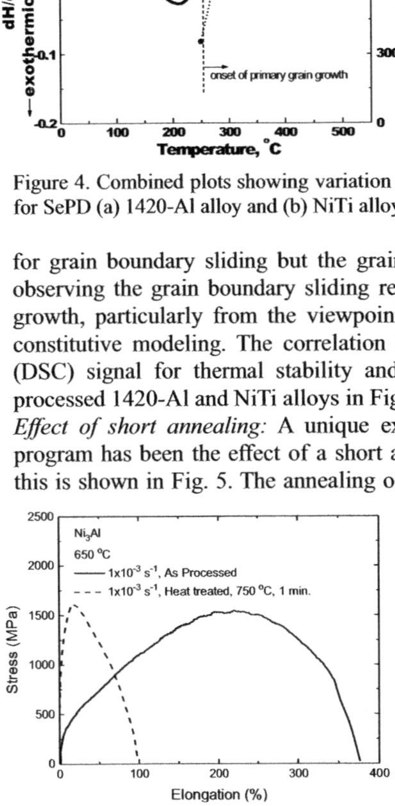

Figure 5. Effect of a short anneal on flow behavior of a nanocrystalline Ni_3Al alloy.

An analysis of experimental data on mechanical behavior of SePD materials under different conditions clearly demonstrates that some microstructural features other than just grain size (grain boundary structure; grain size distribution, etc.) can be also responsible for the material behavior.

Deformation mechanisms in nc materials

All the trends described above indicate that the mechanism of superplasticity in nc materials is fundamentally different from the well established mechanisms in mc materials.

The unique deformation behavior of nc and ufg materials is considered to be caused by suppression of conventional lattice dislocation slip (which dominates in coarse-grained materials) and effective action of alternative deformation mechanisms mostly occurring through motion of grain boundary (GB) defects. An increasing role of grain boundaries in NCM, especially as a source and sink of mobile dislocations,

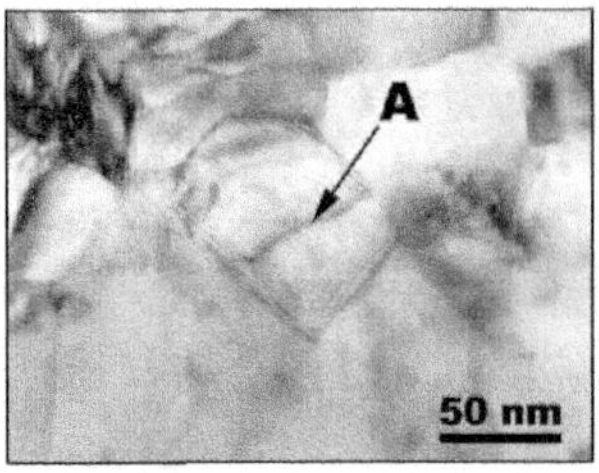

Figure 6. In-situ tensile testing of Ni_3Al at room temperature. Note dislocation "A" in nanosized grain. The dislocation was generated in the grain boundary, moved through the grain interior and disappeared in the opposite grain boundary.

was clearly demonstrated by molecular dynamics simulation [15]. The authors observed grain boundary generation of dislocations in a 45 nm grain size high stacking fault energy metal (12% strain) at room temperature. Such dislocation motion has already been captured by video camera at room temperature in Ni_3Al ordered matrix (Fig.6). Moreover, regardless of the deformation mechanism at elevated temperatures, grain rotation and grain boundary sliding is considered to play a sufficient role in plastic deformation of these materials as the role of dislocations is diminished at smaller and smaller grain sizes. This is, for example, evident from a comparison of the gage surface of the Ti sample before and after deformation at room temperature (Fig.7). It is seen that the surface of the gage before deformation is rather smooth while after tensile test the deformation relief of the shape of separate grains can be observed. This suggests that such types of deformation mechanisms as grain boundary sliding and grain rotation should be activated in this material during tensile testing even at low temperatures. An activation of grain boundary sliding during deformation at relatively low temperatures was also assumed for nanostructured Cu after severe plastic

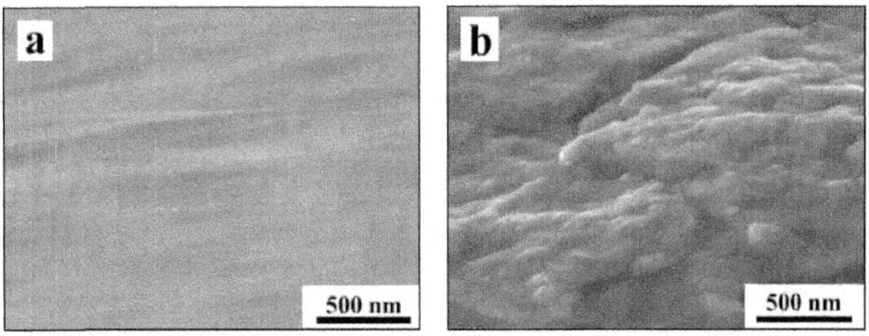

Figure 7. View of the gage surface of HPT Ti before and after deformation at room temperature.

deformation [16]. Other evidence for grain boundary sliding in nc materials is emergence of twin observed in HPT Ti deformed at 600°C (Fig.8). Such twinning can be formed as a result of grain boundary sliding in order to relieve stress concentration on grain boundary. So, during elevated temperature deformation there seems to be a competition between these two mechanisms and both of them can play a critical role in the deformation process.

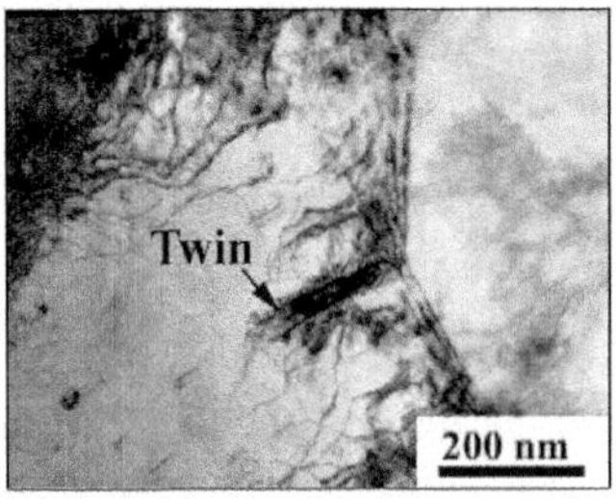

Figure 8. Microstructure of CP Ti produced by HPT and deformed at 600°C.

In real deformation processes, several mechanisms, which act independently from each other, simultaneously may contribute to plastic flow. The suggested representations on the combined actions of GB sliding and rotational mode enhanced by conventional dislocation slip are indirectly supported by the experimentally detected suppression of superplasticity in heat-treated nc materials [8]. Heat treatment results in annihilation of the GB dislocations generated due to conventional dislocation slip in nanocrystalline materials prepared by severe plastic deformation. Therefore, enhanced climb of GB dislocations from triple junctions does not occur causing suppression of the crossover from the GB sliding to the rotational mode. As a result, a nc specimen after heat treatment does not exhibit superplasticity. Recently, clear evidence of grain rotation under applied strain (Fig.9) was detected in Al-1420 alloy produced by HPT during *in-situ* tensile test at elevated temperature [17]. It is worthy to note that grain boundary sliding may concern either individual grains, or on the contrary rigid groups of grains behaving like distinct entities. This has been discussed in the literature in the context of cooperative grain boundary sliding. Experimental evidences supporting the role of cooperative grain boundary sliding in mc materials have been reported by several authors [18-20] In each case these authors observed surfaces of strain localization, separating less deformed zones. These features were explained from the viewpoint of CGBS associated with sliding and rotation of entire grain groups. In fine-grained materials, the respective role of GBS and CGBS is not really known.

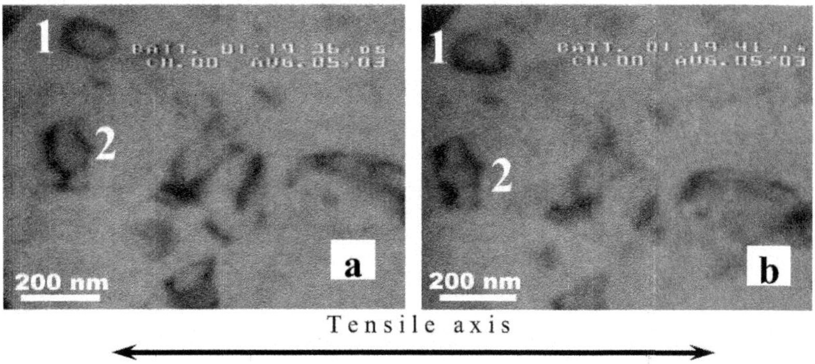

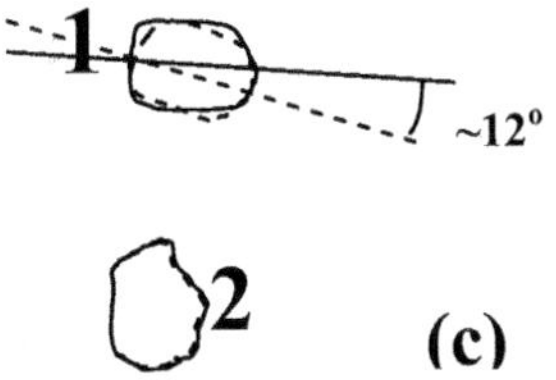

Figure 9. (a) An image of sample microstructure during in-situ tensile test; (b) an image of the same place after 5 sec of straining; (c) overlapping contours of grains 1 and 2 corresponding to their position in (a) and (b). Note a rotation of grain 1.

Concluding remarks

In the most intensively investigated materials produced by SePD, high temperature deformation in the temperature range where the microstructure is stable results in a slight increase in ductility but is far from predicted superplastic behavior. Usually, an enhancement in nanomaterial plasticity was only observed at temperatures for microstructural instability in terms of grain growth. One of the open questions on the micro-mechanism of superplasticity in such materials is the possible mechanism (or mechanisms) for grain boundary sliding accommodation. As noted earlier, the micrograined materials do show significant intragranular dislocation activity and these dislocation features are interpreted as evidence for slip accommodation. On the other hand, deformed nc materials have shown rather limited dislocation activity. Moreover, an analysis of experimental data on mechanical behavior of SePD materials under different conditions clearly demonstrates that some microstructural features other than just grain size (grain boundary structure; grain size distribution, etc.) can also be responsible for the material behavior. Actually, decreasing grain size leads to changes in deformation mechanism of the materials by eliminating some of them (such as classical dislocation glide) and activation of others (grain boundary sliding, grain rotation, triple junction diffusion). The analysis of their effect on material behavior and development of the generalized model which will describe the combined action of different deformation mechanisms, are the major subject of the ongoing theoretical and experimental investigations.

Acknowledgements

This investigation was supported by a grant from the U.S. National Science Foundation (NSF-DMR-0240144).

References

1. Valiev R.Z., Islamgaliev R.K., Alexandrov I.V., Prog. Mater. Sci. 45 (2000), 103-189.
2. Lowe T.C., Valiev R.Z., *Investigations and applications of severe plastic deformation*, Kluwer Academic Publishers, 2000.
3. Salishchev G.A., Valiakhmetov O.R., Valitov V.A., Mukhtarov S.K., Mater. Sci. Forum Vols. 170-172, 1994, p.121.
4. Salishchev G.A., Galeyev R.M., Malisheva S.P., Valiakhmetov O.R., Mater. Sci. Forum Vols. 243-245, 1997, p.585.
5. Bird J.E,, Mukherjee A.K., Dorn J.E., in: *Quantitative Relation Between Properties and Microstructure*, Inter. Conf. Proc., Haifa, Israel University Press, 1969, p.255.
6. Mishra R.S., Valiev R.Z., McFadden S.X., Mukherjee A.K., Mater. Sci. Eng. A252 (1998), 174-178.
7. McFadden S.X., Mishra R. S., Valiev R. Z., Zhilyaev A.P., Mukherjee A. K., Nature, 398 (1999), 684-686.
8. Islamgaliev R. KValiev., R. Z., Mishra R. S., Mukherjee A. K., Mater. Sci. Eng. A304-306 (2001), 206-210.
9. Mishra R.S., Valiev R.Z., Miukherjee A.K., Nanostr. Materials 9, (1997), 473-476.
10. Miukherjee A.K., Mishra R.S., Bieler T.R., Acta Mater. 45, (1997), 561-568.
11. Suri S., Viswanathan G.B., Hou D.-H., Mills M.J., Acta Mater. 47 (1999), 1019-1034.
12. Meier M. L., Lesuer D. R., Mukherjee A. K., Mater. Sci. Eng., A154 (1992), 165-168.
13. A.V. Sergueeva, V.V. Stolyarov, R. Z. Valiev and A.K.Mukherjee, Scripta Mater. 43 (2000), 819-824.
14. Sergueeva A.V. , Mukherjee A.K., in: *Creep Deformation: Fundamentals and Applications*, ed. R.S. Mishra, J.C. Earthman, and S.V. Raj ,TMS, 2002, p.137.
15. Yarmakov V., Wolf D., Philpot S.R., Mukherjee A. K., Gleiter H., Nature Mater. 1 (2002), 45-49.
16. Dubravina A.A., Sergueeva A.V., Alexandrov I.V., Valiev R.Z., in *Nanomaterials by Severe Plastic Deformation: Fundamentals – Processing – Applications*, ed. M.J.Zehetbauer & R.Z.Valiev, J.Wiley VCH Weinheim (Germany), 2004, p.465.
17. Mukherjee A.K., Mara N.A., Sergueeva A.V., in: *Ultrafine Grained Materials III*, ed. Y.T. Zhu, T.G. Langdon, R.Z. Valiev, S.L. Semiatin, D.H. Shin, and T.C. Lowe, TMS (The Minerals, Metals & Materials Society), 2004, p. 603.
18. Zelin M.G., Krasilnikov N.A., Valiev R.Z., Grabski M.W., Yang H.S., Mukherjee A.K., Acta Metall. Mater. 42 (1994), 119-126.
19. Zelin M.G. , Mukherjee A.K., ,Acta Metall. Mater. 43 (1995), 2359-2372.
20. Chen T.R., Huang J.C., Metall. Mater. Trans. A 30 (1999), 53-64.

Features of Creep in Bulk Nanostructured Composite Cu-0.5wt.%Al_2O_3

Yu.R. Kolobov*, G.P. Grabovetskaya

Institute of Strength Physics and Materials Science, SB RAS, Tomsk, Russia

Abstract

The characteristic features of creep in nanostructured materials produced by severe plastic deformation have been studied at temperatures $T < 0.3\ T_{melt}$ using nanostructured the Cu-0.5wt.%Al_2O_3 composite. The role of grain boundary state in plastic deformation development during creep is considered.

Introduction

Processing by severe plastic deformation (SPD) is a promising way to obtain nanostructured (NS)state in metal materials [1,2]. Using the above technique, structure can be formed in bulk billets with grain sizes in the interval of 100-500 nm and specific substructure in the grain bulk, which is related to the presence of dislocations and dislocation sub-boundaries. Transmission electron microscopy (TEM) examination reveals that the grain boundaries in NS materials show a contrast, which suggests the presence of elastic stresses due to the lattice and grain-boundary dislocations. Thus, the microstructure of nanostructured materials has two key features, i.e. small grain size and non-equilibrium high-energy grain boundary state, which might affect significantly, in particular, at elevated temperatures, both the structure evolution during annealing and the mechanical properties. In view of the above, nanostructured materials processed by SPD have a great potential as construction materials. Therefore, in order to establish the creep regularities for these materials at moderate and elevated temperatures, creep studies are called for.

One of the effective ways to enhance the thermal stability of structure and mechanical properties of metals and alloys is dispersion hardening. In recent years, investigations have begun on nanostructured composite development. To achieve these ends, matrix strengthening by high-stability oxide or carbide or some other phase (particle sizes 10-15 nm) is used in conjunction with nanostructure formation (grain sizes 100 - 300 nm) by SPD. The strengthening phases have low solubility and high resistance to coagulation; therefore, nanostructured composites are expected to exhibit high stability of structure and to retain good properties under the simultaneous action of load and temperature.

The available data on the creep regularities in NS composites subjected to SPD are reviewed below with particular emphasis on the effect of grain boundary state on the creep mechanism. microstructure was examined by SEM with high.

Experimental Procedures

Creep tests were carried out in vacuum at 10^{-2} Pa in the temperature interval of 423-923 K in a strain-rate interval of 10^{-6}-10^{-4} s^{-1} under constant load using a PV-3012M test machine. The value of apparent creep activation energy (Q_c) was

* Corresponding Author. kolobovispms@mail.tomsknet.ru

Y.T. Zhu and V. Varyukhin (eds.), Nanostructured Materials by High-Pressure Severe Plastic Deformation, 285-291.

measured by the temperature jump method (ΔT=10 K). Q_c was calculated from the formula reported in [3]:

$$Q_{\Pi} = R\ln(\dot{\varepsilon}_2/\dot{\varepsilon}_1)/(1/T_1 - 1/T_2) \ , \qquad (1)$$

where $\dot{\varepsilon}_1$ and $\dot{\varepsilon}_2$ are the creep rates before and after the temperature change, respectively; R is the gas constant; T_1 and T_2 are the absolute temperatures.

The samples used for mechanical testing were shaped like a double spade and had dimensions 10×2.8×0.5 and 5×1.4×1 mm^3; these were cut by the electric-spark method. Before the tests, a surface layer of approximately 100 μm in thickness was removed by sanding followed by electropolishing.

Results and Discussion

An attempt to account for the anomalously high n values in the stress dependence of creep rate is made by [4]. The workers use a model in which the creep behavior of dispersion strengthened NS materials is controlled by the local climb of dislocations with subsequent thermally activated detachment of dislocations from strengthening particles. Comparison of the calculated and experimental values of $\dot{\varepsilon}$ reveals that by making certain assumption for the threshold stress of dislocation detachment from strengthening particles, one can obtain an experimental dependence of $\dot{\varepsilon}$ on σ for the composite Cu-3vol%TiC at 773 K, which is in satisfactory agreement with the theoretical dependence.

The characteristic features of creep in Ag-5vol%Cr_2O_3 composite (Ag grain sizes varied in the range 50 nm - 80 μm) were investigated by [5]. According to the authors, the steady-state creep rate is unlikely to be controlled by dislocation climb followed by thermal detachment of dislocations from strengthening particles since the threshold stress for the above process ($\sigma \cong 2$ MPa) is much lower than the experimental values of plastic flow stress obtained for NS composites.

The characteristic features of the creep in the NS composite Cu-0.5wt.%Al_2O_3 are examined at the present work. The composite nanostructure is formed by SPD using high-pressure torsion. The investigations were performed for as-received composite (state 1), as well as for the composite processed by SPD (state 2) and by SPD with subsequent anneal at a temperature 373 K, which precludes recrystallization (state 3). For comparison, similar creep tests were performed for pure NS copper processed by high-pressure torsion.

As a result of SPD processing, in the composite there forms ultrafine-grain structure similar to that of pure copper. The average grain size values obtained by dark-field image examination and measured for as-received composite (state 1) are 0.1-0.3 μm and 0.6 μm, respectively. On the grain boundaries and in the bulk of the material in the above three states there occur Al_2O_3 particles having sizes in the range 10 -100 nm. The TEM studies and electrical resistance measurements reveal that the preliminary anneals at temperatures below 673 K cause no changes in the material grain size and in Al_2O_3 particle size; however, the GB defect structure of the copper matrix undergoes relaxation [6]. Recrystallization of the copper matrix of the NS composite begins at 723 K. The nanostructure of pure copper remains stable at temperatures below 473 K.

Fig. 1 demonstrates the creep curves obtained for pure copper and the composite Cu-Al_2O_3 in the above three states at T=423 K and σ=285 MPa ($0.6\sigma_{0.2}$ for the composite). Evidently, the creep curve of pure NS copper shows only an accelerated creep stage, while the curves obtained for the composite exhibit three stages (unsteady-state, steady-state and accelerated creep). It should be noted that the steady-state creep rate of the composite in state 2 is one order of magnitude lower relative to that obtained for state 1, $\dot{\varepsilon} \sim 4{\cdot}10^{-7}$ s^{-1} and $\dot{\varepsilon} = 4.4{\cdot}10^{-6}$ s^{-1}, respectively (curves 1 and 4, respectively).

Thus, nanostructure formation by SPD enables one to reduce the steady-state creep rate of the composite Cu-Al_2O_3 by one order of magnitude, with the time to failure increasing by 4-5 times and deformation to failure decreasing by 25 %.

It can be suggested that a significant enhancement in the creep resistance of the material processed by SPD is related to the distinctive features of its nanostructure, i.e. ultrafine grain size, non-equilibrium GB state, the presence of elastic stresses and high density of defects on the GBS and in near-GB regions. The above suggestion is confirmed by the data on the creep of the composite Cu-Al_2O_3 subjected to anneals in the temperature interval of 353-523 K which precludes recrystallization. As is seen from fig. 1, the creep rate of the composite in a NS state increases with preliminary annealing temperature. After the annealing treatment at 523 K, the creep rate becomes comparable with that of the material in as-received state (after extrusion). The composite in as-received state is subjected to preliminary annealing at 523 K, which causes little or no change in its creep rate, probably due to the material GBS having more equilibrium structure.

Thus, preliminary anneals at temperatures, which preclude recrystallization, cause variation in the steady-state creep rates of the NS metals and of the composite in a NS state, with the material grain size remaining unchanged. This suggests that the state of material GBS plays an important role in the formation of structure-sensitive properties of NS materials.

Figure 2 illustrates the dependencies lgσ-lgε obtained at 423 K for the creep of the Cu-0.5wt.%Al_2O_3 composite in states 1 and 2 (curves 1 and 2, respectively). As is seen from fig. 2, the dependence of ε on σ is described by equations of the type [7] for the both states

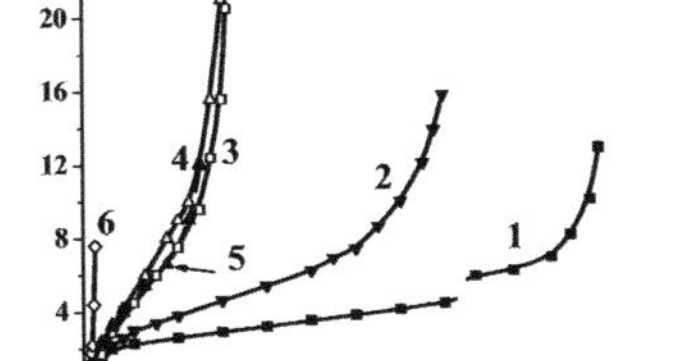

Figure 1. Creep curves of copper (6) and Cu-Al_2O_3 composite in nanostructured (1-3) and as-received (4-5) states. Anneal at 423 (2) and at 523 (3, 5) K. T_{test}=423 K, σ=285 MPa.

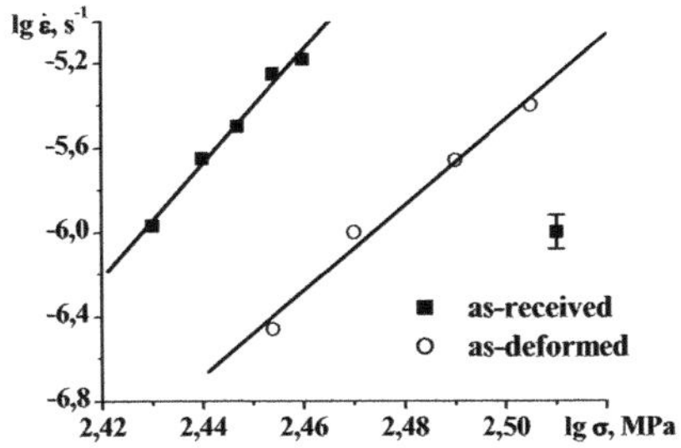

Figure 2. Dependence of creep rate on applied stress obtained for the Cu-Al_2O_3 composite in as-received (1) and nanostructured (2) states. Test temperature is 423 K.

$$\varepsilon = A\sigma^n \exp(Q_c / RT), \qquad (2)$$

where $\dot{\varepsilon}$ is the steady-state creep rate; **n**, the exponent of stress; *A*, a constant; Q_c, the apparent creep activation energy; *R*, the gas constant; *T*, absolute temperature. Equation (1) describes satisfactorily the steady-state creep rate of coarse-grained dispersion strengthened alloys. The values of n and Q_c obtained for the latter materials are, as a rule, higher relative to pure metals (**n** = 3 – 5) and $Q_c = Q_v$ (Q_v is the activation energy of volume self-diffusion). Thus Shilling et al. (1973) have found that the Q_c value obtained for coarse-grained dispersion hardened Cu-Al_2O_3 alloys containing 2.9 to 4.5 vol.% strengthening phase is about 1.5 times that of copper, 286 kJ/mol and 197 kJ/mol, respectively, and the value of **n** > 100.

The **n** value derived for the composite Cu-0.5wt.%Al_2O_3 in states 1 and 2 from the slopes of the curves in fig. 8; these are 30 and 22, respectively, which is significantly higher relative to pure metals. However, the Q_c values obtained for the investigated composite in states 1 and 2 (Table 1) are lower relative to the activation energy of volume self-diffusion for copper. The Q_c values obtained for the composite Cu-0.5wt.%Al_2O_3 after extrusion are matched against those of copper in NS state are found to be virtually the same (71.5 kJ/mol). However, the Q_c value obtained for the composite after extrusion increases to 127 kJ/mol, which is still lower relative to the coarse-grained state.

Table 1. The apparent creep activation energy obtained for NS Cu-0.5 wt. % Al_2O_3 composite.

	Q_c±15, kJ/mol		
	Cu (NS)	Cu-0,5 wt.% Al_2O_3 (as-received)	Cu-0,5 wt.% Al_2O_3 (NS)
423	74	72	127
473	71	148	132
623		194	204

The creep tests conducted for the composite Cu-0.5wt.%Al_2O_3 in the temperature interval of 473-773 K reveal that the steady-state creep rates differ significantly for states 1 and 2 states. At temperatures exceeding 673 K, more specifically, at 773 K the steady-state creep rates and the time to failure obtained for the both states are virtually the same.

Figure 3 shows $\dot{\varepsilon}_1/\dot{\varepsilon}_2$ against temperature ($\dot{\varepsilon}_1$ and $\dot{\varepsilon}_2$ are the steady-state creep rates obtained at the same experimental conditions for the composite Cu-0.5wt.%Al_2O_3 in states 1 and 2, respectively). As is seen from fig. 9, a sharp decrease in the creep resistance of the composite in state 2 occurs in the temperature interval of 673-773 K, in which recrystallization takes place. This may be due to grain size growth and a changeover in the deformation mechanism. The latter suggestion is confirmed by an increase in the apparent creep activation energies of the composite Cu-0.5wt.%Al_2O_3 in states I and 2 up to 194 - 204 kJ/mol (table 2), which is close to the activation energy of volume self-diffusion for copper.

At works [8-910] are shown that a decrease in the Q_c value observed in the temperature interval $T < 0.3T_m$ during the creep of pure metals in NS state may be related to the growing contribution of GBS to the overall deformation. During the creep of coarse-grained metal polycrystals, GBS is generally inhibited by the presence on the GBs of strengthening phase dispersion particles. Therefore, it is reasonable to assume that the process of GBS is hampered during the creep of the composite Cu-0.5wt.%Al_2O_3. An indirect evidence for the above suggestion is provided in the investigations conducted by [11]Kolobov et al. (2000) on the effect of grain boundary aluminum diffusion fluxes on the creep of pure NS copper and the composite Cu-0.5wt.%Al_2O_3. It was found that the effect of increase in creep rate connected with activation of GBS diffusion fluxes of aluminum atoms is observed at creep of nanostructured copper while the creep rate of the composite Cu-0.5wt.%Al_2O_3 remains essentially the same. Therefore, a decrease in the Q_c value observed for the composite Cu-0.5wt.%Al_2O_3 can hardly be attributed to GBS; it is likely to be related to the reduction in the activation energy for the process of dislocation detachment from the strengthening phase particles, which, in turn, might be due to a change in the energy of the particle-matrix interface boundary, which occurs as a result of SPD. This assumption, however, requires special investigation.

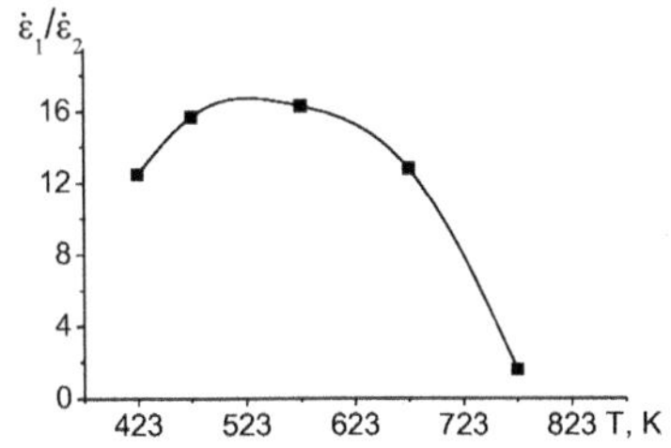

Figure 3. Temperature dependence of the ratio of creep rates of the Cu-0.5 wt. % Al_2O_3 composite in the state 1 to that in the state 2.

At works [12-14] are shown that a decrease in the Q_c value observed in the temperature interval $T < 0.3T_m$ during the creep of pure metals in NS state may be related to the growing contribution of GBS to the overall deformation. During the creep of coarse-grained metal polycrystals, GBS is generally inhibited by the presence on the GBs of strengthening phase dispersion particles. Therefore, it is reasonable to assume that the process of GBS is hampered during the creep of the composite Cu-0.5wt.%Al_2O_3. An indirect evidence for the above suggestion is provided in the investigations conducted by on the effect of grain boundary aluminum diffusion fluxes on the creep of pure NS copper and the composite Cu-0.5wt.%Al_2O_3. It was found that the effect of increase in creep rate connected with activation of GBS diffusion fluxes of aluminum atoms is observed at creep of nanostructured copper while the creep rate of the composite Cu-0.5wt.%Al_2O_3 remains essentially the same. Therefore, a decrease in the Q_c value observed for the composite Cu-0.5wt.%Al_2O_3 can hardly be attributed to GBS; it is likely to be related to the reduction in the activation energy for the process of dislocation detachment from the strengthening phase particles, which, in turn, might be due to a change in the energy of the particle-matrix interface boundary, which occurs as a result of SPD. This assumption, however, requires special investigation.

Thus, the above data suggest that no models have been developed so far in order to account for the reduced apparent creep activation energies of NS composites processed by SPD. It should be emphasized, however, that it is only recently that investigations of regular creep features of these materials were started. Therefore, for lack of experimental evidence, no conclusions can be made about the mechanisms that control the steady-state creep rate.

A characteristic feature of the creep in the investigated NS composite Cu-0.5wt.%Al_2O_3 is the development of localized deformation bands on the meso- and macro-levels. On the meso-level, the deformation localization appears on the sample surface as band-like agglomerates of grains with grain boundary sliding occurring within the agglomerates. Usually such agglomerates measure 1-2 grains in width and up to 10 grains in length. On the macro-level, at the onset of accelerated creep on the sample work part there arise localized deformation bands having width up to 0.5 mm. These extend through the total section of the sample at an angle of ~ 45° to the tensile axis and at an angle of ~ 90° to each other. Several pairs of analogous macro bands have been observed on the surface of the composite studied. Sample failure occurs by cutting along one of the band of localized deformation. During the creep of the samples subjected to preliminary anneals at temperature precluding recrystallization, localized deformation bands develop on their surface. However, in this case neck formation precedes the formation of micro-bands of localized deformation. On the surface of recrystallized samples of the composite Cu-Al_2O_3 no formation of the above bands is found to occur.

Thus, the above experimental evidence suggests that the creep behavior exhibited by NS composite Cu-0.5wt.%Al_2O_3 processed by SPD reveals a number of distinctive regular features relative to the respective coarse-grained counterpart. First, NS composite Cu-0.5wt.%Al_2O_3 processed by SPD show high resistance to creep immediately after the treatment; however, anneals conducted at temperatures precluding recrystallization cause a sharp decrease in their creep resistance with resultant recovery of the non-equilibrium grain boundary structure. Second, NS materials have low apparent activation energies of creep. Third, during the creep of NS materials localization of plastic deformation is observed on meso- and macro-levels.

It has been found that the steady-state creep rate is determined to a significant extent by the grain boundary state of the material irrespective of its grain size. Therefore, by addressing the regular features of creep in NS materials processed by SPD, one has to take into account the above finding.

Acknowledgements

The financial backing of the work came trough the RFBR (grant No. 03-02-16955), ISTC (grants №2398). The authors express their gratitude to Prof. R.Z. Valiev for supplying materials.

References

1. Kolobov, Yu.R., Valiev, R.Z., Grabovetskaya, *Grain boundary diffusion and properties of nanostructured materials*, Novosibirsk: Nauka publ., 2001, p.232.

2. Valiev, R.Z., Alexandrov, I.V., *Nanostructured materials processed by severe plastic deformation*, Moscow: Logos, 2000, p. 272.
3. F, Garofalo. *Fundamentals of creep and creep –rupture in metals* Moscow, Metallurgija, 1967, p. 304.
4. Sauer, C., Weißgärber, T., Dehm G. et al., *Zeitschrift für Metallkunde*. 2 (1998), 119-125.
5. Grundmann, U., Gerner, M., Heilmaier, M. et al, *Mat. Sci. and Eng.* A 234-236 (1997), 505-508.
6. Islamgaliev, R.K., Buchgraber, W., Kolobov, Yu.R. et al., *Mat. Sci. and Eng.* A319-321 (2001), 874-878.
7. Rozenberg, V.M., 1967. Creep of metals. Metallurgija, Moscow.
8. Grabovetskaya, G.P., Ivanov, K.V., Kolobov, Yu.R., *Annales de Chimie Science des Matériaux*. 3(2002), 89-98.
9. Grabovetskaya, G.P., Kolobov, Yu.R., Ivanov, K.V., Girsova N.V., *The Physics of Metals and Metallography*.2 (2002.), 37-44.
10. Kolobov, Yu.R., Grabovetskaya, G.P., Ivanov, K.V., Girsova, N.V., *The Physics of Metals and Metallography*. 5 (2001), 532-537.
11. Kolobov, Yu.R., Ivanov, K.V., Grabovetskaya, G.P., ad,: *Investigation and Applications of Severe Plastic Deformation*. Kluwer Academic Publishers, Boston, 2000, p.339-344.
12. Grabovetskaya, G.P., Ivanov, K.V., Kolobov, Yu.R., *Annales de Chimie Science des Matériaux*. 3(2002), 89-98.
13. Grabovetskaya, G.P., Kolobov, Yu.R., Ivanov, K.V., Girsova N.V., *The Physics of Metals and Metallography*.2 (2002.), 37-44.
14. Kolobov, Yu.R., Grabovetskaya, G.P., Ivanov, K.V., Girsova, N.V., *The Physics of Metals and Metallography*. 5 (2001), 532-537.

Effect of Second Phase Particles on Annealing Ultrafine Grained Aluminium Alloys Deformed by ECAE

M. Berta[1], P.B. Pragnell[1] and P.J. Apps[1*]

[1]Manchester Materials Science Centre, UMIST, Grosvenor St., Manchester M1 7HS, UK

Abstract

The annealing behaviour of Al-alloys containing fine dispersoids (Al-0.2wt.%Sc) and coarse particles (AA8079) have been compared to a single phase alloy (Al-0.13Mg) after deformation by ECAE to a strain of ten. The fraction of HAGB area and average boundary misorientations did not change significantly during annealing in any of the materials, whereas, the width of the grain size distributions was found to be affected by second phase particles. At high temperatures abnormal grain growth was found to occur with both stable and unstable pinning particles. Good agreement was found with the conditions required for abnormal growth, predicted by Humphreys model of cellular stability.

Introduction

The past few years have seen considerable interest in the development of submicron grained (SMG) alloys by severe deformation deformation [1]. A relatively simple method of producing ultrafine grained metals is Equal Channel Angular Extrusion (ECAE), in which a material is subjected to severe plastic deformation by simple shear. Due to the large stored energy associated with severely deformed microstructures, there is a high driving force for instability of the material during subsequent thermal exposure [1,2]. It has been known for some time that heavily deformed metals containing a high fraction of high angle grain boundaries (HAGBs) can continuously recrystallize on annealing [2]. Humphreys et al. have shown that a cellular structure will become resistant to discontinuous growth once 70% of the boundaries are high angle in character, as the cell boundaries will then all have similar energies and mobilities [3]. However, other phenomena such as enhanced grain growth rates at low temperature and abnormal grain growth have been reported [4]. This range of observations can probably be partly attributed to the influence of the as-deformed microstructure, which is rarely homogeneous and dependent on the processing method, but material parameters, such as second phase particle, are also important. To date, few of the recrystallization phenomena reported in severely deformed alloys have been systematically related to the material variables. The present work compares statistical microstructural parameters measured by EBSD, during the isochronal annealing of two model materials with coarse and fine homogeneously dispersed particles, to that of a single-phase alloy after severe deformation. The results are discussed in the context of the behaviour of more conventional alloys and compared to Humphreys' model on the effect of second-phase particles on the stability of an ideal grain assembly during grain growth [5].

* Corresponding Author: m.berta@postgrad.umist.ac.uk.

Y.T. Zhu and V. Varyukhin (eds.), Nanostructured Materials by High-Pressure Severe Plastic Deformation, 293-298.

Experimental Procedures

The annealing behavior of three alloys were compared in this investigation.; i) AA8079, (Al-1.3%Fe-0.09%Si), chosen to give a uniform distribution of coarse second-phase particles with no fine dispersoids and a minimal solute level in the matrix; ii) a high purity Al-0.2wt%Sc alloy, selected to provide a high density of fine dispersoids, and iii) a high purity single phase Al-0.13%Mg alloy, using previously obtained data [2]. All the alloys were processed to give a similar recrystallized starting grain size of ~ 500 μm. The alloy AA8079 contained 2.5 vol.% of ~ 1-2 μm $Al_{13}Fe_4$ particles after pre-processing. It was homogenized for 8.5 h at 610°C to spheroidise the particles, hot rolled, and subsequently recrystallized at 500°C for 1 hr. The Al-0.2wt%Sc alloy was cast, solutionised for 72 hours at 620°C and cold rolled 50%, followed by recrystallization for 30 minutes above the solvus temperature (at 620°C), using rapid heating to retain the scandium in solution. A heat treatment of 16 hrs at 400°C was then applied precipitating a homogeneous dispersion of 20 nm dispersoids, with a spacing of ~ 100 nm. 15 mm diameter rods were machined from the materials, in the rolling direction, and processed by ECAE, at 20°C to 15 ECAE passes without billet rotation (Route A), using a 120° die - giving a von Mises' effective strain per pass of $\varepsilon_{vm} \sim 0.7$ [4]. The processed billets were annealed for one hour at a range of temperatures and analyzed using a high resolution FEGSEM-EBSD system in the plane of symmetry of the die. Average microstructural parameters were determined using orientation maps containing at least 1500 grains, with a minimum misorientation of 15° defining a high angle grain boundary (HAGB).

Results and Discussion

Microstructure of the deformed materials after ECAE processing,

Before considering the annealing data of the three materials, it is interesting to summarize their microstructures after severe deformation, which, if significantly different, could influence their recrystallisation behaviour. From the EBSD data in Table 1, it can be seen that on average all the materials contain submicron grain fragments and have elongated deformation structures with a high fraction of HAGB area. Such grain structure are comprised of lamellar HAGBs aligned in the direction of deformation with the low angle boundaries being orientated mainly in the transverse direction [2]. The coarse particle containing alloy shows a slightly lower grain aspect ratio and higher HAGB fraction than the single phase alloy, whereas the dispersoid containing alloy has a significantly lower fraction of HAGB area and average boundary misorientation distribution and larger grain size and aspect ratio. Previous work has demonstrated that coarse second phase particles accelerates the rate of grain refinement, and slightly refine the SMG structure at ultra-high strains, by giving rise to particle stimulated deformation zones at low strains, and disrupting the lamellar deformation structures that form at high strains [2]. In comparison the presence of fine non-shearable dispersoids reduces the rate of grain subdivision by homogenizing slip and inhibiting the formation of the micro-shear bands, which normally give rise to the majority of new HAGBs formed during the initial stages of severe deformation. At high strains this results in a slightly less refined structure, containing a lower fraction of HAGBs, with more potential for exhibiting a discontinuous recrystallization behavior on annealing.

Tablel. Comparison of the average EBSD data for the particle containing and single-phase Al-0.13%Mg alloys, obtained from the EBSD maps after deformation to a strain of ~ 10 (15 ECAE passes).

	Al 0.13%Mg	Al 0.2%Sc	AA 8079
Grain Width (μm)	0.45	0.7	0.4
Grain Lenght (μm)	1.01	2.2	0.6
Aspect Ratio	2.3	3	1.6
Mean Misorientation	33°	27°	34°
Fraction HAGBs (%)	79	64	82

Grain structure evolution during isochronal annealing

In order to investigate the effect of different types of particles on the materials grain structure evolution during annealing, the EBSD orientation maps were post processed to reconstruct the grains and fit Gaussian curves to logarithmic grain size distributions, for directions of maximum and minimum average grain size (i.e. parallel to and normal to the lamellar boundary structures produced after deformation), using fitted ellipsoids of equivalent area and weighted aspect ratio. Example EBSD maps, showing typical grain size distributions as inserts, are given in Fig. 1 for annealing at 150°C. In general a good fit was obtained with all materials in both the as deformed condition and after annealing to a log-normal distribution, except after high temperatures in the particle containing alloys (see below). The average grain size, aspect ratio, fraction of HAGB area and boundary misorientation are plotted in Fig. 2 as a function of temperature and the normalised standard deviation of the fitted Gaussian distributions are shown in Fig. 3.

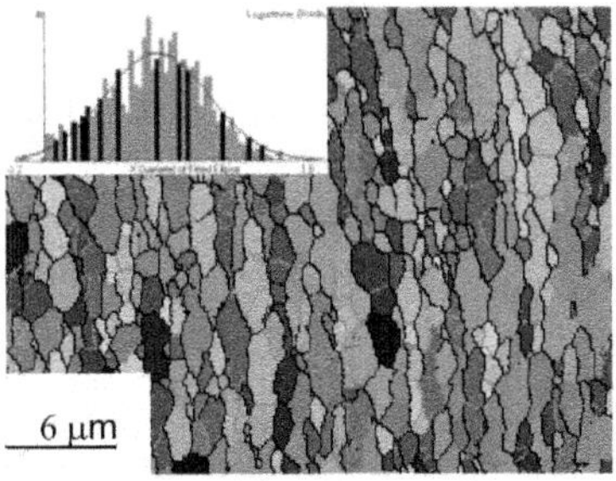

a)

b)

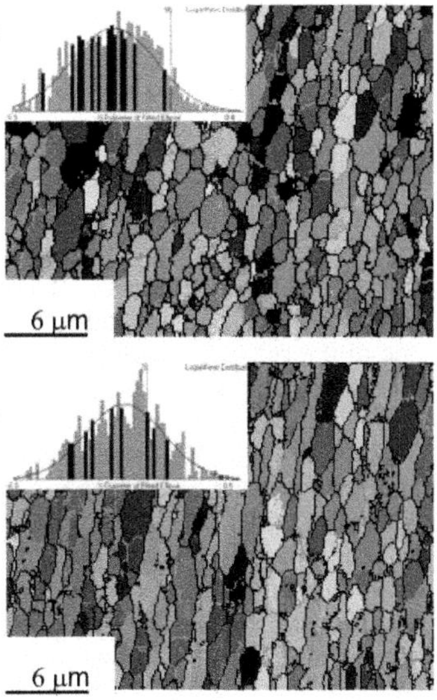

c)

Figure 1. EBSD Orientation maps and relative grain size distribution on logarithmic scale for a) Al 0.13%Mg, b) AA8079 and c) Al0.2%Sc after 1h annealing at 150°C. Note in (b) the black areas are second phase particles.

None of the materials exhibited a significant discontinuity in the fraction of HAGB area and average boundary misorientations, which increased only very slightly with annealing temperature. This is indicative of a continuous recrystallisation behaviour [3], but does exclude the possibility of abnormal grain growth. In the high purity Al-0.13% single phase alloy and the large particle containing 8079 material the grain structures evolved very similarly at low temperatures below 400°C, prior to the grain size approaching the inter-particle spacing in the 8079 alloy, although coarsening occurred at different rates due to the inhibiting effect of Fe on boundary mobility in 8079. In both alloys the grain structures rapidly spheroidised, driven mainly by the equilibration of surface tensions at triple points. However, as has been previously demonstrated [2], larger high aspect ratio grain fragments retained in the deformed state also subdivided along their length due to imbalances in surface tension from the transverse LAGBs, causing the lamellar HAGBs to pinch-off. This phenomenon occurred less in the course particle containing alloy as the Al-matrix had a lower average grain aspect ratio and contained fewer high aspect ratio grain fragments after deformation.

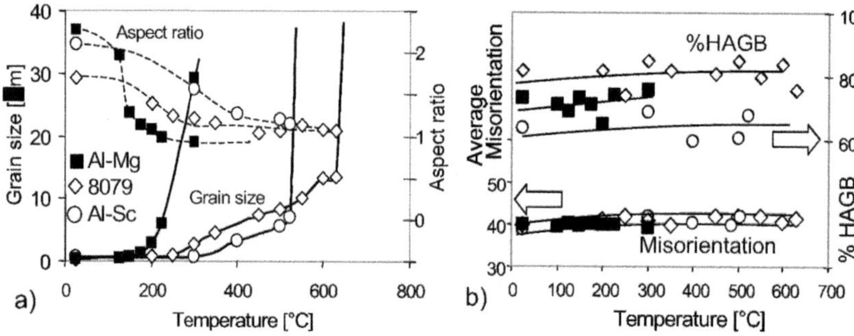

Figure 2. a) Average grain size and aspect ratio and b) percentage of high angle grain boundary area and average boundary misorientation with annealing temperature, for one hour heat treatments.

It can be seen in Fig. 3 that the normalised spread of the grain size distributions in grain length (σ_y/d_y) are initially reduced in all materials as the grain size increase and becomes more equiaxed. In comparison the distribution spread is very similar in and initially increases identically with grain coarsening in the grain width direction, due to the relatively uniform spacing of the high angle lamellar boundaries formed during processing.

A more detailed study was performed to see if the large particles had any local effect on the SMG matrix during annealing of the 8079 alloy. The EBSD orientation maps were analysed in areas 2 particle diameters in radius from individual particles and compared to measurements made 2 particle diameters away from any particle. Although because of the small areas measured, problems of 2D sectioning and the high particle volume fraction, the results should be treated with caution, this indicated that the grain size and aspect ratio is slightly smaller near large particles after deformation and this difference is retained until the average grain size has approximately trebled (by ~ 200°C) whereupon the grains become large relative to the particle size. Overall, the presence of coarse particles therefore appears to have only a very small effect on the early grain coarsening behaviour in severely deformed alloys when the grain size is smaller than the inter-particle spacing.

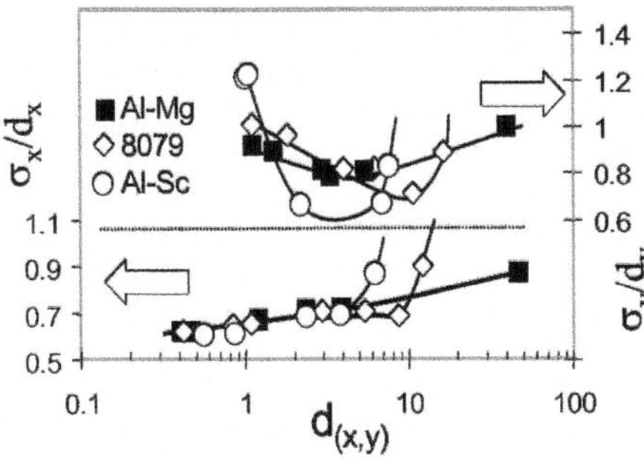

Figure 3. Standard deviation (σ) of the log-normal grain size distributions normalised with respect to the distribution mean, in the grain width (x) and length (y) directions plotted against mean grain size for the three alloys. (note the scales for the x and y directions are shifted for clarity).

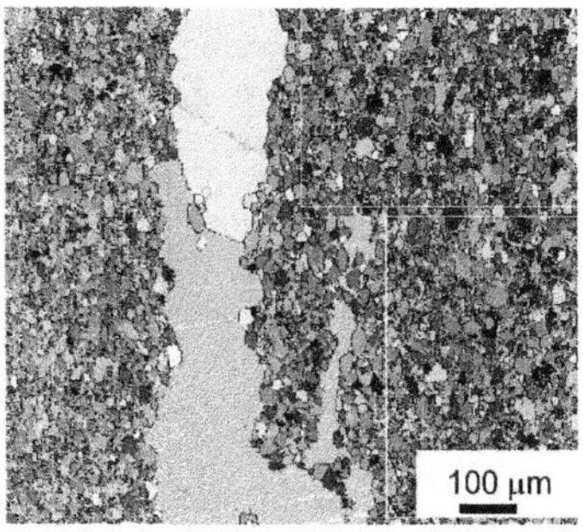

Figure 4. An example of abnormal grain growth in the 8079 alloy after annealing at 620 °C for 1 hr.

In comparison to the other two materials, the Al-Sc alloy showed little change in grain size on annealing below 300°C, although there was a small reduction in grain aspect ratio (Fig. 2a). The high stability of the as deformed structure in this alloy is due to very effective Zener pinning pressure, resulting from the high density of coherent Al_3Sc dispersoids present in the material, and is well known [6]. Gradual stable grain growth was observed above 300°C until 450°C when the grain size was found to dramatically increase. In the 8079 alloy the initial grain size was much finer than the inter-particle spacing and was not found to interfere greatly with grain growth at low annealing temperatures. However, as the grain size increased and approached the Zener limit (see below) there was evidence that the grain growth rate started to reduce and stagnate (Fig. 2a). There was then a dramatic increase in grain size at 620°C. In both particle containing alloys this rapid increase in grain size coincided with the onset of abnormal grain growth, an example of which can be seen in Fig. 4, which resulted in mm scale grains being produced. This trend is also reflected in the spread in the grain size distributions shown, in Fig. 3, which narrows as grain stagnation occurs in the initially more heterogeneous grain length direction and then rapidly widens as abnormal grain growth occurs.

Comparison with predictions of instability during grain growth

Fig. 5 shows the experimental data for the AA8079 and the Al 0.2%Sc alloys compared to the theoretical model developed by Humphreys [5], for predicting the stability of cellular microstructures during coarsening in the presence of incoherent stable second phase particles. In the coarse particle containing alloy the particle pinning pressure (proportional to the volume fraction to particle diameter ratio; measured as $F_v/d_p = 0.025$) was not found to change with annealing temperature and the model predicts that normal grain growth will occur until a grain size of 10 μm is reached, whereupon it will stagnate at the Zener limiting grain size. Abnormal grain growth may then occur if a sufficiently large grain is present within the distribution. In this alloy abnormal grain growth began when the average matrix grain size stagnated at ~ 13 μm, in close agreement with the model. In comparison, the Al_3Sc dispersoids in the Al-0.2%Sc alloy are unstable at high temperatures and their F_v/d_p ratio was found to reduce dramatically at above 400°C, from $F_v/d_p \sim 0.2$ at room temperature, to $F_v/d = 0$ at ~ 585°C, when the particles go into solution. The particles are also coherent at low temperatures and lose their coherency on annealing for 1 hr. above ~ 500°C [6], becoming semicoherent. For an initial grain size of ~ 0.7μm the Al-0.2Sc data initially approximately tracks the Zener limiting grain size curve, as the grain size increases simultaneously with particle coarsening, but would in reality lie further to the right within the range where abnormal growth is possible, because for coherent particles the pinning pressure is double that for incoherent, effectively doubling F_v/d [6]. Abnormal grain growth then readily occurs when the temperature reaches the point where the dispersoid density rapidly reduces (~ 400°C). This may occur inhomogeneously on a local scale allowing some grains to develop a growth advantage, nucleating abnormal growth.

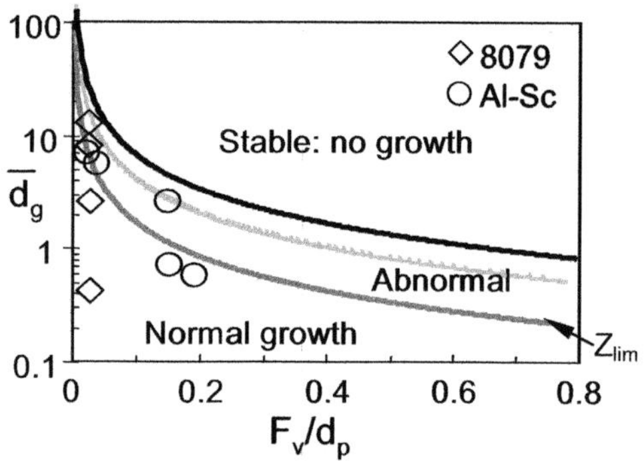

Figure 5. Comparison of experimental measurements of F_v/d_p and average grain size (d_g) with the zone of instability (or abnormal grain growth) predicted by Humphreys' model for stability of cellular microstructures during grain coarsening

Conclusion

Particles affect the rate of grain refinement during ECAE processing, with coarse and fine particles slightly increasing and reducing the level of grain refinement seen at ultra-high strains, respectively. The fraction of HAGB area and average boundary misorientations were not found to change significantly during annealing in any of the materials, whereas, the evolution of the grain size distributions was found to be affected by second phase particles. While Al_3Sc dispersoids are very effective in maintaining thermal stability of the deformed microstructure at moderate temperatures, coarse particles do not greatly alter the matrix grain growth behaviour until the grain size approaches the inter-particle spacing. At high temperatures abnormal grain growth was found, both in the presence of stable and unstable pinning particles. Good agreement was seen with predictions using Humphreys model of cellular stability for the conditions required for abnormal growth.

References

1. Valiev, R.Z A.V. Korznikov, and I.V. Alexandrov, *Annales de Chimie Science des Materiaux*, 2002. 27(3): 1-2.
2. Prangnell, P.B., et al., *Acta Materialia*, 2004. 52(11): 3193-3206.
3. Humphreys, F.J., et al., in *Philos.Trans.R.Soc.* 1999: London. p. 1663.
4. Apps, P.J., J.R. Bowen, and P.B. Prangnell, *Acta Materialia*, 2003. 51(10): 2811-2822.
5. Humphreys, F.J., *Acta Materialia*, 1997. 45(12): p. 5031-5039.
6. Jones, M.J. and F.J. Humphreys, *Acta Materialia*, 2003. 51(8): p. 2149-2159.

The Influence of the SPD Regimes on Superplastic Behavior of an Aluminum Alloy

R.K. Islamgaliev[1], N.F. Yunusova[2], R.Z. Valiev[2*]

[1]Institute of Mechanics Russian Academy of Sciences, Ufa, Russia, 450000
[2]Institute of Physics of Advanced Materials, Ufa State Aviation Technical University, Ufa, Russia 450000

Abstract

Recent studies have demonstrated that ultrafine-grained alloys processed by severe plastic deformation (SPD) techniques, i.e., equal channel angular pressing (ECAP) or high pressure torsion (HPT), can exhibit enhanced superplacticity at relatively low temperature and/or high strain rates. In this paper we consider the influence of the SPD regimes on microstructure and superplastic behavior of commercial 1421 aluminum alloy. Special attention is paid to processing guidelines for enhancement of superplastic properties through SPD techniques.

Introduction

Recent development of SPD techniques for processing bulk ultrafine-grained (UFG) metals and alloys [1-3] has provided new opportunities in investigations of superplasticity in metallic materials. Processing of UFG structures by SPD techniques made it possible to begin systematic experimental studies in this field. These investigations have been used both the bulk samples with submicrocrystalline structure with a grain size of less than 1 μm fabricated by equal channel angular pressing [2-4] and nanostructured specimens processed by high pressure torsion [5].

The present work was initiated to investigate an important factor that may influence the nature of the enhanced superplasticity in UFG aluminum alloys. In particularly, to evaluate the significance of the SPD temperature since it is known that the SPD temperature affects the nature of the morphology of second phases [6]. The investigation was conducted using a commercial 1421 aluminum alloy where this alloy was selected because it is anticipated that the presence of a minor amount of scandium (0.2%) will lead to high thermal stability and, in addition, the studies have shown that ECAP of this alloy produces mean grain sizes of <1 μm and a consequent enhancement in the superplastic properties [6].

Corresponding Author. rzvaliev@mail.rb.ru

Y.T. Zhu and V. Varyukhin (eds.), Nanostructured Materials by High-Pressure Severe Plastic Deformation, 299-304.

Experimental Procedures

The Russian commercial aluminum alloy 1421 (Al-5.5%Mg-2.2%Li-0.12%Zr-0.2%Sc) has been selected for this investigation. The UFG structure was formed by two SPD techniques, namely, high pressure torsion and equal-channel angular pressing. Before straining, the billets were annealed at 470°C for 30 min, and water quenched to prevent the decomposition of a solid solution. The HPT samples, having a diameter of 13 mm and a thickness of 0.3 mm, were processed by torsion at different temperatures (20-370°C) to a true logarithmic strain of ~ 6 under a pressure of 5 GPa. Further details of the HPT procedure were given earlier [2]. For ECAP we used a specially developed die-set which principal scheme was discussed in our previous works [2,3]. As the initial billets, 100 mm long cylinders with a diameter of 20 mm were used. Then, the specimens were ECA-pressed using two channels with the same cross section and a diameter of 20mm intersecting at an angle of 90°. To form different structural and phase states in the alloy, four regimes of ECAP were used (see below). It is known that a single passage through the ECAP die corresponds to an equivalent strain of ~ 1. In the present investigations, samples were pressed for 10 passes, thereby giving a total strain of ~ 10.

The structure was studied by transmission electron microscopy (TEM) using Philips EM430 microscope at an accelerating voltage of 300 kV equipped with a Gatan goniometric attachment.

Tensile tests were performed on the as-pressed samples at selected elevated temperatures using specimens cut parallel to the pressing direction with gauge lengths of 1 mm and cross-sectional areas of ~0.4 mm^2. The variation in stress with strain was recorded using a computer-controlled tensile testing machine operating at a constant rate of crosshead displacement.

Results and Discussion

Careful inspection of the as-pressed samples by TEM showed that the optimal microstructure, as represented by the minimum grain size, was achieved after pressing at 370°C where the grain size was ~0.3-0.4 μm. This microstructure is illustrated in Fig. 1(a) and the ultrafine grain size is attributed to the presence of extensive precipitation and consequent enhanced thermal stability. EDS analysis [6,7] has shown precipitates have variable composition and can be designated $Al(Mg,Sc,Zr)_x$.

By contrast, coarser microstructures were obtained after ECAP at the other temperatures [6], where the measured grain sizes are in the range of ~0.8-1.0 μm. There were also significant differences in the size and volume fractions of the $Al(Mg,Sc,Zr)_x$ precipitates: thus, precipitates were essentially absent after ECAP at the lowest temperature of 340°C but they had a mean size of ~200 nm and represented ~10-20% of the volume fraction when pressing at 370°C, and at the higher pressing temperatures of 390 and 410°C the particles coarsened to ~400-500 nm.

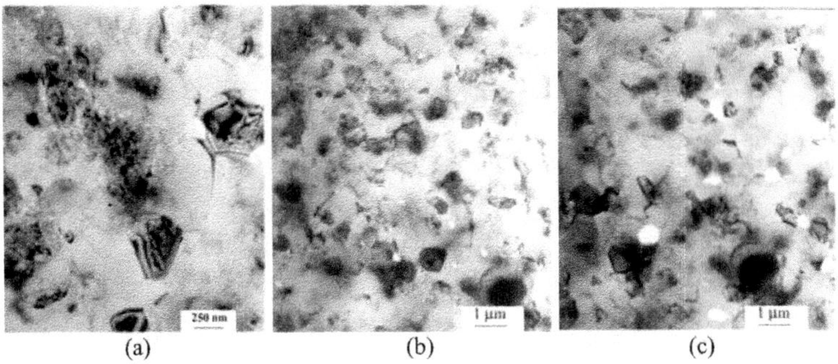

Figure 1. Microstructures of the 1421 alloy after ECA pressing at temperatures of 370°C: micrographs are shown after pressing (a) and after *in situ* heating in the TEM at 270°C (b) and 400°C (c).

During *in situ* heating in the TEM, no additional precipitation was revealed after ECAP at 340°C but there was evidence of additional precipitation in the vicinities of grain boundaries and triple junctions for the samples pressed at the other three temperatures with particle sizes of ~100 and ~300-400 nm after heating at 270 and 400°C, respectively: examples of this effect are visible in the center and right columns of Fig. 1. A similar effect has been reported in the 1420 alloy [7] where these particles were identified as the AlLi phase which is often observed in aluminum-lithium alloys at elevated temperatures [8].

Plots of true stress against true strain are shown in Fig. 2 for the specimens tested in tension at 400°C: in Fig. 2(a) results are given for specimens tested at an initial strain rate of 10^{-2} s^{-1} for samples prepared by ECAP at the four different pressing temperatures whereas in Fig. 2(b) the curves relate to tests conducted at initial strain rates of 10^{-1} and 10^{-2} s^{-1} for the specimens pressed at the optimal temperature of 370°C. Two conclusions may be drawn from these curves. First, all specimens exhibit superplastic characteristics but the ductility of the alloy is limited after ECAP at 340°C when no precipitates are available to minimize grain growth. Second, high tensile ductility is observed after ECAP at temperatures from 370 to 410°C but a pressing temperature of 370°C leads not only to the smallest as-pressed grain size but also to the highest tensile ductility. For this material, the highest elongation to failure was ~1500% when testing at an initial strain rate of 10^{-1} s^{-1} and a comparable elongation of ~1300% was achieved at an initial strain rate of 10^{-2} s^{-1}. By abruptly changing the strain rate during tests conducted at strain rates in the vicinity of ~10^{-1} and ~10^{-2} s^{-1}, the values of the strain rate sensitivities, *m*, were determined as ~0.22 and ~0.34 at these two strain rates, respectively. These values of *m* are lower than in conventional superplasticity where, typically, $m \approx 0.5$ but the values are comparable to those reported after ECAP in Al-0.2% Sc alloys containing 0.5% and 1% Mg [9].

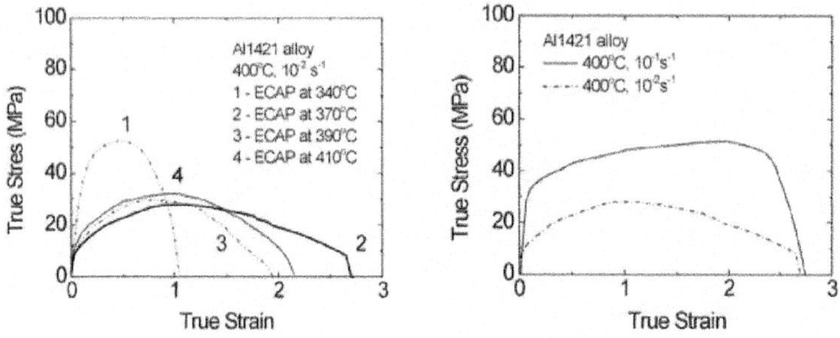

(a) (b)

Figure 2. True stress versus true strain for the as-pressed 1421 alloy: (a) after pressing at different temperatures and testing at 400°C with an initial strain rate of 10^{-2} s^{-1} and (b) after pressing at 370°C and testing at 400°C at two different strain rates.

Table. Microstructural and superplastic characteristics of the 1421 alloy processed by HPT at various temperatures.

HPT temperature, °C	Microstructural characteristics				Elongation to failure at strain rate $10^{-2}c^{-1}$, (%)	
	particle size, nm / volume fraction, %					
	d, nm	$Al(Mg,Sc,Zr)_x$	Al_3Sc	AlLi	375°C	400°C
20	80	-	-	-	595	593
120	40- 500	-	-	-	554	534
300	100	60-70 / 5	20 / 2	40-50 / 15	655	670
340	80	-	20 / 3	30-40 / 10	633	533
370	70	-	30 / 3	30-40 / 10	623	590

The results demonstrate also the beneficial influence of scandium in retaining a very small grain size and producing high tensile ductility. Thus, published data for the 1420 alloy, which has a similar composition but without the addition of scandium, show an as-pressed grain size of ~0.8 μm and a maximum elongation of ~1200% when testing at 400°C with an initial strain rate of $10^{-1}s^{-1}$: this contrasts with a grain size of ~0.3-0.4 μm and a maximum elongation of ~1500% achieved in the 1421 alloy in the present investigation under the same testing conditions.

Microstructural and superplastic characteristics of nanostructured 1421 alloy processed by HPT at various temperatures are presented in Table.

From Table one can see that there is similarity in matrix grain size and strong difference in morphology of precipitates for microstructure in the 1421 alloy processed at various HPT temperatures. For instance, no precipitates have been found in specimens after HPT at room temperature (RT) and 120°C. At the same time there are various types of precipitates in specimens processed by HPT at temperatures (300-370°C). Though we observed uniform microstructure with a grain size of about 80 nm in specimens fabricated by HPT at RT and 300-370°C, however after HPT at 120°C the grain structure was non-uniform. In this case grain size have been distributed from 40 nm to 500 nm.

Intensive grain growth was observed during *in-situ* heating in the TEM of thin foils from the 1421 alloy processed by HPT at RT. Grain size has been achieved 0.5 μm and 1 μm after heating up to 300°C and 400°C, respectively. It should be noted that no additional precipitation was revealed in these specimens, which could lead to stabilization of grain structure and prevent grain growth at elevated temperature of superplastic deformation.

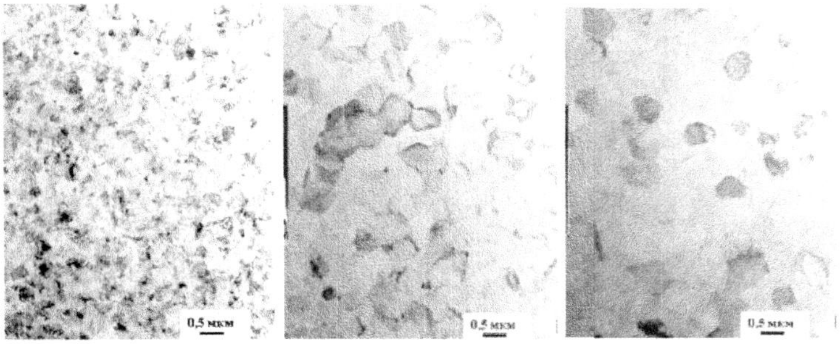

20°C 300°C 400°C

Figure. 3. Typical microstructures observed during *in-situ* heating of the 1421 alloy processed by HPT at RT.

The 1421 samples after HPT were pulled to failure at various testing temperatures in the interval of 250-400°C. The samples demonstrated the fundamental characteristics of superplastic flow: uniform deformation within the gauge length and absence of visible localized necking. The optimum level of superplasticity was achied at a rather low temperature of 375°C and high strain rate of 10^{-2} c^{-1}. The maximum elongation 670% was observed in the alloy processed by HPT at 300°C. Elongation changes slightly with various HPT temperatures although it remains quite high for specimens processed by HPT at all temperatures 20-370°C (Table).

Conclusion

The applications of SPD techniques for processing of UFG structures have provided an opportunity to attain enhanced superplastic properties, namely, low temperature and high strain rate superplasticity, in the 1421 aluminum alloy. At the same time, from the results obtained in present work we can conclude that a very important pre-requisite of the enhanced superplasticity in the UFG materials, in addition to the strong refinement of microstructure by SPD, is the enhanced thermal stability of precipitates preventing grain growth at elevated temperatures. On the other hand, these microstructural parameters depend strongly on the processing routes and regimes, which should be controlled carefully, in order to attain enhanced superplastic properties.

As a result of the SPD processing the following development in superplastic properties of the 1421 aluminium alloy have been achieved:

1. The grain size of the 1421 alloy was reduced to the submicrometer level through ECA pressing. A minimum grain size of ~0.3-0.4 mm was recorded when pressing at 370°C due to presence of thermal stable particles.
2. Under the optimal ECAP conditions, a superplastic elongation of ~1500% was achieved at 400°C using an initial strain rate of 10^{-1} s^{-1}.
3. Specimens of the 1421 alloy processed by HPT at various temperatures 20-370°C demonstrated both the small grain size of about 70-100 nm and low temperature superplasticity with elongation to failure d=550-670% at temperature 375-400°C and strain rate $10^{-2}c^{-1}$.

References

[1] Valiev R.Z., *Nature Materials.* 3, August (2004), 511-516.
[2] Valiev R.Z., Islamgaliev R.K., Alexandrov.I.V., *Prog. Mater. Sci.* 45 (2000), 103-189.
[3] Langdon T.G., Furukawa M., Nemoto M., Horita Z., *JOM.* 52, 4 (2000), 30-33.
[4] Lee S., Utsonomia A., Alamatsu H., Neishi K., Furukawa M., Horita Z., Langdon T.G., *Acta Mater.* 50 (2002), 553-564
[5] Mishra R.S., Valiev R.Z., McFadden S.X., Islamgaliev R.K., Mukherjee A.K., *Phil. Mag.* A81, 1 (2001), 37-48.
[6] Islamgaliev R.K., Yunusova N.F., Valiev R.Z., Tsenev N.K., Perevezentsev V.N., Langdon T.G., *Scripta Mater.* 49, 5 (2003), 467-472.
[7] Islamgaliev R.K., Yunusova N.F., Valiev R.Z., *Phys. Met. Metallogr.* 94, 6 (2002), 606-616.
[8] Papazian J.M., Sigli C., Sanchez J.M., *Scripta Metallurgica.* 20 (1986), 201-206.
[9] Furukawa M, Utsunomiya A, Matsubara K, Horita Z, Langdon T.G., *Acta Mater.* 49 (2001), 3829-3838.

Workshop Participants

Prof. Alexey Babushkin
Ural State University
Russia
Aelxey.Babushkin@usu.ru

Mrs. Julia Bazhenova
Ufa State Aviation Technical University
Russia
RZValiev@mail.rb.ru

Dr. Nikolai Belousov
Donetsk Phys.&Tech. Institute of the NAS of Ukraine
Ukraine
bil@hpress.dipt.donetsk.ua

Prof. Vladimir Bengus
B. Verkin Institute for Low Temperature Physics & Engineering
Ukraine
bengus@ilt.kharkov.ua

Mr. Berta, Marco
University of Manchester/UMIST
UK
m.berta@postgrad.umist.ac.uk

Dr.Yan Beygelzimer
Donetsk Phys.&Tech. Institute of the NAS of Ukraine
Ukraine
tean@an.dn.ua

Prof. Irina Brodova
Institute of metal physics, Russian Academy of science, Ural branch
Russia
brodova@imp.uran.ru

Dr. Marcello Cabibbo
Dipartimento di Meccanica - Universita' Politecnica delle Marche
Italy
marcello.cabibbo@univpm.it

Dr. Vladimir Chuvildeev
Nizhny Novgorod State University
Russia
chuvildeev@nifti.unn.ru

Prof. Sergey Dobatkin
A.A. Baikov Institute of Metallurgy and Materials Science
Russia
dobatkin@ultra.imet.ac.ru

Dr. Boris Efros
Donetsk Phys.&Tech. Institute of the NAS of Ukraine
Ukraine;
efros@ukr.net

Dr. Nariman Enikeev
Ufa State Aviation Technical University
Russia
carabus@mail.rb.ru

Prof. Sergey Firstov
I.M. Frantcevych Institute for Problems of Materials Science of the NAS of Ukraine
Ukraine
fsa@materials.kiev.ua

Mrs. Gahlert, Corinna
University of Technology Chemnitz
Germany
corinna.gahlert@mb.tu-chemnitz.de

Dr. Ralph Hellmig
Institut fur Werkstoffkunde und Werkstofftechnik
Germany
ralph.hellmig@tu-clausthal.de

Mr. Hockauf, Matthias
University of Technology Chemnitz
Germany
matthias.hockauf@mb.tu-chemnitz.de

Prof. Zenji Horita
Kyushu University
Japan
horita@zaiko.kyushu-u.ac.jp

Dr. Rinat Islamgaliev
Institute of Mechanics Russian Academy of Sciences
Russia
saturn@mail.rb.ru

Prof. Hyoung S. Kim
Chungnam National University
Korea
hskim@cnu.ac.kr

Prof. Yuriy Kolobov
Institute of Strength Physics and Materials Science SB RAS
Russia
kolobovispms@mail.tomsknet.ru

Dr. Tatjana Konstantinova
Donetsk Phys.&Tech. Institute of the NAS of Ukraine
Ukraine
tatjana@konstant.fti.ac.donetsk.ua

Dr. Vladimir Kopylov
Physico-Technical Institute of the National Academy of Sciences of Belarus
Belarus
kpl@hmti.ac.by

Dr. Alexander Korshunov
RFNC-VNIIEF
Russia
korshun1@sar.ru

Prof. Terence Langdon
University of Southern California
USA
langdon@usc.edu

Mrs. Larisa Loladze
Donetsk Phys.&Tech. Institute of the NAS of Ukraine
Ukraine
larisa@ukr.net

Dr. Terry Lowe
Los Alamos National Laboratory
USA
tlowe@lanl.gov

Mr. Makarychev, Konstantin
Ufa State Aviation Technical University
Russia

Mrs. Malysheva, Svetlana
Institute for Metals Superplasticity Problems of the RAS
Russia
svufa@imsp.da.ru

Dr. Suveen Mathaudhu
Texas A&M University
USA
suveen@tamu.edu

Prof. Zenon Mroz
Institute of Fundamental Technological Research
Poland
zmroz@ippt.gov.pl

Prof. Hael Mughrabi
Universität Erlangen-Nürnberg
Germany
mughrabi@ww.uni-erlangen.de

Prof. Amiya Mukherjee
University of California
USA
akmukherjee@ucdavis.edu

Prof. Nina Noskova
Institute of metal physics, Russian Academy of science, Ural branch
Russia
noskova@imp.uran.ru

Dr. Dmitry Orlov
Donetsk Phys.&Tech. Institute of the NAS of Ukraine
Ukraine
orlov@donec.net

Prof. Yuriy Podrezov
I.M. Frantcevych Institute for Problems of Materials Science of the NAS of Ukraine
Ukraine
podrezov@materials.kiev.ua

Ms. Oksana Prokofyeva
Donetsk Phys.&Tech. Institute of the NAS of Ukraine
Ukraine
prokofok@land.ru

Mr. Alexey Reshetov
Donetsk Phys.&Tech. Institute of the NAS of Ukraine
Ukraine

Dr. Tatyana Ryumshyna
Donetsk Phys.&Tech. Institute of the NAS of Ukraine
Ukraine
Tatjana@konstant.fti.ac.donetsk.ua

Ms. Lilya Saitova
Ufa State Aviation Technical University
Russia
Saitova-LR@yandex.ru

Prof. Gennady Salishchev
Institute for Metals Superplasticity Problems of the RAS
Russia
gensal@imsp.da.ru

Dr. Irina Semenova
Ufa State Aviation Technical University
Russia
irina-s@mail.rb.ru

Mr. Marat Shagiev
Institute for Metals Superplasticity Problems of the RAS
Russia
marat@imsp.da.ru

Mr. Sitdikov, Vil
Ufa State Aviation Technical University
Russia

Dr. Mikhail Shtern
I.M. Frantcevych Institute for Problems of Materials Science of the NAS of Ukraine
Ukraine
mbsh@i.com.ua

Prof. Vladimir Stolyarov
Ufa State Aviation Technical University
Russia
vlst@mail.rb.ru

Dr. Sergey Synkov
Donetsk Phys.&Tech. Institute of the NAS of Ukraine
Ukraine
reshetov@pisem.net

Dr. Elena Tabachnikova
B. Verkin Institute for Low Temperature Physics & Engineering
Ukraine
tabachnikova@ilt.kharkov.ua

Mr. Tirel, Johan
Toyota Motor Engineering and Manufacturing nv
Belgium
Johan.tirel@toyota-europe.com

Prof. Nobuhiro Tsuji
Osaka University
Japan
tsuji@ams.eng.osaka-u.ac.jp

Prof. Ruslan Valiev
Ufa State Aviation Technical University
Russia
RZValiev@mail.rb.ru

Prof. Viktor Varyukhin
Donetsk Phys.&Tech. Institute of the NAS of Ukraine
Ukraine
var@hpress.dipt.donetsk.ua

Mrs. Vildanova, Nina
Institute of metal physics, Russian Academy of science, Ural branch
Russia

Mr. Guney Yapici
Texas A&M University
USA
ggyapici@neo.tamu.edu

Dr. Alexandra Yurkova
National Technical University of Ukraine
Ukraine
byakova@vic.com.ua

Mrs. Zakharova, Natalia
I.M. Frantcevych Institute for Problems of Materials Science of the NAS of Ukraine
Ukraine
sirko@ipms.kiev.ua

Prof. Michael Zehetbauer
University of Vienna
Austria
zehet@ap.univie.ac.at

Dr. Alexander Zhilyaev
Naval Postgraduate School
USA
apzhilya@nps.edu

Dr. Yuntian T. Zhu
Los Alamos National Laboratory
USA
yzhu@lanl.gov

Subject Index

GPSR Compliance

The European Union's (EU) General Product Safety Regulation (GPSR) is a set of rules that requires consumer products to be safe and our obligations to ensure this.

If you have any concerns about our products, you can contact us on ProductSafety@springernature.com

In case Publisher is established outside the EU, the EU authorized representative is:

Springer Nature Customer Service Center GmbH
Europaplatz 3
69115 Heidelberg, Germany

Batch number: 10755284

Printed by Printforce, the Netherlands